Product Design Modeling using CAD/CAE

*This book, the first of the four Computer-Aided Engineering Design series,
is dedicated to my mentors, Professor Kyung K. Choi, Professor Edward J. Haug,
and Professor Vijay K. Goel, who have taught me so much,
inspiration and influence for a lifetime.*

Product Design Modeling using CAD/CAE

The Computer Aided Engineering Design Series

Kuang-Hua Chang

AMSTERDAM • BOSTON • HEIDELBERG • LONDON
NEW YORK • OXFORD • PARIS • SIGNAPORE • SAN DIEGO
SAN FRANCISCO • SYDNEY • TOKYO
Academic Press is an Imprint of Elsevier

Academic Press is an imprint of Elsevier
The Boulevard, Langford Lane, Kidlington, Oxford, OX5 1GB, UK
225 Wyman Street, Waltham, MA 02451, USA

First published 2014

Notices
Knowledge and best practice in this field are constantly changing. As new research and experience broaden
our understanding, changes in research methods, professional practices, or medical treatment may become
necessary.

Practitioners and researchers must always rely on their own experience and knowledge in evaluating and using
any information, methods, compounds, or experiments described herein. In using such information or methods
they should be mindful of their own safety and the safety of others, including parties for whom they have
a professional responsibility.

To the fullest extent of the law, neither the Publisher nor the authors, contributors, or editors, assume any liability
for any injury and/or damage to persons or property as a matter of products liability, negligence or otherwise, or
from any use or operation of any methods, products, instructions, or ideas contained in the material herein.

British Library Cataloguing in Publication Data
A catalogue record for this book is available from the British Library

Library of Congress Cataloguing in Publication Data
A catalog record for this book is available from the Library of Congress

ISBN: 978-0-12-398513-2

For information on all Academic Press publications
visit our website at **store.elsevier.com**

Contents

Preface

The conventional product development process employs a design-build-test philosophy. The sequentially executed process often results in a prolonged lead time and an elevated product cost. The e-Design paradigm presented in the Computer Aided Engineering Design series employs IT-enabled technology, including computer-aided design, engineering, and manufacturing (CAD/CAE/CAM) tools, as well as advanced prototyping technology to support product design from concept to detailed designs, and ultimately manufacturing. This e-Design approach employs virtual proto-typing technology to support a cross-functional team in analyzing product performance, reliability, and manufacturing costs early in the product development stage and in conducting quantitative trade-offs for design decision making. Physical prototypes of the product design are then produced using rapid prototyping (RP) technique mainly for design verification. The e-Design approach holds potential for shortening the overall product development cycle, improving product quality, and reducing product cost. The Computer Aided Engineering Design series intends to provide readers with a comprehensive coverage of essential elements for understanding and practicing the e-Design paradigm in support of product design, including design method and process, and computer-based tools and technology. The book series consists of four books: *Product Design Modeling using CAD/CAE*, *Product Performance Evaluation using CAD/CAE*, *Product Manufacturing and Cost Estimating using CAD/CAE*, and *Design Theory and Methodology using CAD/CAE*. *Product Design Modeling using CAD/CAE* discusses virtual mockup of the product that is first created in the CAD environment. The critical design parameterization that converts the product solid model into parametric representation, enabling the search for better designs, is an indispensable element of practicing the e-Design paradigm, especially in the detailed design stage. The second book, *Product Performance Evaluation using CAD/CAE*, focuses on applying numerous CAE technologies and software tools to support evaluation of product performance, including structural analysis, fatigue and fracture, rigid body kinematics and dynamics, and failure probability prediction and reliability analysis. The third book, *Product Manufacturing and Cost Estimating using CAD/CAE*, introduces CAM technology to support manufacturing simulations and process planning, RP technology, and computer numerical control machining for fast product prototyping, as well as manufacturing cost estimate that can be incorporated into product cost calculations. The product performance, reliability, and cost calculated can then be brought together to the cross-functional team for design trade-offs based on quantitative engineering data obtained from simulations. Design trade-off is one of the key topics included in the fourth book, *Design Theory and Methodology using CAD/CAE*. In addition to conventional design optimization methods, the fourth book discusses decision theory, utility theory, and decision-based design. Simple examples are included to help readers understand the fundamentals of concepts and methods introduced in this book series.

In addition to the discussion on design principles, methods, and processes, this book series offers reviews on the commercial off-the-shelf software tools for the support of modeling, simulations, manufacturing, and product data management and data exchanges. Tutorial style lessons on using commercial software tools are provided together with project-based exercises. Two suites of engineering software are included: they are Pro/ENGINEER-based, including Pro/MECHANICA Structure, Pro/ENGINEER Mechanism Design, and Pro/MFG; and SolidWorks-based, including

SolidWorks Simulation, SolidWorks Motion, and CAMWorks. These tutorial lessons are designed to help readers gain hands-on experiences to practice the e-Design paradigm.

The book you are reading, *Product Design Modeling* using CAD/CAE, is the first book of the *Computer Aided Engineering Design* series. The objective of *Product Design Modeling* is to provide readers with fundamental understanding in product modeling principles and modern engineering tools for solid and assembly modeling, and apply the principles and software tools to support practical design applications. In Chapter 1, a brief introduction to the e-Design paradigm and tool environment is given. Following this introduction, important topics in product design modeling, including geometric and solid modeling, assembly modeling, design parameterization, and product data management and data exchange are discussed.

Chapter 2 focuses on geometric modeling, in which general geometric modeling techniques and methods commonly employed in CAD are discussed. Fundamentals in geometric modeling, such as mathematic representation of parametric curves and surfaces, continuity, and geometric trans-formations are presented to provide readers a basic understanding in geometric modeling. The goal of this chapter is to help readers understand how geometric entities, such as curves and surfaces are created in CAD, which is critical to understand the theories and methods that support part modeling in CAD.

Chapter 3 offers basic knowledge on the theories of solid modeling in CAD. Basic solid modeling theories, including constructive solid geometry and boundary representation (B-Rep), are briefly presented. The goal of this chapter is to help readers understand how solid parts are created in CAD and the theories and methods that support part modeling in CAD.

Chapter 4 provides a brief discussion on product assembly in CAD, which involves both modeling and analysis of the articulated assemblies for support of product design. In CAD, an assembly is created by defining relative position and orientation of parts, whereas a kinematic model is created by specifying kinematic constraints between parts. Both are important for engineers to create functional assemblies in CAD to support product design. The goal of this chapter is to help readers understand how solid parts are put together that perform desired functions in CAD and the theories and methods that do the tricks.

Chapter 5 is the key chapter of this book, in which design parameterization concept and method are discussed for the support of capturing design intents in the parts and assembly of the product model. A set of guidelines are presented for the designers to parameterize solid models at sketch, part, and assembly levels in order to properly capture design intents. The goal of the chapter is to provide design parameterization concept, methods, and guidelines that support designers to explore product design alternatives in the context of e-Design paradigm.

After learning how parts and assemblies are created in CAD, we discuss how to manage product data to support product design in Chapter 6. In addition, data exchange between CAD systems, which is one of the major issues encountered in product design using e-design paradigm, is discussed to offer readers practical approaches in dealing with such issues.

In addition to theories and methods, two companion projects are included: *Project S1 Solid Modeling with SolidWorks* and *Project P1 Solid Modeling with Pro/ENGINEER*. These projects offer tutorial lessons that should help readers to learn and be able to use the respective software tools for support of solid modeling, assembly modeling, design parameterization, and model translations for practical applications. Example files needed for going through the tutorial lessons are available for download at the book's companion site. The goal of the projects is to help readers become confident and competent in using CAD tools for creating adequate product models to support product design.

Product Design Modeling should serve well for a half semester (8 weeks) instruction in engineering colleges of general universities. Typically, a 3-h lecture and 1-h laboratory exercise per week are desired. This book (and the book series) aims at providing engineering senior and first-year graduate students a comprehensive reference to learn advanced technology in support of engineering design using IT-enabled technology. Typical engineering courses that the book serves include computer-aided design, engineering design, integrated product and process development, concurrent engineering, design and manufacturing, modern product design, computer-aided engineering, as well as senior capstone design. In addition to classroom instruction, this book should support practicing engineers who wish to learn more about the e-Design paradigm at their own pace.

About the Author

Dr Kuang-Hua Chang is a David Ross Boyd Professor and Williams Companies Foundation Presidential Professor at the University of Oklahoma (OU), Norman, OK, USA. He received his PhD in mechanical engineering from the University of Iowa in 1990. Since then, he joined the Center for Computer-Aided Design at Iowa as a research scientist and Computer-Aided Engineering technical area manager. In 1996, he joined Northern Illinois University as an assistant professor. In 1997, he joined OU. He teaches mechanical design and manufacturing, in addition to conducting research in computer-aided modeling and simulation for design and manufacturing of mechanical systems. He has worked with aerospace and automotive industries and served as technical consultant to US industry and foreign companies. His work has been published in six books and more than 140 articles in international journals and conference proceedings.

He has received numerous awards for his teaching and research, including the Presidential Professorship in 2005 for meeting the highest standards of excellence in scholarship and teaching, OU Regents Award for Superior Accomplishment in Research and Creative Activity in 2004, OU BP AMOCO Foundation Good Teaching Award in 2002, and OU Regents Award for Superior Teaching in 2010. He is a five-time recipient of the CoE Alumni Teaching Award, given to top teachers in the College of Engineering at OU. His research paper was given a Best Paper Award at the iNEER Conference for Engineering Education and Research in 2005 (iCEER-2005). In 2006, he was awarded a Ralph R. Teetor Educational Award by the Society of Automotive Engineers in recognition of significant contributions to teaching, research, and student development. He was honored by the Oklahoma City Mayor's Committee on Disability Concerns with the 2009 Don Davis Award, which is the highest honor granted in public recognition of extraordinarily meritorious service which has substantially advanced opportunities for people with disabilities by removing social, attitudinal, and environmental barriers in the greater Oklahoma City area. In 2013, Dr Chang was named David Ross Boyd Professor, one of the highest honors, at the University of Oklahoma, for having consistently demonstrated outstanding teaching, guidance, and leadership for students in an academic discipline or in an interdisciplinary program within the University.

About the Cover

The cover page shows the solid model of a single-piston engine in computer that is commonly found in light airplane such as the one in the background. The computer model of the single-piston engine is constructed in computer-aided design (CAD) and is fully parameterized, in which the bore diameter of the engine case is defined as the design variable. When the diameter is changed, the engine case is regenerated first in CAD by properly updating solid features that are affected by the change. At the same time, the change propagates to other parts in the assembly, including the piston, piston pin, cylinder head, cylinder sleeve, cylinder fins, and crankshaft. More important, the parts stay intact, maintaining adequate assembly placement constraints, and the change does not induce interference nor leave excessive gaps between parts. With such parametric models, designers are given tremendous freedom to explore design alternatives efficiently and accurately.

Acknowledgement

I would like to first thank Mr. Joseph P. Hayton for recognizing the need for an engineering design book series that offers knowledge in modern engineering design principles, methods, and tools to mechanical engineering students. His enthusiasm in moving the book idea forward and eventually publishing the book series is highly appreciated. Mr. Hayton's colleagues at Elsevier, Ms. Lisa Jones, Ms. Chelsea Johnson, Ms. Fiona Geraghty, Ms. Marilyn Rash, and the entire production team, have made significant contributions in transforming the original manuscripts into well-organized and professionally-polished books that are suitable and presentable to our readers.

I am also thankful to Mr. Yunxiang Wang, PhD student of Mechanical Engineering at the University of Oklahoma, for his help in preparing part of the manuscripts. His contribution to this book, especially Tutorial Project P1, is highly valuable. I am grateful to my former graduate students Dr. Mangesh Edke, Dr. Qunli Sun, Dr. Sung-Hwan Joo, Dr. Xiaoming Yu, Dr. Hsiu-Ying Hwang, Mr. Trey Wheeler, Mr. Yunxiang Wang, Mr. Iulian Grindeanu, Mr. Tyler Bunting, Mr. David Gibson, Mr. Chienchih Chen, Mr. Tim Long, Mr. Poh-Soong Tang, and Mr. Javier Silver, for their excellent efforts in conducting research on numerous aspects of engineering design. Ideas and results that came out of their research have been largely incorporated into this book. Their dedication to the research in developing computer-aided approaches for support of product design modeling is acknowledged and is highly appreciated.

Introduction to e-Design

1

CHAPTER OUTLINE

Conventional product development employs a design-build-test philosophy. The sequentially executed development process often results in prolonged lead times and elevated product costs. The proposed e-Design paradigm employs IT-enabled technology for product design, including virtual prototyping (VP) to support a cross-functional team in analyzing product performance, reliability, and manufacturing costs early in product development, and in making quantitative trade-offs for design decision making. Physical prototypes of the product design are then produced using the rapid prototyping (RP) technique and computer numerical control (CNC) to support design verification and functional prototyping, respectively.

e-Design holds potential for shortening the overall product development cycle, improving product quality, and reducing product costs. It offers three concepts and methods for product development:

- Bringing product performance, quality, and manufacturing costs together early in design for consideration.
- Supporting design decision making based on quantitative product performance data.
- Incorporating physical prototyping techniques to support design verification and functional prototyping.

1.1 **Introduction**

A conventional product development process that is usually conducted sequentially suffers the problem of the *design paradox* (Ullman 1992). This refers to the dichotomy or mismatch between the design engineer's knowledge about the product and the number of decisions to be made (flexibility) throughout the product development cycle (see Figure 1.1). Major design decisions are usually made in the early design stage when the product is not very well understood. Consequently, engineering changes are frequently requested in later product development stages, when product design evolves and is better understood, to correct decisions made earlier.

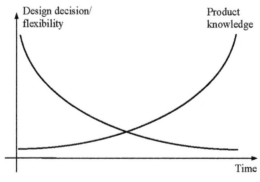

FIGURE 1.1 The Design Paradox.

Conventional product development is a design-build-test process. Product performance and reliability assessments depend heavily on physical tests, which involve fabricating functional prototypes of the product and usually lengthy and expensive physical tests. Fabricating prototypes usually involves manufacturing process planning and fixtures and tooling for a very small amount of production. The process can be expensive and lengthy, especially when a design change is requested to correct problems found in physical tests.

In conventional product development, design and manufacturing tend to be disjointed. Often, manufacturability of a product is not considered in design. Manufacturing issues usually appear when the design is finalized and tests are completed. Design defects related to manufacturing in process planning or production are usually found too late to be corrected. Consequently, more manufacturing procedures are necessary for production, resulting in elevated product cost.

With this highly structured and sequential process, the product development cycle tends to be extended, cost is elevated, and product quality is often compromised to avoid further delay. Costs and the number of engineering change requests (ECRs) throughout the product development cycle are often proportional according to the pattern shown in Figure 1.2. It is reported that only 8% of the total product budget is spent for design; however, in the early stage, design determines 80% of the lifetime cost of the product (Anderson 1990). Realistically, today's industries will not survive worldwide competition unless they introduce new products of better quality, at lower cost, and with shorter lead times. Many approaches and concepts have been proposed over the years, all with a common goal—to shorten the product development cycle, improve product quality, and reduce product cost.

A number of proposed approaches are along the lines of virtual prototyping (Lee 1999), which is a simulation-based method that helps engineers understand product behavior and make design decisions in a virtual environment. The virtual environment is a computational framework in which the geometric and physical properties of products are accurately simulated and represented. A number of successful virtual prototypes have been reported, such as Boeing's 777 jetliner, General Motors' locomotive engine, Chrysler's automotive interior design, and the Stockholm Metro's Car 2000 (Lee 1999). In addition to virtual prototyping, the concurrent engineering (CE) concept and methodology have been studied and developed with emphasis on subjects such as product life cycle design, design for X-abilities (DFX), integrated product and process development (IPPD), and Six Sigma (Prasad 1996).

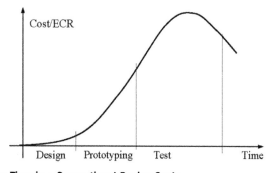

FIGURE 1.2 Cost/ECR versus Time in a Conventional Design Cycle.

Although significant research has been conducted in improving the product development process and successful stories have been reported, industry at large is not taking advantage of new product development paradigms. The main reason is that small and mid-size companies cannot afford to develop an in-house computer tool environment like those of Boeing and the Big-Three automakers. On the other hand, commercial software tools are not tailored to meet the specific needs of individual companies; they often lack proper engineering capabilities to support specific product development needs, and most of them are not properly integrated. Therefore, companies are using commercial tools to support segments of their product development without employing the new design paradigms to their full advantage.

The e-Design paradigm does not supersede any of the approaches discussed. Rather, it is simply a realization of concurrent engineering through virtual and physical prototyping with a systematic and quantitative method for design decision making. Moreover, e-Design specializes in performance and reliability assessment and improvement of complex, large-scale, compute-intensive mechanical systems. The paradigm also uses design for manufacturability (DFM), design for manufacturing and assembly (DFMA), and manufacturing cost estimates through virtual manufacturing process planning and simulation for design considerations.

The objective of this chapter is to present an overview of the e-Design paradigm and the sample tool environment that supports a cross-functional team in simulating and designing mechanical products concurrently in the early design stage. In turn, better-quality products can be designed and manufactured at lower cost. With intensive knowledge of the product gained from simulations, better design decisions can be made, breaking the aforementioned design paradox. With the advancement of computer simulations, more hardware tests can be replaced by computer simulations, thus reducing cost and shortening product development time. The desirable cost and ECR distributions throughout the product development cycle shown in Figure 1.3 can be achieved through the e-Design paradigm.

A typical e-Design software environment can be built using a combination of existing computer-aided design (CAD), computer-aided engineering (CAE), and computer-aided manufacturing (CAM) as the base, and integrating discipline-specific software tools that are commercially available for specific simulation tasks. The main technique in building the e-Design environment is tool integration. Tool integration techniques, including product data models, wrappers, engineering views, and design process management, have been developed (Tsai et al. 1995) and are described in *Design Theory and Methods using CAD/CAE*, a book in The Computer Aided Engineering Design Series. This integrated e-Design tool environment allows small and mid-size companies to conduct efficient product development using the e-Design paradigm. The tool environment is flexible so that additional engineering tools can be incorporated with a lesser effort.

In addition, the basis for tool integration, such as product data management (PDM), is well established in commercial CAD tools and so no wheel needs to be reinvented. The e-Design paradigm employs three main concepts and methods for product development:

- Bringing product performance, quality, and manufacturing cost for design considerations in the early design stage through virtual prototyping.
- Supporting design decision making through a quantitative approach for both concept and detail designs.
- Incorporating product physical prototypes for design verification and functional tests via rapid prototyping and CNC machining, respectively.

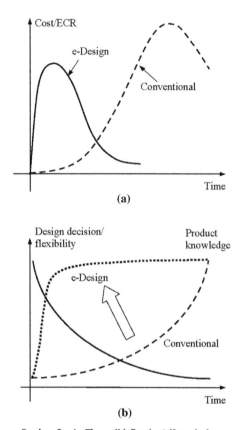

FIGURE 1.3 (a) Cost/ECR versus e-Design Cycle Time; (b) Product Knowledge versus e-Design Cycle Time.

In this chapter, the e-Design paradigm is introduced. Then components that make up the paradigm, including knowledge-based engineering (KBE) (Gonzalez and Dankel 1993), virtual prototyping, and physical prototyping, are briefly presented. Designs of a simple airplane engine and a high-mobility multipurpose wheeled vehicle (HMMWV) are briefly discussed to illustrate the e-Design paradigm. Details of modeling and simulation are provided in later chapters.

1.2 The e-Design paradigm

As shown in Figure 1.4, in e-Design, a product design concept is first realized in solid model form by design engineers using CAD tools. The initial product is often established based on the designer's experience and legacy data of previous product lines. It is highly desirable to capture and organize designer experience and legacy data to support decision making in a discrete form so as to realize an initial concept. The KBE (Gonzalez and Dankel 1993) that computerizes knowledge about specific product domains to support design engineers in arriving at a solution to a design problem supports

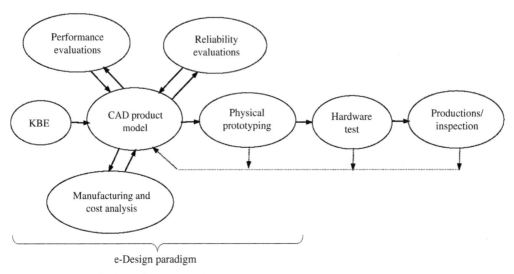

FIGURE 1.4 The e-Design Paradigm.

the concept design. In addition, a KBE system integrated with a CAD tool may directly generate a solid model of the concept design that directly serves downstream design and manufacturing simulations.

With the product solid model represented in CAD, simulations for product performance, reliability, and manufacturing can be conducted. The product development tasks and the cross-functional team are organized according to engineering disciplines and expertise. Based on a centralized computer-aided design product model, simulation models can be derived with proper simplifications and assumptions. However, a one-way mapping that governs changes from CAD models to simulation models must be established for rapid simulation model updates (Chang et al. 1998). The mapping maintains consistency between CAD and simulation models throughout the product development cycle.

Product performance, reliability, and manufacturing can then be simulated concurrently. Performance, quality, and costs obtained from multidisciplinary simulations are brought together for review by the cross-functional team. Design variables—including geometric dimensions and material properties of the product CAD models that significantly influence performance, quality, and cost—can be identified by the cross-functional team in the CAD product model. These key performance, quality, and cost measures, as well as design variables, constitute a product design model. With such a model, a systematic design approach, including a parametric study for concept design and a trade-off study for detail design, can be conducted to improve the product with a minimum number of design iterations.

The product designed in the virtual environment can then be fabricated using rapid prototyping machines for physical prototypes directly from product CAD solid models, without tooling and process planning. The physical prototypes support the cross-functional team for design verification and assembly checking. Change requests that are made at this point can be accommodated in the virtual environment without high cost and delay.

The physics-based simulation technology potentially minimizes the need for product hardware tests. Because substantial modeling and simulations are performed, unexpected design defects encountered during the hardware tests are reduced, thus minimizing the feedback loop for design modifications. Moreover, the production process is smooth since the manufacturing process has been planned and simulated. Potential manufacturing-related problems will have been largely addressed in earlier stages.

A number of commercial CAD systems provide a suite of integrated CAD/CAE/CAM capabilities (e.g., Pro/ENGINEER and SolidWorks®). Other CAD systems, including CATIA® and NX, support one or more aspects of the engineering analysis. In addition, third-party software companies have made significant efforts in connecting their capabilities to CAD systems. As a representative example, CAE and CAM software companies worked with SolidWorks and integrated their software into SolidWorks environments such as CAMWorks®. Each individual tool is seamlessly integrated into SolidWorks.

In this book, Pro/ENGINEER and SolidWorks, with a built-in suite of CAE/CAM modules, are employed as the base for the e-Design environment. In addition to their superior solid modeling capability based on parametric technology (Zeid 1991), Pro/MECHANICA® and SolidWorks Simulation support simulations of nominal engineering, including structural and thermal problems. Mechanism Design of Pro/ENGINEER and SolidWorks Motion support motion simulation of mechanical systems. Moreover, CAM capabilities implemented in CAD, such as Pro/MFG (Parametric Technology Corp., www.ptc.com), and CAMWorks, provide an excellent basis for manufacturing process planning and simulations. Additional CAD/CAE/CAM tools introduced to support modeling and simulation of broader engineering problems encountered in general mechanical systems can be developed and added to the tool environment as needed.

1.3 **Virtual prototyping**

Virtual prototyping is the backbone of the e-Design paradigm. As presented in this chapter, VP consists of constructing a parametric product model in CAD, conducting product performance simulations and reliability evaluations using CAE software, and carrying out manufacturing simulations and cost estimates using CAM software. Product modeling and simulations using integrated CAD/CAE/CAM software are the basic and common activities involved in virtual prototyping. However, a systematic design method, including parametric study and design trade-offs, is indispensable for design decision making.

1.3.1 **Parameterized CAD product model**

A parametric product model in CAD is essential to the e-Design paradigm. The product model evolves to a higher-fidelity level from concept to detail design stages (Chang et al. 1998). In the concept design stage, a considerable portion of the product may contain non-CAD data. For example, when the gross motion of the mechanical system is sought, the non-CAD data may include engine, tires, or transmission if a ground vehicle is being designed. Engineering characteristics of the non-CAD parts and assemblies are usually described by engineering parameters, physics laws, or mathematical equations. This non-CAD representation is often added to the product model in the concept design stage for a complete product model. As the design evolves, non-CAD parts and assemblies are refined into solid-model forms for subsystem and component designs as well as for manufacturing process planning.

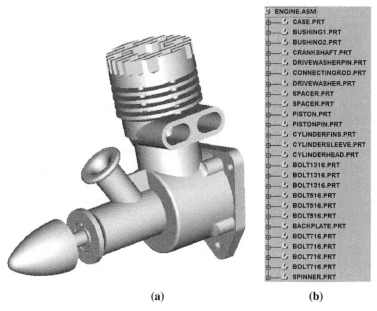

ENGINE.ASM
CASE.PRT
BUSHING1.PRT
BUSHING2.PRT
CRANKSHAFT.PRT
DRIVEWASHERPIN.PRT
CONNECTINGROD.PRT
DRIVEWASHER.PRT
SPACER.PRT
SPACER.PRT
PISTON.PRT
PISTONPIN.PRT
CYLINDERFINS.PRT
CYLINDERSLEEVE.PRT
CYLINDERHEAD.PRT
BOLT1316.PRT
BOLT1316.PRT
BOLT1316.PRT
BOLT516.PRT
BOLT516.PRT
BOLT516.PRT
BACKPLATE.PRT
BOLT716.PRT
BOLT716.PRT
BOLT716.PRT
BOLT716.PRT
SPINNER.PRT

(a)   (b)

FIGURE 1.5 Airplane Engine Model: (a) CAD Model and (b) Model Tree.

A primary challenge in conducting product performance simulations is generating simulation models and maintaining consistency between CAD and simulation models through mapping. Challenges involved in model generation and in structural and dynamic simulations are discussed next, in which an airplane engine model in the detail design stage, as shown in Figure 1.5, is used for illustration.

1.3.1.1 Parameterized product model

A parameterized product model defined in CAD allows design engineers to conveniently explore design alternatives for support of product design. The CAD product model is parameterized by defining dimensions that govern the geometry of parts through geometric features and by establishing relations between dimensions within and across parts. Through dimensions and relations, changes can be made simply by modifying a few dimensional values. Changes are propagated automatically throughout the mechanical product following the dimensions and relations. A single-piston airplane engine with a change in its bore diameter is shown in Figure 1.6, so as illustrating change propagation through parametric dimensions and relationships. More in-depth discussion of the modeling and parameterization of the engine example can be found in *Product Design Modeling using CAD/CAE*, a book in The Computer Aided Engineering Design Series.

1.3.1.2 Analysis models

For product structural analysis, finite element analysis (FEA) is often employed. In addition to structural geometry, loads, boundary conditions, and material properties can be conveniently defined in the CAD model. Most CAD tools are equipped with fully automatic mesh generation capability. This

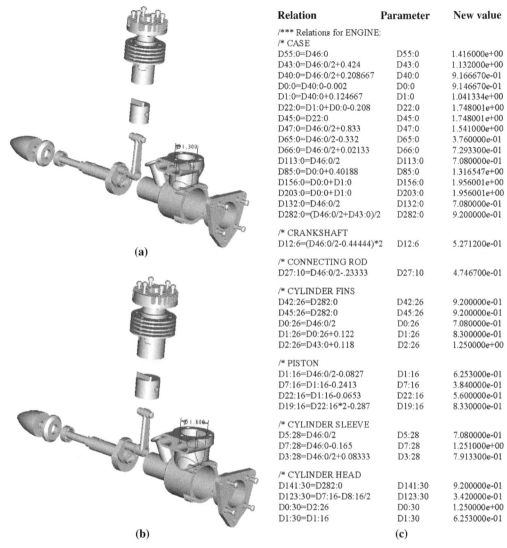

Relation	Parameter	New value
/*** Relations for ENGINE:		
/* CASE		
D55:0=D46:0	D55:0	1.416000e+00
D43:0=D46:0/2+0.424	D43:0	1.132000e+00
D40:0=D46:0/2+0.208667	D40:0	9.166670e-01
D0:0=D40:0-0.002	D0:0	9.146670e-01
D1:0=D40:0+0.124667	D1:0	1.041334e+00
D22:0=D1:0+D0:0-0.208	D22:0	1.748001e+00
D45:0=D22:0	D45:0	1.748001e+00
D47:0=D46:0/2+0.833	D47:0	1.541000e+00
D65:0=D46:0/2-0.332	D65:0	3.760000e-01
D66:0=D46:0/2+0.02133	D66:0	7.293300e-01
D113:0=D46:0/2	D113:0	7.080000e-01
D85:0=D0:0+0.40188	D85:0	1.316547e+00
D156:0=D0:0+D1:0	D156:0	1.956001e+00
D203:0=D0:0+D1:0	D203:0	1.956001e+00
D132:0=D46:0/2	D132:0	7.080000e-01
D282:0=(D46:0/2+D43:0)/2	D282:0	9.200000e-01
/* CRANKSHAFT		
D12:6=(D46:0/2-0.44444)*2	D12:6	5.271200e-01
/* CONNECTING ROD		
D27:10=D46:0/2-.23333	D27:10	4.746700e-01
/* CYLINDER FINS		
D42:26=D282:0	D42:26	9.200000e-01
D45:26=D282:0	D45:26	9.200000e-01
D0:26=D46:0/2	D0:26	7.080000e-01
D1:26=D0:26+0.122	D1:26	8.300000e-01
D2:26=D43:0+0.118	D2:26	1.250000e+00
/* PISTON		
D1:16=D46:0/2-0.0827	D1:16	6.253000e-01
D7:16=D1:16-0.2413	D7:16	3.840000e-01
D22:16=D1:16-0.0653	D22:16	5.600000e-01
D19:16=D22:16*2-0.287	D19:16	8.330000e-01
/* CYLINDER SLEEVE		
D5:28=D46:0/2	D5:28	7.080000e-01
D7:28=D46:0-0.165	D7:28	1.251000e+00
D3:28=D46:0/2+0.08333	D3:28	7.913300e-01
/* CYLINDER HEAD		
D141:30=D282:0	D141:30	9.200000e-01
D123:30=D7:16-D8:16/2	D123:30	3.420000e-01
D0:30=D2:26	D0:30	1.250000e+00
D1:30=D1:16	D1:30	6.253000e-01

    (a)   **(b)**   **(c)**

FIGURE 1.6 Design Change Propagation: (a) Bore Diameter = 1.3 in.; (b) Bore Diameter Changed to 1.6 in.; (c) Relations of Geometric Dimensions.

capability is convenient but often leads to large FEA models with some geometric discrepancy at the part boundary. Plus, triangular and tetrahedral elements are often the only elements supported. An engine connecting rod example meshed using Pro/MESH (part of Pro/MECHANICA) with default mesh parameters is shown in Figure 1.7. The FEA model consists of 1,270 nodes and 4,800 tetrahedron elements, yet it still reveals discrepancy to the true CAD geometry. Moreover, mesh distortion due to large deformation of the structure, such as hyperelastic problems, often causes FEA to abort

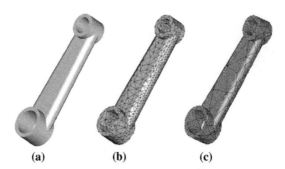

(a)   (b)   (c)

FIGURE 1.7 Finite Element Meshes of a Connecting Rod: (a) CAD Solid Model, (b) h-version Finite Element Mesh, and (c) p-version Finite Element Mesh.

prematurely. Semiautomatic mesh generation is more realistic; therefore, tools such as MSC/Patran® (MacNeal-Schwendler Corp., www.mscsoftware.com) and HyperMesh® (Altair® Engineering, Inc., www.altair.com) are essential to support the e-Design environment for mesh generation.

In general, p-version FEA (Szabó and Babuška 1991) is more suitable for structural analysis in terms of minimizing the gap in geometry between CAD and finite element models, and in lessening the tendency toward mesh distortion. It also offers capability in convergence analysis that is superior to regular h-version FEA. As shown in Figure 1.7c, the same connecting rod is meshed with 568 tetrahedron p-elements, using Pro/MECHANICA with a default setting. A one-way mapping between changes in CAD geometric dimensions and finite element mesh for both h- and p-version FEAs can be established through a design velocity field (Haug et al. 1986), which allows direct and automatic generation of the finite element mesh of new designs.

Another issue worth considering is the simplification of 3D solid models to surface (shell) or curve (beam) models for analysis. Capabilities that semiautomatically convert 3D thin-shell solids to surface models are available in, for example, Pro/MECHANICA and SolidWorks Simulation.

1.3.1.3 Motion simulation models

Generating motion simulation models involves regrouping parts and subassemblies of the mechanical system in CAD as bodies and often introducing non-CAD components to support a multibody dynamic simulation (Haug 1989). Engineers must define the joints or force connections between bodies, including joint type and reference coordinates. Mass properties of each body are computed by CAD with the material properties specified. Integration between Mechanism Design and Pro/ENGINEER, as well as between SolidWorks Motion (Chang 2008) and SolidWorks, is seamless. Design changes made in geometric dimensions propagate to the motion model directly. In addition, simulation tools, such as Dynamic Analysis and Design Systems (DADS) (LMS, www.lmsintl.com/DADS) and communication and data systems integration, are also integrated with CAD with proper parametric mapping from CAD to simulation models that support parametric study. As an example, the motion inside an airplane engine is modeled as a slider-crank mechanism in Mechanism Design, as shown in Figure 1.8.

A common mistake made in creating motion simulation models is selecting improper joints to connect bodies. Introducing improper joints creates an invalid or inaccurate model that does not

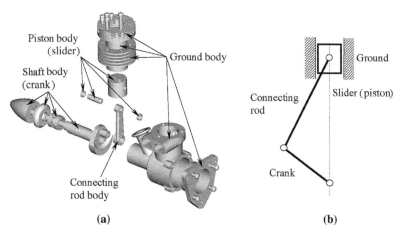

Piston body
(slider)

Shaft body
(crank)

Ground body

Ground

Connecting
rod

Slider (piston)

Crank

Connecting
rod body

(a)

(b)

FIGURE 1.8 Engine Motion Model: (a) Definition and (b) Schematic View.

simulate the true behavior of the mechanical system. Intelligent modeling capability that automatically specifies joints in accordance with assembly relations defined between parts and subassemblies in solid models is available in, for example, SolidWorks Motion.

1.3.2 Product performance analysis

As mentioned earlier, product performance evaluation using physics-based simulation in the computer environment is usually called, in a narrow sense, virtual prototyping, or VP. With the advancement of simulation technology, more engineering questions can be answered realistically through simulations, thus minimizing the needs for physical tests. However, some key questions cannot be answered for sophisticated engineering problems—for example, the crashworthiness of ground vehicles. Although VP will probably never replace hardware tests completely, the savings it achieves for less sophisticated problems is significant and beneficial.

1.3.2.1 Motion analysis

System motion simulations include workspace analysis (kinematics), rigid- and flexible-body dynamics, and inverse dynamic analysis. Mechanism Design and SolidWorks Motion, based on theoretical work (Kane and Levinson 1985), mainly support kinematics and rigid-body simulations for mechanical systems. They do not properly support mechanical system simulation such as a vehicle moving on a user-defined terrain. General-purpose dynamic simulation tools, such as DADS (www.lsmintl.com) or Adams® (www.mscsoftware.com), are more desirable for simulation of general mechanical systems.

1.3.2.2 Structural analysis

Pro/MECHANICA supports linear static, vibration, buckling, fatigue, and other such analyses, using p-version FEA. General-purpose finite element codes, such as MSC/Nastran® (MacNeal-Schwendler Corp., www.mscsoftware.com) and ANSYS® (ANSYS Analysis Systems, Inc., www.ansys.com) are ideal for the e-Design environment to support FEA for a broad range of structural problems—for example, nonlinear, plasticity, and transient dynamics. Meshless methods developed in recent years (for example, Chen et al. 1997) hold promise for avoiding finite element mesh distortion in large-deformation problems. Multiphase problems (e.g., acoustic and aero-structural) are well supported by specialized tools such as LMS® SYSNOISE (Numerical Integration Technologies 1998). LS-DYNA® (Hallquist 2006) is currently one of the best codes for nonlinear, plastic, dynamics, friction-contact, and crashworthiness problems. These special codes provide excellent engineering analysis capabilities that complement those provided in CAD systems.

1.3.2.3 Fatigue and fracture analysis

Fatigue and fracture problems are commonly encountered in mechanical components because of repeated mechanical or thermal loads. MSC Fatigue® (MacNeal-Schwendler Corp., www.mscsoftware.com), with an underlying computational engine developed by nCode® (www.ncode.com) is one of the leading fatigue and fracture analysis tools. It offers both high- and low-cycle fatigue analyses. A critical plane approach is available in MSC Fatigue for prediction of fatigue life due to general multiaxial loads.

Note that the recently developed extended finite element method (XFEM) supports fracture propagation without re-meshing (Moës et al. 2002). XFEM was recently integrated in ABAQUS®. Also note that additional capabilities, such as thermal analysis, computational fluid dynamics (CFD) and combustion, can be added to meet specific needs in analyzing mechanical products. Integration of additional engineering disciplines are briefly discussed in Section 1.3.4.

1.3.2.4 Product reliability evaluations

Product reliability evaluations in the e-Design environment focus on the probability of specific failure events (or failure mode). The failure event corresponds to a product performance measure, such as the fatigue life of a mechanical component. For the reliability analysis of a single failure event, the failure event or failure function is defined as (Madsen et al. 1986)

$$g(X) = \psi^u - \psi(X) \tag{1.1}$$

where

ψ is a product performance measure
ψ^u is the upper bound (or design requirement) of the product performance
X is a vector of random variables.

When product performance does not meet the requirement—that is, when $\psi^u \leq \psi(X)$, the event fails. Therefore, the probability of failure P_f of the particular event $g(X) \leq 0$ is

$$P_f = P[g(X) \leq 0] \tag{1.2}$$

where $P[\bullet]$ is the probability of event $\bullet$.

Given the joint probability density function $f_X(x)$ of the random variables X, the probability of failure for a single event of a mechanical component can be expressed as

$$P_f = P[g(X) \leq 0] = \int \int_{g(X) \leq 0} \cdots \int f_X(x) dx. \tag{1.3}$$

The probability of failure in Eq. 1.3 is commonly evaluated using the Monte Carlo method or the first- or second-order reliability method (FORM or SORM) (Wu and Wirsching 1984, Yu et al. 1998).

Once the probabilities of several failure events in subsystems or components are computed, system reliability can be obtained by, for example, fault-tree analysis (Ertas and Jones 1993). No general-purpose software tool for reliability analysis of general mechanical systems is commercially available yet. Numerical evaluation of stochastic structures under stress (NESSUS®) (www.nessus.swri. org), which is currently in development can be a good candidate for incorporation into the e-Design environment. With the probability of failure, critical quality design criteria, such as mean time between failure (MTBF), can be computed (Ertas and Jones 1993).

Two main challenges exist in reliability analysis: One, realistic distribution data are difficult to acquire and often are not available in the early stage; and two, failure probability computations are often expensive. The first challenge may be alleviated by employing legacy data from previous product lines. Approximation techniques (e.g., Yu et al. 1998) can be employed to make the computation affordable even for an individual failure event within a mechanical component.

1.3.3 Product virtual manufacturing

Virtual manufacturing addresses issues of design for manufacturability (DFM) (Prasad 1996) and design for manufacturing and assembly (DFMA) (Boothroyd et al. 1994) early in product development. In the e-Design paradigm, DFM and DFMA are performed by conducting virtual manufacturing and assembly using, for example, Pro/MFG. DFM and DFMA of the product are verified through animations of the virtual manufacturing and assembly process.

Pro/MFG is a Pro/ENGINEER module supporting the virtual machining process, including milling, drilling, and turning. By incorporating part design and also defining workpieces, workcells, fixtures, cutting tools, and cutting parameters, Pro/MFG automatically generates a toolpath (see Figure 1.9a), which simulates the machining process (Figure 1.9b), calculates machining time, and produces cutter location (CL) data. The CL data can be post-processed for CNC codes. In addition, casting, sheet metal, molding, and welding can be simulated using Pro/CASTING, Pro/SHEETMETAL, Pro/MOLD, and Pro/WELDING, respectively.

With such virtual manufacturing process planning and animation, manufacturability of the product design can, to some extent, be verified. The DFMA tool (Boothroyd et al. 1994) developed by Boothroyd Dewhurst, Inc., assists the cross-functional team in quantifying product assembly time and labor costs. It also challenges the team to simplify product structure, thereby reducing product as well as assembly costs.

One of the limitations in using virtual manufacturing tools (e.g., Pro/MFG) is that chip formation (Fang and Jawahir 1996), a primary consideration in computer numerical control (CNC), is not incorporated into the simulation. In addition, machining parameters, such as power consumption, machining temperature, and tool life, which contribute to manufacturing costs are not yet simulated.

1.3.4 Tool integration

Techniques developed to support tool integration (Chang et al. 1998) include parameterized product data models, engineering views, tool wrappers, and design process management. Parameterized product data models represent engineering data that are needed for conducting virtual prototyping of

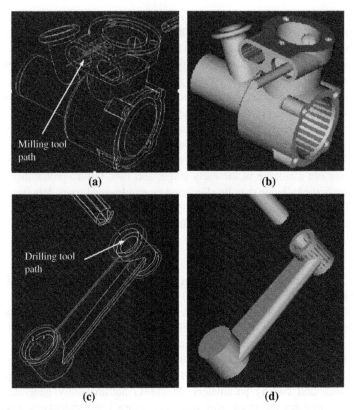

FIGURE 1.9 Virtual Machining Process: (a) Engine Case—Milling Toolpath; (b) Milling Simulation; (c) Connecting Rod—Drilling Toolpath; (d) Drilling Simulation.

the mechanical system. The main sources of the product data model are CAD and non-CAD models. The product data model evolves throughout the product development cycle as illustrated in Figure 1.10.

Engineering views allow engineers from various disciplines to view the product from their own technical perspectives. Through engineering views, engineers create simulation models that are consistent with the product model by simplifying the CAD representation, as needed adding non-CAD product representation and mapping. Tool wrappers provide two-way data translation and transmission between engineering tools and the product data model. Design process management provides the team leader with a tool to monitor and manage the design process. When a new tool of an existing discipline, for example ANSYS for structural FEA, is to be integrated, a wrapper for it must be developed. Three main tasks must be carried out when a new engineering discipline, say computational fluid dynamics (CFD), is added to the environment. First, the product data model must be extended to include engineering data needed to support CFD. Second, engineering views must be added to allow design engineers to generate CFD models. Finally, wrappers must be developed for specific CFD tools.

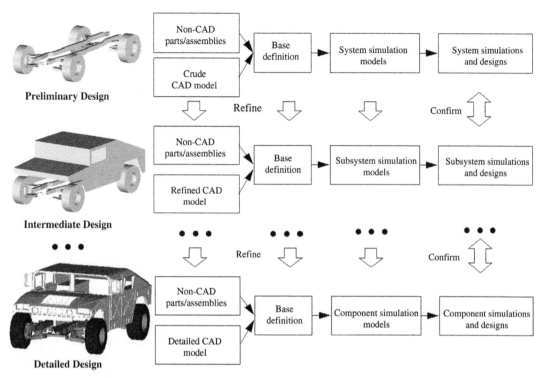

FIGURE 1.10 Hierarchical Product Models Evolved Through the e-Design Process.

1.3.5 Design decision making

Product performance, reliability, and manufacturing cost that are evaluated using simulations can be brought to the cross-functional team for review. Product performance and reliability are checked against product specifications that have been defined and have evolved from the beginning of the product development process. Manufacturing cost derived from the virtual manufacturing simulations can be added to product cost. The cross-functional team must address areas of concern identified in product performance, reliability, and manufacturability, and it must identify a set of design variables that influence these areas. Design modifications can then be conducted. In the past, quality functional deployment (QFD) (Ertas and Jones 1993) was largely employed in design modification to assign qualitative weighting factors to product performance and design changes. e-Design employs a systematic and quantitative approach to design modifications (for example, Yu et al. 1997).

1.3.5.1 Design problem formulation

Before a design can be improved, design problems must be defined. A design problem is often presented in a mathematical form, typically as

$$\text{Minimize } \varphi(\boldsymbol{b}) \tag{1.4a}$$

Subject to

$$\psi_i(\boldsymbol{b}) \le \psi_i^u \quad i = 1, \, m \tag{1.4b}$$

$$P_{f_j}(\boldsymbol{b}) \le P_{f_j}^u \quad j = 1, \, n \tag{1.4c}$$

$$b_k^l \le b_k \le b_k^u \quad k = 1, \, p \tag{1.4d}$$

where

$\varphi(\boldsymbol{b})$ is the objective (or cost) function to be minimized
$\psi_i(\boldsymbol{b})$ is the ith constraint function that must be no greater than its upper bound ψ_i^u
$P_{f_j}(\boldsymbol{b})$ is the jth failure probability index that must be no greater than its upper bound $P_{f_j}^u$
$\boldsymbol{b}$ is the vector of design variables
b_k^l and b_k^u are the lower and upper bounds of the design variable b_k, respectively.

Note that in e-Design design variables are usually associated with dimensions of geometric features and part material properties in the parameterized CAD models. The feature-based design parameters serve as the common language to support the cross-functional team while conducting parametric study and design trade-offs.

1.3.5.2 Design sensitivity analysis
Before quantitative design decisions can be made, there must be a design sensitivity analysis (DSA) that computes derivatives of performance measures, including product performance, failure proba-bility, and manufacturing cost, with respect to design variables. Dependence of performance measures on design variables is usually implicit. How to express product performance in terms of design var-iables in a mathematical form is not straightforward. Analytical DSA methods combined with nu-merical computations have been developed mainly for structural responses (Haug et al. 1986) and fatigue and fracture (Chang et al. 1997). DSA for failure probability with respect to both deterministic and random variables has also been developed (Yu et al. 1997). In addition, DSA and optimization using meshless methods have been developed for large-deformation problems (Grindeanu et al. 1999). More details about the analytical DSA for structural responses also referred to Haug et al. (1985).

For problems such as motion and manufacturing cost, where premature or no analytical DSA capability is available, the finite difference method is the only choice. The finite difference method is expressed in the following equation:

$$\frac{\partial \psi}{\partial b_j} \approx \frac{\psi(\boldsymbol{b} + \Delta b_j) - \psi(\boldsymbol{b})}{\Delta b_j} \tag{1.5}$$

where Δb_j is a perturbation in the jth design variable. With sensitivity information, parametric study and design trade-offs can be conducted for design improvements at the concept and detail stages, respectively.

1.3.5.3 Parametric study
A parametric study that perturbs design variables in the product design model to explore various design alternatives can effectively support product concept designs. The parametric study is simple and easy to perform as long as the mapping between CAD and simulation models has been established. The mapping

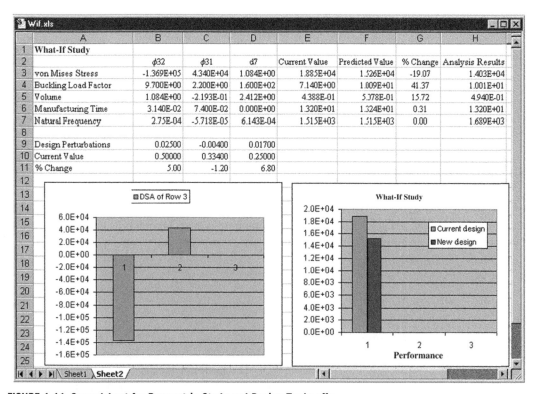

	A	B	C	D	E	F	G	H
1	What-If Study							
2		φ32	φ31	d7	Current Value	Predicted Value	% Change	Analysis Results
3	von Mises Stress	-1.369E+05	4.340E+04	1.084E+00	1.885E+04	1.526E+04	-19.07	1.403E+04
4	Buckling Load Factor	9.700E+00	2.200E+00	1.600E+02	7.140E+00	1.009E+01	41.37	1.001E+01
5	Volume	1.084E+00	-2.193E-01	2.412E+00	4.388E-01	5.378E-01	15.72	4.940E-01
6	Manufacturing Time	3.140E-02	7.400E-02	0.000E+00	1.320E+01	1.324E+01	0.31	1.320E+01
7	Natural Frequency	2.75E-04	-5.718E-05	6.143E-04	1.515E+03	1.515E+03	0.00	1.689E+03
8								
9	Design Perturbations	0.02500	-0.00400	0.01700				
10	Current Value	0.50000	0.33400	0.25000				
11	% Change	5.00	-1.20	6.80				

FIGURE 1.11 Spreadsheet for Parametric Study and Design Trade-offs.

supports fast simulation model generation for performance analyses. It also supports DSA using the finite difference method. The parametric study is possible for concept design because the number of design variables to perturb is usually small. A spreadsheet with a proper formula defined among cells is well suited to support the parametric study. The use of Microsoft Excel is illustrated in Figure 1.11.

1.3.5.4 Design trade-off analysis

With design trade-off analysis, the design engineer can find the most appropriate design search direction for the design problem formulated in Eq. 1.4, using four possible algorithms:

- Reduce cost.
- Correct constraint neglecting cost.
- Correct constraint with a constant cost.
- Correct constraint with a cost increment.

As a general rule, the first algorithm, reduce cost, can be chosen when the design is feasible; in other words, all constraint functions are within the desired limits. When the design is infeasible, generally one may start with the third algorithm, correct constraint with a constant cost. If the design remains infeasible, the fourth algorithm, correct constraint with a cost increment—say 10%—may

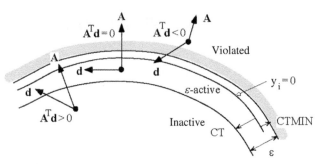

FIGURE 1.12 ε**-Active Constraint Strategy.**

be appropriate. If a feasible design is still not found, the second algorithm, correct constraint neglecting cost, can be selected. A quadratic programming (QP) subproblem can be formulated to numerically find the search direction that corresponds to the algorithm selected.

An ε-active constraint strategy (Arora 1989), shown in Figure 1.12, can be employed to support design trade-offs. The constraint functions in Eq. 1.4 are normalized by

$$y_i = \frac{\psi_i}{\psi_i^u} - 1 \leq 0, \quad i = 1, m \tag{1.6}$$

When y_i is between *CT* (usually 0.03) and *CTMIN* (usually 0.005), it is active—that is, $\varepsilon = |CT| + CTMIN$, as illustrated in Figure 1.12. When y_i is less than *CT*, the constraint function is inactive or feasible. When y_i is larger than *CTMIN*, the constraint function is violated. A QP subproblem can be formulated to find the search direction numerically corresponding to the option selected. For example, the QP subproblem for the first algorithm (cost reduction) can be formulated as

$$\begin{aligned} \text{Minimize} \quad & \mathbf{c}^T\mathbf{d} + 0.5\,\mathbf{d}^T\mathbf{d} \\ \text{Subject} \quad & \mathbf{A}^T\mathbf{d} \leq \mathbf{y} \\ & \mathbf{b}^L - \mathbf{b}^{(k)} \leq \mathbf{d} \leq \mathbf{b}^U - \mathbf{b}^{(k)} \end{aligned} \tag{1.7}$$

where

$$\mathbf{c} = [c_1, c_2, \ldots, c_{n1+n2}]^T, \quad c_i = \partial\varphi/\partial b_i$$

d is the search direction to be determined.

$$A_{ij} = \partial P_{y_i}/\partial b_j; \quad \mathbf{b} = [b_1, b_2, \ldots b_n]^T$$

k is the current design iteration.

The objective of the design trade-off algorithm is to find the optimal search direction *d* under a given circumstance. Details are discussed in *Design Theory and Methods using CAD/CAE*, a book in The Computer Aided Engineering Design Series.

1.3.5.5 What-if study

After the search direction *d* is found, a number of step sizes α can be used to perturb the design along the direction *d*. Objective and constraint function values, represented as ψ_i, at a perturbed design

$b + \alpha d$ can be approximated using the first-order sensitivity information of the functions by Taylor series expansion about the current design b without going through simulations; that is,

$$\psi_i(b + \alpha d) \approx \psi_i(b) + \frac{\partial \psi_i}{\partial b} \alpha d. \tag{1.8}$$

Note that since there is no analysis involved, the what-if study can be carried out very efficiently. This allows the design engineer to explore design alternatives more effectively.

Once a satisfactory design is identified, after trying out different step sizes α in an approximation sense, the design model can be updated to the new design and then simulations of the new design can be conducted. Equation 1.8 also supports parametric study, in which the design perturbation δb is determined by engineers based on sensitivity information. To ensure a reasonably accurate function prediction using Eq. 1.8, the step sizes must be small so that the perturbation $\partial \psi_i / (\partial b)(\alpha d)$ is, as a rule of thumb, less than 10% of the function value $\psi_i(b)$.

1.4 **Physical prototyping**

In general, two techniques are suitable for fabricating physical prototypes of the product in the design process: rapid prototyping (RP) and computer numerical control (CNC) machining. RP systems, based on solid freeform fabrication (SFF) technology (Jacobs 1994), fabricate physical prototypes of the structure for design verification. The CNC machining fabricates functional parts as well as the mold or die for mass production of the product.

1.4.1 **Rapid prototyping**

The Solid Freeform Fabrication (SFF) technology, also called Rapid Prototyping (RP), is an additive process that employs a layer-building technique based on horizontal cross-sectional data from a 3D CAD model. Beginning with the bottom-most cross-section of the CAD model, the rapid prototyping machine creates a thin layer of material by slicing the model into so-called 2½ D layers. The system then creates an additional layer on top of the first based on the next higher cross-section. The process repeats until the part is completely built. It is illustrated using an engine case in the example shown in Figure 1.13. Rapid prototyping systems are capable of creating parts with small internal cavities and complex geometry.

Most important, SFF follows the same layering process for any given 3D CAD models, so it requires neither tooling nor manufacturing process planning for prototyping, as required by conventional manufacturing methods. Based on CAD solid models, the SFF technique fabricates physical prototypes of the product in a short turnaround time for design verification. It also supports tooling for product manufacturing, such as mold or die fabrications, through, for example, investment casting (Kalpakjian 1992).

Note that there are various types of SFF systems commercially available, such as the SLA® 7000 and Sinterstation® by 3D Systems (Figures 1.14a and 1.14b). In this chapter, the Dimension 1200 sst® machine (www.stratasys.com), as shown in Figure 1.14c, is presented. More details about it as well as other RP systems will be discussed in *Product Manufacturing and Cost Estimating using CAD/CAE*, a book in The Computer Aided Engineering Design Series.

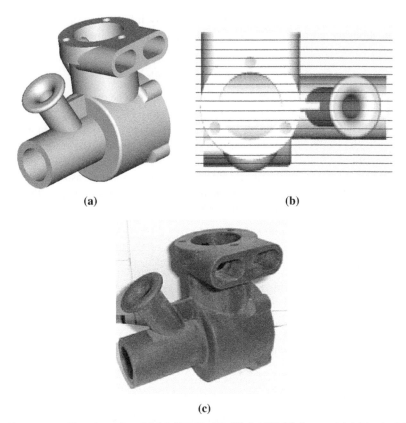

FIGURE 1.13 SFF: Layered Manufacturing: (a) 3D CAD Model, (b) 2-1/2D Slicing, and (c) Physical Model.

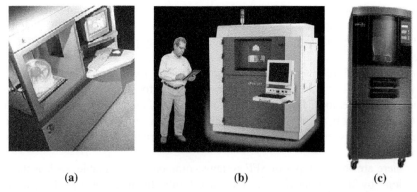

FIGURE 1.14 Commercial RP systems: (a) 3D Systems' SLA 7000, (b) SinterStation 2500, (c) Stratasys Inc.'s Dimension 1200 sst.

Sources: (b) 3D Systems Corporation, USA; (c) Stratasys Ltd.

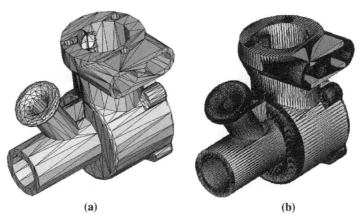

(a)   (b)

FIGURE 1.15 STL Engine Case Models: (a) Coarse and (b) Refined.

The CAD solid model of the product is first converted into a stereolithographic (STL) format (Chua and Leong 1998), which is a faceted boundary representation uniformly accepted by the industry. Both the coarse and refined STL models of an engine case are shown in Figure 1.15. Even though the STL model is an approximation of the true CAD geometry, increasing the number of triangles can minimize the geometric error effectively. This can be achieved by specifying a smaller chord length, which is defined as the maximum distance between the true geometric boundary and the neighboring edge of the triangle. The faceted representation is then sliced into a series of 2D sections along a prespecified direction. The slicing software is SFF-system dependent.

The Dimension 1200 sst employs fused deposition manufacturing (FDM) technology. Acrylonitrile butadiene styrene (ABS) materials are softened (by elevating temperature), squeezed through a nozzle on the print heads, and laid on the substrate as build and support materials, respectively, following the 2D contours sliced from the 3D solid model (Figure 1.16). Note that various crosshatch options are

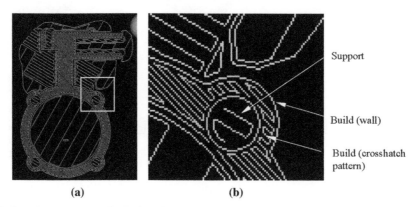

Support

Build (wall)

Build (crosshatch pattern)

(a)   (b)

FIGURE 1.16 Crosshatch Pattern of a Typical Cut-out Layer: (a) Overall and (b) Enlarged.

available in CatalystEX® software (www.dimensionprinting.com), which comes with the rapid prototyping system.

The physical prototypes are mainly for the cross-functional team to verify the product design and check the assembly. However, they can also be used for discussion with marketing personnel to develop marketing ideas. In addition, the prototypes can be given to potential customers for feedback, thus bringing customers into the design loop early in product development.

1.4.2 CNC machining

The machining operations of virtual manufacturing, such as milling, turning, and drilling, allow designers to plan the machining process, generate the machining toolpath, visualize and simulate machining operations, and estimate machining time. Moreover, the toolpath generated can be converted into CNC codes (M-codes and G-codes) (Chang et al. 1998, McMahon and Browne 1998) to fabricate functional parts as well as a die or mold for production.

For example, the cover die of a mechanical part is machined from an 8 in. × 5.25 in. × 2 in. steel block, as shown in Figure 1.17a. The cutter location data files generated from virtual machining are post-processed into machine control data (MCD)—that is, G- and M-codes, for CNC machining, using post-processor UNCX01.P11 in Pro/MFG. In addition to volume milling and contour surface milling, drilling operations are conducted to create the waterlines. A 3-axis CNC mill, HAAS VF-series (HAAS Automation, Inc. 1996), is employed for fabricating the die for casting the mechanical part (Figure 1.17b).

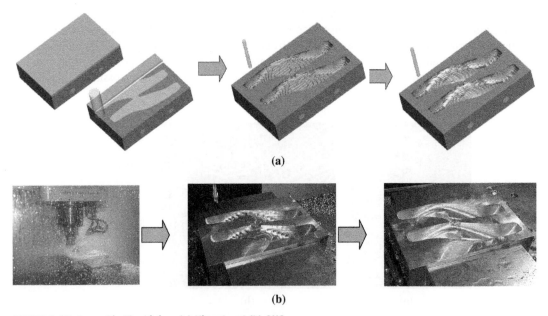

(a)

(b)

FIGURE 1.17 Cover Die Machining: (a) Virtual and (b) CNC.

1.5 Example: simple airplane engine

A single-piston, two-stroke, spark-ignition airplane engine (shown in Figure 1.5) is employed to illustrate the e-Design paradigm and tool environment. The cross-functional team is asked to develop a new model of the engine with a 30% increment in both maximum torque and horsepower at 1,215 rpm. The design of the new engine will be carried out at two interrelated levels: system and component. At the system level, the performance measure is the power output; at the component level, the structural integrity and manufacturing cost of each component are analyzed for improvement. Note that only a very brief discussion is provided in this introductory chapter. The computation and modeling details are discussed in later chapters and *Product Design Modeling using CAD/CAE*, a book in The Computer Aided Engineering Design Series.

1.5.1 System-level design

Power is proportional to the rotational speed of the crankshaft (N), the swept volume (V_s), and the brake mean effective pressure (P_b) (Taylor 1985):

$$W_b = P_b \, V_s \, N. \tag{1.9}$$

The effective pressure P_b applied on top of the piston depends on, among other factors, the swept volume and the rotational speed of the crankshaft. The pressure is limited by the integrity of the engine structure.

Design variables at the system level include bore diameter (d46:0) and stroke, defined as the distance between the top face of the piston at the bottom and top dead-center positions. In the CAD model, the stroke is defined as the sum of the crank offset length (d6:6) and the connecting rod length (d0:10), as shown in Figure 1.18. To achieve the requirement for system performance, these three

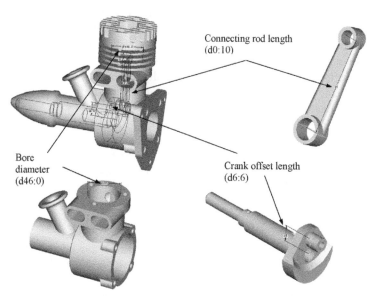

FIGURE 1.18 Engine Assembly with Design Variables at the System Level.

Table 1.1 Changes in Design Variables at the System Level

Design Variable	Current Value (in.)	New Value (in.)	Change (in.)	% Change
Bore diameter (d46:0)	1.416	1.6	0.164	11.6
Crank length (d6:6)	0.5833	0.72	0.1567	26.9
Connecting rod length (d0:10)	2.25	2.49	0.24	10.7

design variables are modified as listed in Table 1.1. The design variable values were calculated following theory and practice for internal combustion engines (Taylor 1985). Details of the computation can be found in Silva (2000).

The solid models of the entire engine are automatically updated and properly assembled using the parametric relations established earlier (refer to Figure 1.6b). The change causes P_b to increase from 140 to 180 lbs, so the peak load increases from 400 to 600 lbs. The load magnitude and path applied to the major load-carrying components, such as the connecting rod and crankshaft, are therefore altered. Results from motion analysis show that the system performs well kinematically. Reaction forces applied to the major load-carrying components are computed—for example, for the connecting rod shown in Figure 1.19. The change also affects manufacturing time for some components.

1.5.2 Component-level design

Structural performance is evaluated and redesigned to meet the requirements. In addition, virtual manufacturing is conducted for components with significant design changes. Build materials (volume) and manufacturing times constitute a significant portion of the product cost. In this section, the design of the connecting rod is presented to demonstrate the design decision-making method discussed.

Because of the increased load transmitted through the piston and the increased stroke length, the connecting rod can experience buckling failure during combustion. In addition, because changes in stroke length, stiffness, and mass vary, the natural frequency of the rod may be different. Moreover, load is repeatedly applied to the connecting rod, potentially leading to fatigue failure. Structural FEA are conducted to evaluate performance. In addition, virtual manufacturing is carried out to determine the machining cost of the rod.

Because of the increment of the connecting rod length (d0:10) and the magnitude of the external load applied (see Figure 1.20), the rod's maximum von Mises stress increases from 13,600 to 18,850 psi and the buckling load factor decreases from 33 to 7. The first natural frequency is 1,515 Hz. The machining time estimated for the connecting rod is 13.2 minutes using hole-drilling and face-milling operations (shown earlier in Figure 1.9d).

1.5.3 Design trade-off

The design trade-off method discussed in Section 1.3.5 is applied to the components, with significant changes resulting from the system-level design. Only the design trade-off conducted for the connecting rod is discussed.

Performance measures for the connecting rod, including buckling load factor, fatigue life, natural frequency, volume, and machining costs (time), are brought together for design trade-off. Three design

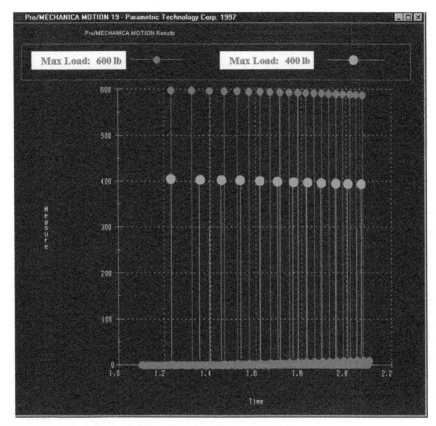

FIGURE 1.19 Dynamic Load Applied to the Connecting Rod.

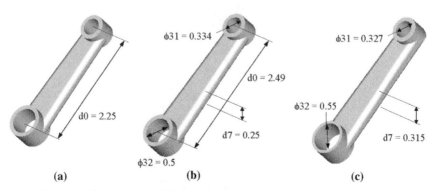

FIGURE 1.20 Engine Connecting Rod: (a) Original Design; (b) Changes at the System Level; (c) Changes at the Component Level.

variables, $\phi32$, $\phi31$, and d7, are identified, as shown in Figure 1.20b. The objective is to minimize volume and manufacturing time subject to maximum allowable von Mises stress, operating frequency, and minimum allowable buckling load factor. The engine is designed to work at 21 kHz, and the minimum allowable buckling load factor for the connecting rod is assumed to be 10.

Sensitivity coefficients for performance and cost measures with respect to design variables are calculated (refer to Figure 1.11) using the finite difference method. Design trade-offs are conducted followed by a what-if study. When a satisfactory design is found, the solid model of the rod is updated for performance evaluation and virtual manufacturing. This process is repeated twice when all the requirements are met. The design change is summarized in Tables 1.2 and 1.3, which show that the machining time is maintained and a small volume increment is needed to achieve the required performance.

Table 1.2 Changes in Design Variables at the Component Level

Design Variable	Current Value (in.)	New Value (in.)	% Change
Diameter of the large hole ($\phi32$)	0.50	0.55	10
Diameter of the small hole ($\phi31$)	0.334	0.32728	−2.01
Thickness (d7)	0.25	0.31484	25.9

Table 1.3 Changes in Performance Measures at the Component Level

Performance Measure	Current Value	New Value	% Change
VM stress	18.9 ksi	10.5 ksi	−44.4
Buckling load factor	7.1	14.2	100
Volume	0.438813 in.^{3}	0.5488 in.^{3}	25.1
Machining time	13.2 min	13.2 min	0
Natural frequency	1515 Hz	1840 Hz	21.5

FIGURE 1.21 Physical Prototypes of Engine Parts.

1.5.4 **Rapid prototyping**

When the design is finalized through virtual prototyping, rapid prototyping is used to fabricate a physical prototype of the engine, as shown in Figure 1.21. The prototype can be used for design verification as well as tolerance and assembly checking.

1.6 **Example: High-mobility multipurpose wheeled vehicle**

The overall objective of the high-mobility multipurpose wheeled vehicle (HMMWV) design is to ensure that the vehicle's suspension is durable and reliable after accommodating an additional armor loading of 2,900 lb. A design scenario using a hierarchical product model (see Figure 1.10) that evolves during the design process is presented in this section.

In the preliminary design stage, vehicle motion is simulated and design changes are performed to improve the vehicle's gross motion. At this stage, the dynamic behavior of the HMMWV's suspension is simulated and designed. The specific objectives of the preliminary design are to avoid the problem of metal-to-metal contact in the shock absorber due to added armor load, and to improve the driver's comfort by reducing vertical acceleration at the HMMWV driver's seat.

By modifying the spring constant to improve the HMMWV suspension design at the preliminary design stage, the load path generated in HMMWV dynamics simulation is affected in the suspension unit. In the detail design stage, the objective is to assess and redesign the durability, reliability, and structural performance of selected suspension components affected by the added armor load that result in changes in load path and load magnitude.

Note that only a very brief discussion is provided in this introductory chapter. The computation and modeling details are discussed in later chapters.

1.6.1 **Hierarchical product model**

In this particular case, a hierarchical product model is employed to support the HMMWV's design. In all models, nonsuspension parts, such as instrument panel, seats, and lights, are not modeled. Important vehicle components, such as engine and transmission, are modeled using engineering parameters without depending on CAD representation. A low-fidelity CAD model consisting of 18 parts (Figure 1.22) is created using Pro/ENGINEER to support the preliminary design. This model has accurate joint definition and fairly accurate mass property, but less accurate geometry. The goal of the

FIGURE 1.22 HMMWV CAD Model for Preliminary Design.

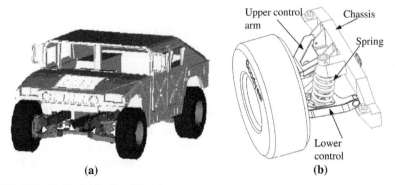

(a)   (b)

FIGURE 1.23 HMMWV CAD Model for Detail Design.

low-fidelity model is to support vehicle dynamic simulation. It is created using substantially less effort compared to that required for the detailed model.

The detailed product model, consisting of more than 200 parts and assemblies (Figure 1.23), is created to support the detail design of suspension components. The detailed model is derived from the preliminary model by (1) breaking an entity into more parts and assemblies (e.g., the gear hub assembly, shown in Figure 1.24) to simulate and design detailed parts, and (2) refining the geometry of mechanical components to support structural FEA (e.g., the lower control arm, shown in Figure 1.25).

1.6.2 Preliminary design

The HMMWV is driven repeatedly on a virtual proving ground, as shown in Figure 1.26, with a constant speed of 20 MPH for a period of 23 seconds. A dynamic simulation model, shown in Figure 1.27, is first derived from the low-fidelity CAD solid model of the HMMWV (refer to Figure 1.22). A more in-depth discussion of the HMMWV vehicle dynamic model is provided in Chapter 3.

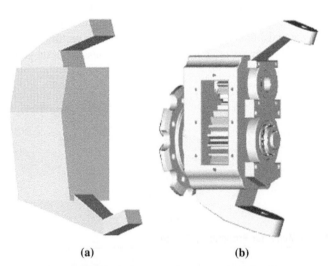

(a)   (b)

FIGURE 1.24 HMMWV gear hub Assembly Models: (a) Preliminary and (b) Detailed.

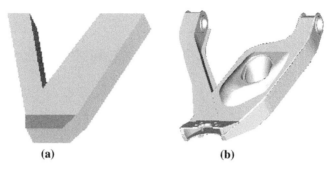

(a)   (b)

FIGURE 1.25 HMMWV Lower Control Arm Models: (a) Preliminary and (b) Detailed.

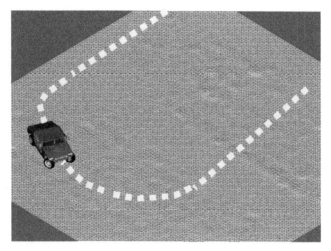

FIGURE 1.26 HMMWV Dynamic Simulation.

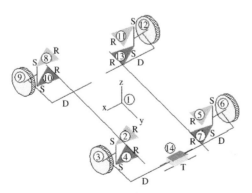

Body
1 Chassis
2 Right front upper control arm
3 Right front wheel spindle
4 Right front lower control arm
5 Left front upper control arm
6 Left front wheel spindle
7 Left front lower control arm
8 Right rear upper control arm
9 Right rear wheel spindle
10 Right rear lower control arm
11 Left rear upper control arm
12 Left rear wheel spindle
13 Left rear lower control arm
14 Rack

Joint types

R: Revolute joint
T: Translational joint
S: Spherical joint
D: Distance constraint

FIGURE 1.27 HMMWV Dynamic Model.

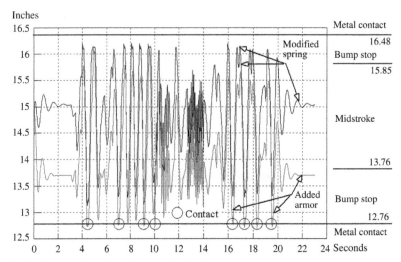

FIGURE 1.28 Shock Absorber Operation Distance (in inches).

Using DADS, severe metal-to-metal contact is identified within the shock absorber, caused by the added armor load and rough driving conditions, as shown in Figure 1.28. The spring constant is adjusted to avoid any contact problems; it is increased in proportion to the mass increment of the added armor to maintain the vehicle's natural frequency. This design change not only eliminates the contact problem (see Figure 1.28) but also reduces the amplitude of vertical acceleration at the driver's seat, which improves driving comfort (see Figure 1.29). However, the change alters the load path in the components of the suspension subsystem—for example, the shock absorber force acting on the control arm increases about 75%, as shown in Figure 1.30.

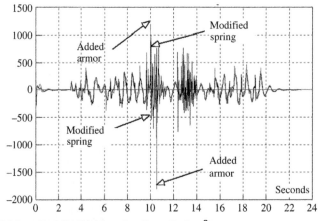

FIGURE 1.29 HMMWV Driver Seat Vertical Accelerations (in./sec²).

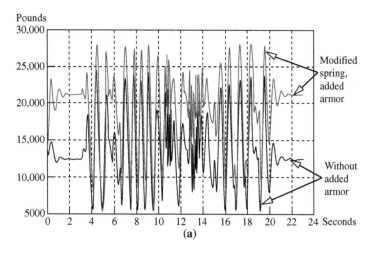

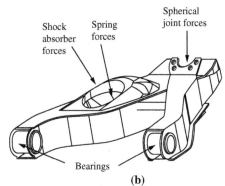

FIGURE 1.30 History of Shock Absorber Forces (lbs): (a) Force History with and without Added Armor Load, (b) Locations of Force Application.

1.6.3 Detail design

Simulations are carried out for fatigue, vibration, and buckling of the lower control arm (Figure 1.30); reliability of gears in the gear hub assembly (refer to Figure 1.24b); the spring of the shock absorber (see Figure 1.23); and the bearings of the control arm (see Figure 1.30).

Using ANSYS, the first natural frequency of the lower control arm is obtained as 64 Hz, which is far away from vehicle vibration frequency, eliminating concern about resonance. The buckling load factor is analyzed using the peak load at time 10.05 seconds in the 23-second simulation period. The result shows that the control arm will not buckle even under the most severe load. Therefore, the current design is acceptable as far as buckling and resonance of the lower control arm are concerned.

Results obtained from fatigue analyses show that fatigue life (crack initiation) of the lower control arm degrades significantly—for example, from 6.61E+09 to 1.79E+07 blocks (one

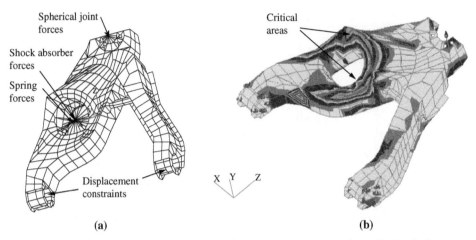

Spherical joint forces

Shock absorber forces

Spring forces

Displacement constraints

Critical areas

X Y Z

(a)   (b)

FIGURE 1.31 HMMWV Lower Control Arm Models: (a) Finite Element and (b) Fatigue Life Prediction.

block is 20 seconds) at critical areas (see Figure 1.31b)—because of the additional armor load and change of load path. Therefore, the design must be altered to improve control arm durability. Reliability of the bearing, gear, and spring at a 99% fatigue failure rate is 2.18E+07, 3.36E+06, and 1.27E+02 blocks, respectively. Note that the fatigue life of the spring at the required reliability is not desirable.

1.6.4 Design trade-off

Eleven design parameters, including geometric dimensions (d1 and d2 in Figure 1.32a), material property (cyclic strength coefficient K' of the lower control arm), and thickness of the control arm sheet metal (t1 to t7 in Figure 1.32b) are defined to support design modification.

A global design trade-off that involves changes in more than one component is conducted first. Geometric design parameters d1 and d2 are modified to reduce loads applied to the control arm, bearing, spring, and gears in the gear hub so that the durability and reliability of these components can be improved. Changes in d1 and d2 affect not only the lower control arm but also the upper control arm and the chassis frame. Sensitivity coefficients of loads at discretized time steps (a total of 10 selected time steps) with respect to parameters d1 and d2 are calculated using a finite difference method. Sensitivity coefficients can be displayed in bar charts (see Figure 1.33a) to guide design modifications. A what-if study is carried out with a design perturbation of 0.6 and 0.3 in. for d1 and d2, respectively, to obtain a reduction in loads. An example of the what-if results is shown in Figure 1.33b.

A local design trade-off that involves design parameters of a single component is carried out for the lower control arm. Thickness design parameters t1 to t7 and the material design parameter K' are modified to increase the control arm's fatigue life. Fatigue life at ten nodes of its finite element model in the critical area is measured. Sensitivity coefficients of control arm fatigue

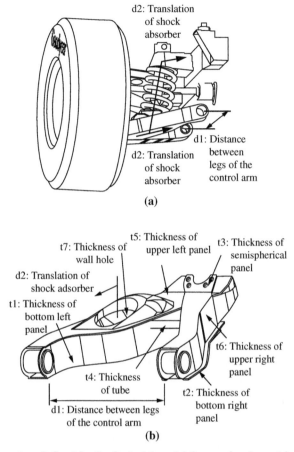

FIGURE 1.32 Design Parameters Defined for the Control Arm: (a) Suspension Geometric Dimensions and (b) Thickness Dimensions.

life at these nodes with respect to the thickness and material parameters are calculated. A design trade-off method using a QP algorithm is employed because of the large number of design parameters and performance measures involved. An improved design obtained shows that with a 0.6% weight increment, fatigue life at the critical area increases about ten times: from 1.79 E+07 to 1.68 E+08 blocks.

A dynamic simulation is performed again with the detailed model and modified design to ensure that the metal contact problem, encountered in the preliminary design stage, is eliminated as a result of model refinement and design changes in the detail design stage. The global design trade-off reduces the load applied to the shock absorber spring. This reduction significantly increases the spring fatigue life to the desired level.

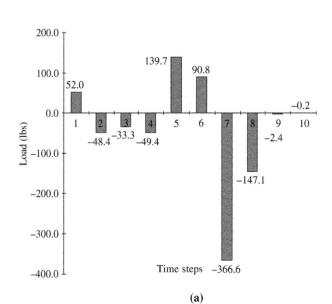

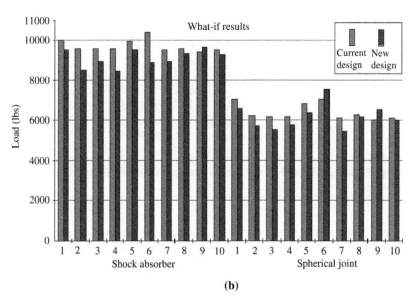

FIGURE 1.33 Sensitivity of Load on the Spherical Joint of Control Arm w.r.t d2 at 10 Time Steps (a) Design Sensitivity Display and (b) What-if Study.

1.7 Summary

In this chapter, the e-Design paradigm and software tool environment were discussed. The e-Design paradigm employs virtual prototyping for product design and rapid prototyping and computer numerical control (CNC) for fabricating physical prototypes of a design for design verification and functional tests. The e-Design paradigm offers three unique features:

- The VP technique, which simulates product performance, reliability, and manufacturing costs; and brings these measures to design.
- A systematic and quantitative method for design decision making for the parameterized product in solid model forms.
- RP and CNC for fabricating prototypes of the design that verify product design and bring marketing personnel and potential customers into the design loop.

The e-Design approach holds potential for shortening the overall product development cycle, improving product quality, and reducing product costs. With intensive knowledge of the product gained from simulations, better design decisions can be made, thereby overcoming what is known as the *design paradox*. With the advancement of computer simulations, more hardware tests can be replaced by them, reducing cost and shortening product development time. Manufacturing-related issues can be largely addressed through virtual manufacturing in early design stages. Moreover, manufacturing process planning conducted in virtual manufacturing streamlines the production process.

Questions and exercises

1.1. In this assignment, you are asked to search and review articles (such as in *Mechanical Engineering* magazine) that document successful stories in industry that involve employing the e-Design paradigm and/or employing CAD/CAE/CAM technology for product design.
  - Briefly summarize the company's history and its main products.
  - Briefly summarize the approach and process that the company adopted for product development in the past.
  - Why must the company make changes? List a few factors.
  - Which approach and process does the company currently employ?
  - What is the impact of the changes to the company?
  - In which journal, magazine, or website was the article published?
1.2. In this chapter we briefly discussed rapid prototyping technology and the Dimension 1200 sst machine. The sst uses fused deposition manufacturing technology for support of layer manufacturing. Search and review articles to understand the FDM technology and machines that employ such technology other than the Dimension series.

References

Anderson, D.M., 1990. Design for Manufacturability: Optimizing Cost, Quality, and Time to Market. CIM Press.
Arora, J.S., 1989. Introduction to Optimal Design. McGraw-Hill.

Boothroyd, G., Dewhurst, P., Knight, W., 1994. Product Design for Manufacturing and Assembly. Marcel Dekker.

Chang, K.H., Choi, K.K., Wang, J., Tsai, C.S., Hardee, E., 1998. A multi-level product model for simulation-based design of mechanical systems. Concurrent Engineering Research and Application (CERA) Journal 6 (2), 131–144.

Chang, K.H., Yu, X., Choi, K.K., 1997. Shape design sensitivity analysis and optimization for structural durability. International Journal of Numerical Methods in Engineering 40, 1719–1743.

Chang, T.C., Wysk, R.A., Wang, H.P., 1998. Computer-Aided Manufacturing, second ed. Prentice Hall.

Chen, J.S., Pan, C., Wu, T.C., 1997. Large deformation analysis of rubber based on a reproducing kernel particle method. Computational Mechanics 19, 153–168.

Chua, C.K., Leong, K.F., 1998. Rapid Prototyping: Principles and Applications in Manufacturing. John Wiley.

Ertas, A., Jones, J.C., 1993. The Engineering Design Process. John Wiley.

Fang, X.D., Jawahir, I.S., 1996. A hybrid algorithm for predicting chip form/chip breakability in machining. International Journal of Machine Tools and Manufacture 36 (10), 1093–1107.

Gonzalez, A.J., Dankel, D.D., 1993. The Engineering of Knowledge-Based Systems, Theory and Practice. Prentice Hall.

Grindeanu, I., Choi, K.K., Chen, J.S., Chang, K.H., 1999. Design sensitivity analysis and optimization of hyperelastic structures using a meshless method. AIAA Journal 37 (8), 990–997.

HAAS Automation Inc., 1996. VF Series Operations Manual.

Hallquist, J.O., 2006. LS-DYNA3D Theory Manual. Livermore Software Technology Corp.

Haug, E.J., Choi, K.K., Komkov, V., 1986. Design Sensitivity Analysis of Structural Systems. Academic Press.

Haug, E.J., 1989. Computer-Aided Kinematics and Dynamics of Mechanical Systems, vol. I: Basic Methods. Allyn and Bacon.

Jacobs, P.F., 1994. StereoLithography and Other RP&M Technologies. ASME Press.

Kalpakjian, S., 1992. Manufacturing Engineering and Technology, second ed. Addison-Wesley.

Kane, T.R., Levinson, D.A., 1985. Dynamics: Theory and Applications. McGraw-Hill.

Lee, W., 1999. Principles of CAD/CAM/CAE Systems. Addison-Wesley Longman.

Madsen, H.O., Krenk, S., Lind, N.C., 1986. Methods of Structural Safety. Prentice Hall.

McMahon, C., Browne, J., 1998. CADCAM, second ed. Addison-Wesley.

Moës, N., Gravouil, A., Belytschko, T., 2002. Nonplanar 3D crack growth by the extended finite element and level sets. Part I: Mechanical model. International Journal for Numerical Methods in Engineering 53 (11), 2549–2568.

Numerical Integration Technologies, 1998. SYSNOISE 5.0.

Prasad, B., 1996. Concurrent Engineering Fundamentals, vols. I and II: Integrated Product and Process Organization. Prentice Hall.

Silva, J., 2000. Concurrent Design and Manufacturing for Mechanical Systems. MS thesis. University of Oklahoma.

Szabó, B., Babuška, I., 1991. Finite Element Analysis. John Wiley.

Taylor, C., 1985. The Thermal-Combustion Engine in Theory and Practice, vol. I: Thermodynamics, Fluid Flow, Performance, second ed. MIT Press.

Tsai, C.S., Chang, K.H., Wang, J., 1995. Integration infrastructure for a simulation-based design environment, Proceedings, Computers in Engineering Conference and the Engineering Data Symposium. ASME Design Theory and Methodology Conference.

Ullman, D.G., 1992. The Mechanical Design Process. McGraw-Hill.

Wu, Y.T., Wirsching, P.H., 1984. Advanced reliability method for fatigue analysis. Journal of Engineering Mechanics 110, 536–563.

Yu, X., Chang, K.H., Choi, K.K., 1998. Probabilistic structural durability prediction. AIAA Journal 36 (4), 628–637.

Yu, X., Choi, K.K., Chang, K.H., 1997. A mixed design approach for probabilistic structural durability. Journal of Structural Optimization 14, 81–90.

Zeid, I., 1991. CAD/CAM Theory and Practice. McGraw-Hill.

Sources

Adams: www.mscsoftware.com
ANSYS: www.ansys.com
CAMWorks: www.camworks.com
CatalystEX: www.dimensionprinting.com
LMS DADS: http://lsmintl.com
Dimension sst: www.stratasys.com
HAAS VF-Series: www.haascnc.com
HyperMesh: www.altairhyperworks.com
LS-DYNA: www.lstc.com
MSC/Nastran, MSC/Patran: www.mscsoftware.com
nCode: www.ncode.com
NESSUS: www.nessus.swri.org
Pro/ENGINEER, Pro/MECHANICA, Pro/MFG, Pro/SHEETMETAL, Pro/WELDING, etc.:
www.ptc.com
SLA-7000, Sinterstation: www.3dsystems.com
SolidWorks Motion, SolidWorks Simulation: www.solidworks.com
SYSNOISE 5.0: www.lmsintl.com/SYSNOISE

Geometric Modeling

CHAPTER OUTLINE

Virtual prototyping is becoming a cornerstone in modern product development. In e-Design, product design is first realized in computer-aided design (CAD) solid model form as parts and assemblies. The CAD solid model describes a geometric shape and physical properties that are essential for support of the product design, particularly product performance evaluation, virtual manufacturing, and cost estimating. The CAD solid model must be also properly parameterized in order for the design team to explore design alternatives for better product performance and hopefully lesser cost.

Most CAD software employs a geometric modeling kernel, such as Parasolid or ACIS, which is the library of core mathematical functions that define and store three-dimensional (3D) solid objects, for support of product modeling. In solid modeling, geometry is formed as a combination of constituent solid objects (more specifically solid features), which are created mostly by sketching a two-dimensional (2D) profile, composed of line or curve entities and protruding the profile for a solid object. While protruding the profile for a solid object, the trace of the line or curve entities forms boundary surfaces that wrap the solid object.

Therefore, before getting into the solid modeling and CAD theories, it is indispensable for readers to acquire a fundamental knowledge in curves and surfaces, which are often referred to as geometric modeling. This chapter focuses on introducing basic topics in geometric modeling, including curves, surfaces, and geometric transformations that are required for transforming geometric entities to meet specific needs. We assume readers who have used CAD software for creating solid models, but have no or little background in geometric modeling (and CAD theory). Therefore, instead of focusing on the use of CAD software, we focus more on understanding the selected topics in geometric modeling (and CAD theories in the next chapter). These topics are essential and relevant for readers to gain more in-depth understanding in behind-the-scenes operations while using CAD software. For a more comprehensive discussion on geometric modeling, readers are referred to excellent books, such as Mortenson (2006).

In this chapter, we provide fairly thorough discussions on parametric representations for the basic curves and surfaces that are widely employed in geometric modeling. Such curves and surfaces include Hermit curve, Coons patch, Bézier curves and surfaces, B-spline curves and surfaces, and nonuniform rational B-spline (NURB) curves and surfaces that are considered to be the most versatile and general form for representing geometric entities. We also discuss surfaces generated by protruding sketch profiles in CAD, such as cylindrical, ruled, revolved, sweep, and loft. In addition, we discuss geometric transformations, including scaling, translation, and rotation, which are commonly employed to manipulate geometric entities to meet specific modeling needs.

This chapter is essentially a prelude to the subjects that are more directly relevant to product design modeling, such as CAD theories, parameterization, product data exchanges, and so forth, which are discussed next in this book. Overall, the objective of this chapter is to provide an introduction to the parametric representations for curves and surfaces that help readers understand how the geometric models are defined mathematically.

2.1 **Introduction**

Geometric modeling is in general considered as a branch of applied mathematics and computational geometry that studies methods and algorithms for the mathematical description of shapes that represent geometry of objects. Geometric modeling has been an important and interesting subject for many years from the purely mathematical and computer science viewpoint, and also from the standpoint of engineering and various other applications, such as CAD/CAM (computer-aided manufacturing), entertainment, animation, and multimedia. Our interest is certainly in CAD/CAM, especially product modeling in CAD, in which objects are constructed by first creating curves and surfaces before reaching a solid model. Geometric modeling is indeed the backbone of a CAD system. It deserves our attention because understanding geometric modeling technique is a key step in learning CAD, especially for understanding its behind-the-scenes operations. With a solid background in geometric modeling, it should be easier for you to learn to use CAD software, avoid potential pitfalls, and be able to diagnose problems encountered in solid modeling.

Before getting into the discussion, a few points must be kept in mind. First, geometric modeling is only a means, not the goal, in engineering. In engineering design, analysis, and manufacturing, product geometry with an adequate level of details must be available. This is especially true in the e-Design paradigm, in which product design is refined with significant geometric details as the development process gets into later stages. Geometric modeling provides the fundamental means in representing design with the needed level of details that supports engineering design in various stages.

This chapter attempts to help you understand how modeling is carried out on a computer. To support creating geometric models on a computer, computational algorithms must be implemented on computer systems. Traditional mathematical methods learned in high school mathematical courses are based upon continuous functions, and computer systems do not generally work this way; they are discrete beasts. Thus, in the early 1970s, it was recognized that we could not represent curves by a general continuous function but must represent them as discrete entities. It is the reduction of these continuously defined mathematical objects to a more discrete representation that has motivated the field of geometric modeling. In any case, the mathematical representations of curves and surfaces are essential, to say the least.

This chapter focuses on the discussion of mathematical representations of curves and surfaces, as well as the transformation of these geometric entities for various modeling purposes, such as scaling, rotation, and translation. We provide fairly thorough discussions on parametric representations for the most basic and popular curves and surfaces in Sections 2.2 and 2.3, respectively. Such curves and surfaces include the Hermit curve, Coons patch, Bézier curves and surfaces, B-spline curves and surfaces, and NURB curves and surfaces, which are considered to be the most versatile and general forms for representing geometric entities. In addition to the basic surfaces, we include surfaces generated by CAD in Section 2.4. In CAD, we sketch a profile and protrude it for a surface (or solid). The protrusion capabilities commonly available in CAD include extrusion, blend (or loft), revolve, and sweep. The mathematical representations of these surfaces are presented together with Matlab scripts that graph them for visualization. We also include the discussion of geometric transformations in Section 2.5. Finally, in Section 2.6, we include case studies that showcase the applications of geometric modeling to practical applications, which include curve fitting and surface skinning techniques, and applications of the techniques to practical modeling examples.

2.2 Parametric curves

Typically, in high school mathematics or trigonometry, a curve is presented as a graph of a function f(x), as shown in Figure 2.1a. As x is varied, $y = f(x)$ is computed by the function f, and the pair of coordinates (x,y) sweeps out the curve. This is called the *explicit* form of the curve representation.

In addition to the explicit form, a geometric curve can be presented in an *implicit* form as $F(x,y) = 0$. For example, a circle of radius r and center at (a,b) in a Cartesian coordinate system x–y, shown in Figure 2.1b, can be written as

$$F(x, y) = (x - a)^2 + (y - b)^2 - r^2 = 0. \tag{2.1}$$

In addition to a circle, a number of basic Conic curves shown in Figure 2.2 are commonly found in describing geometry of mechanical parts. They are all quadratic functions in two variables represented in the Cartesian coordinate system as an implicit form. As shown in Figure 2.2, the graph of them is always a conic section.

Is CAD using such mathematical forms, either the explicit form $y = f(x)$ or the implicit form $F(x,y) = 0$, to represent curves (or surfaces) and carry out computations internally? The answer is generally no.

From a design perspective, such forms are inadequate in several ways. If we take the circle shown in Figure 2.1b as an example, the implicit form is inconvenient for computing points on the curve. For example, consider a circle defined as $F(x,y) = (x - 1)^2 + (y - 2)^2 - 1^2 = 0$. If one chooses $x = 3$, then $(y - 2)^2 + 3 = 0$ does not have a real solution for y, implying that the vertical line $x = 3$ does not intersect with the circle. Also, the implicit form may not be a single-valued function when there is a solution. As shown in Figure 2.3a, the curve is not single-valued along lines that are inside the circle

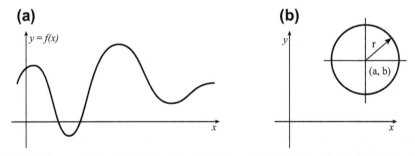

FIGURE 2.1 A Curve Representation. (a) An Explicit Function $y = f(x)$. (b) An Implicit Form $F(x,y) = 0$.

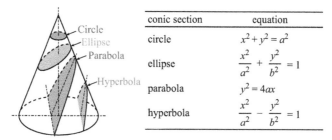

conic section	equation
circle	$x^2 + y^2 = a^2$
ellipse	$\dfrac{x^2}{a^2} + \dfrac{y^2}{b^2} = 1$
parabola	$y^2 = 4ax$
hyperbola	$\dfrac{x^2}{a^2} - \dfrac{y^2}{b^2} = 1$

FIGURE 2.2 Conic Curve Representations.

and are parallel to the y axis. Also, it is cumbersome to transform, such as to rotate geometric curves represented in such forms. For example, as shown in Figure 2.3b, writing an explicit or implicit function for the 90°-circular arc that is rotated a −α angle along the y-axis is not straightforward, to say the least.

CAD relies on parametric forms to describe curves and surfaces. What is a parametric curve? Is the equation we are familiar with, $(x - a)^2 + (y - b)^2 = r^2$, representing a circle shown in Figure 2.1b, a parametric curve? The answer is no. What about representing the same circle in a polar coordinate system? For example, a circle of radius r and center point (a,b) can be written in a polar coordinate form as the following:

$$x = a + r\cos\theta$$
$$y = b + r\sin\theta, \quad \text{and} \quad \theta \in [0, 2\pi] \tag{2.2}$$

Is Eqn (2.2) parametric? Yes, it is one of the parametric forms, in which x and y are decoupled and are separately represented in respective trigonometric functions in terms of the common parameter θ. In this case, the angle varying between 0 and 2π.

In general, a parametric curve that lies on an x–y plane is defined by two functions, x(u) and y(u), which use the parameter u. x(u) and y(u) are coordinate functions since because their values

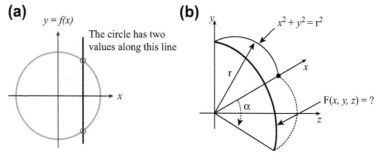

FIGURE 2.3 A Circle of Radius r and Center at (a,b). (a) In a Cartesian Coordinate System. (b) In a Polar Coordinate System.

represent the coordinates of points on the curve. As u varies, the coordinates (x(u), y(u)) sweep out the curve.

In general, CAD deals primarily with polynomial or rational functions (made by dividing one polynomial by another) and less on trigonometric functions like those in Eqn (2.2). For example, the circle can also be given by allowing u to vary from $-\infty$ to $+\infty$ in the following functions:

$$x(u) = \frac{2u}{1+u^2}$$

$$y(u) = \frac{1-u^2}{1+u^2}$$

(2.3)

Both Eqns (2.2) and (2.3) yield circles, so how do they differ? It is the curve parameterization. The motion of the point (x(u), y(u)) is different, even if the paths (the circles) are the same.

A good physical model for parametric curves is that of a moving particle. The parameter u represents time. At any time u, the position of the particle is (x(u), y(u)). Two paths (or parametric curves) may be identical even though the motion (or parameterization) is different.

In general, a planar or spatial curve $\mathbf{P}(u)$ can be represented in parametric forms, respectively, as follows:

$$\mathbf{P}(u) = \left[P_x(u), \ P_y(u)\right]_{1\times2}, \quad u \in [0, 1], \ \text{and}$$

$$\mathbf{P}(u) = \left[P_x(u), \ P_y(u), \ P_z(u)\right]_{1\times3}, \quad u \in [0, 1]$$

(2.4)

where u is the parameter, usually in [0,1]. Parametric curves are suitable for modeling curves in CAD. As shown in Eqn (2.4), coordinates of the curves x and y (and z) are decoupled and represented independently by their respective functions, usually explicit, in terms of a single parameter u. When a u value is specified, the coordinates of the curve can always be evaluated using Eqn (2.4). Also, the curve transformations can be easily taken care of by transforming characteristic points of the curve instead of the functions, which will be discussed more in Section 2.5.

In this section, we discuss parametric curves, both polynomial and rational. We will start with a simple straight line, quadratic curves, cubic curves, B-splines, and then rational curves. We assume spatial curves. Planar curves can be easily obtained by removing $P_z(u)$ (the z-component) of the curve equations. Some of the detailed derivations are either left as exercises or presented in appendices.

2.2.1 Straight line

A straight line shown in Figure 2.4 can be defined in a parametric form as a linear function of u as

$$\mathbf{P}(u) = \left[P_x(u), \ P_y(u), \ P_z(u)\right]_{1\times3} = (1-u)\mathbf{P}_0 + u\mathbf{P}_1, \quad u \in [0, 1] \tag{2.5}$$

where $\mathbf{P}_0 = [P_{0x}, P_{0y}, P_{0z}]_{1\times3}$ and $\mathbf{P}_1 = [P_{1x}, P_{1y}, P_{1z}]_{1\times3}$ are the start and end points of the line, respectively. When u is 0, $\mathbf{P}(u) = \mathbf{P}(0) = \mathbf{P}_0$, and when $u = 1$, $\mathbf{P}(u) = \mathbf{P}(1) = \mathbf{P}_1$.

Note that Eqn (2.5) can be derived in a more formal way. For a straight line, its coordinates can be represented by a liner function in u, in which $u \in [0,1]$:

$$\begin{aligned}
P_x(u) &= a_{1x}u + a_{0x} \\
P_y(u) &= a_{1y}u + a_{0y} \\
P_z(u) &= a_{1z}u + a_{0z}
\end{aligned}$$

(2.6)

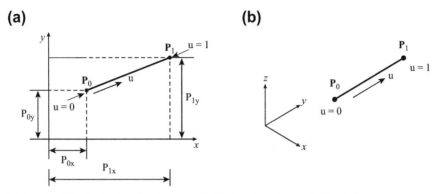

FIGURE 2.4 A Straight Line Defined by Its Start and End Points. (a) Planar. (b) Spatial.

where a_{1x}, a_{1y}, a_{1z}, a_{0x}, a_{0y}, and a_{0z} are unknown coefficients to be determined by the locations of start and end points $\mathbf{P}_0$ and $\mathbf{P}_1$. Rewriting Eqn (2.6) in a matrix form, we have

$$\mathbf{P}(u) = [u \quad 1]_{1 \times 2} \begin{bmatrix} a_{1x} & a_{1y} & a_{1z} \\ a_{0x} & a_{0y} & a_{0z} \end{bmatrix}_{2 \times 3} = \mathbf{U}_{1 \times 2} \mathbf{A}_{2 \times 3} \tag{2.7}$$

where matrix $\mathbf{A}$ contains the unknown coefficients.

By plugging $u = 0$ and $u = 1$ into Eqn (2.7), we have respectively $\mathbf{P}(0) = \mathbf{P}_0 = [0\ 1]\ \mathbf{A}$, and $\mathbf{P}(1) = \mathbf{P}_1 = [1\ 1]\ \mathbf{A}$.

By rewriting the above equations in a matrix form, we have

$$\begin{bmatrix} \mathbf{P}(0) \\ \mathbf{P}(1) \end{bmatrix}_{2 \times 3} = \begin{bmatrix} \mathbf{P}_0 \\ \mathbf{P}_1 \end{bmatrix}_{2 \times 3} = \begin{bmatrix} 0 & 1 \\ 1 & 1 \end{bmatrix}_{2 \times 2} \mathbf{A}_{2 \times 3} \tag{2.8}$$

where $\mathbf{P}_0$ and $\mathbf{P}_1$ must be known in order to define the straight line. From Eqn (2.8), the matrix $\mathbf{A}$ can be obtained by

$$\mathbf{A} = \begin{bmatrix} 0 & 1 \\ 1 & 1 \end{bmatrix}^{-1} \begin{bmatrix} \mathbf{P}_0 \\ \mathbf{P}_1 \end{bmatrix} = \begin{bmatrix} -1 & 1 \\ 1 & 0 \end{bmatrix} \begin{bmatrix} \mathbf{P}_0 \\ \mathbf{P}_1 \end{bmatrix}. \tag{2.9}$$

Hence, from Eqn (2.7), we have

$$\mathbf{P}(u) = \mathbf{UA} = [u \quad 1] \begin{bmatrix} -1 & 1 \\ 1 & 0 \end{bmatrix} \begin{bmatrix} \mathbf{P}_0 \\ \mathbf{P}_1 \end{bmatrix} = (1 - u)\mathbf{P}_0 + u\mathbf{P}_1 \tag{2.10}$$

where $(1 - u)$ and u are so-called basis functions associated with the characteristic points (in this case, the start and end points $\mathbf{P}_0$ and $\mathbf{P}_1$) of the straight line. The parametric curve equations for other polynomials can be derived in the same way.

2.2.2 Quadratic curves

A quadratic curve can be created by three distinct points—$\mathbf{P}_0$, $\mathbf{P}_1$, and $\mathbf{P}_2$, as shown in Figure 2.5a. Such a curve is called *spline curve*. In addition to a spline curve, a quadratic curve can be defined by two end points and a vector (Figure 2.5b), and by three control points forming a control polygon that encloses a Bézier curve (shown in Figure 2.5c), among others.

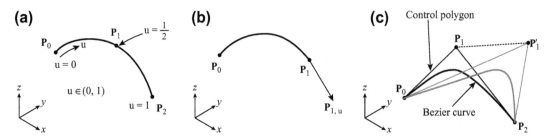

FIGURE 2.5 Quadratic Curves Defined by (a) Three Distinct Points (Spline Curve), (b) Two End Points and a Vector, and (c) Three Control points (Bézier Curve).

Similar to Eqn (2.7), a quadratic curve can be written in the following parametric form:

$$\mathbf{P}(u) = \begin{bmatrix} u^2 & u & 1 \end{bmatrix}_{1\times 3} \begin{bmatrix} a_{2x} & a_{2y} & a_{2z} \\ a_{1x} & a_{1y} & a_{1z} \\ a_{0x} & a_{0y} & a_{0z} \end{bmatrix}_{3\times 3} = \mathbf{U}_{1\times 3}\mathbf{A}_{3\times 3} \tag{2.11}$$

where matrix $\mathbf{A}$ contains 9 (3×3) unknown coefficients.

2.2.2.1 Spline curve—three points

For a quadratic spline curve, we assume the three distinct points are $\mathbf{P}_0 = \mathbf{P}(0)$, $\mathbf{P}_1 = \mathbf{P}(\frac{1}{2})$, and $\mathbf{P}_2 = \mathbf{P}(1)$. Note that $\mathbf{P}_1$ does not have to be located at $u = \frac{1}{2}$.

By plugging $u = 0$, $u = \frac{1}{2}$, and $u = 1$ into Eqn (2.11), we have respectively $\mathbf{P}(0) = \mathbf{P}_0 = [0\ 0\ 1]\ \mathbf{A}$, $\mathbf{P}(\frac{1}{2}) = \mathbf{P}_1 = [\frac{1}{4}\ \frac{1}{2}\ 1]\ \mathbf{A}$, and $\mathbf{P}(1) = \mathbf{P}_2 = [1\ 1\ 1]\ \mathbf{A}$.

By rewriting the above equations in a matrix form, we have

$$\begin{bmatrix} \mathbf{P}(0) \\ \mathbf{P}(\frac{1}{2}) \\ \mathbf{P}(1) \end{bmatrix}_{3\times 3} = \begin{bmatrix} \mathbf{P}_0 \\ \mathbf{P}_1 \\ \mathbf{P}_2 \end{bmatrix}_{3\times 3} = \begin{bmatrix} 0 & 0 & 1 \\ \frac{1}{4} & \frac{1}{2} & 1 \\ 1 & 1 & 1 \end{bmatrix} \mathbf{A}_{3\times 3} \tag{2.12}$$

where $\mathbf{P}_0$, $\mathbf{P}_1$, and $\mathbf{P}_2$ are known. The matrix $\mathbf{A}$ can be obtained by

$$\mathbf{A} = \begin{bmatrix} 0 & 0 & 1 \\ \frac{1}{4} & \frac{1}{2} & 1 \\ 1 & 1 & 1 \end{bmatrix}^{-1} \begin{bmatrix} \mathbf{P}_0 \\ \mathbf{P}_1 \\ \mathbf{P}_2 \end{bmatrix} = \begin{bmatrix} 2 & -4 & 2 \\ -3 & 4 & -1 \\ 1 & 0 & 0 \end{bmatrix} \begin{bmatrix} \mathbf{P}_0 \\ \mathbf{P}_1 \\ \mathbf{P}_2 \end{bmatrix}. \tag{2.13}$$

Hence from Eqn (2.11), we have

$$\mathbf{P}(u) = \mathbf{UA} = \begin{bmatrix} u^2 & u & 1 \end{bmatrix} \begin{bmatrix} 2 & -4 & 2 \\ -3 & 4 & -1 \\ 1 & 0 & 0 \end{bmatrix} \begin{bmatrix} \mathbf{P}_0 \\ \mathbf{P}_1 \\ \mathbf{P}_2 \end{bmatrix} = \mathbf{U}\mathbf{N}^s\mathbf{G}^s = \mathbf{B}^s\mathbf{G}^s \tag{2.14}$$

where $\mathbf{N}^s$ is a constant 3×3 matrix for any given quadratic spline curve for which $\mathbf{P}_1$ is at $u = \frac{1}{2}$, and $\mathbf{B}^s$ is the 1×3 vector of the basis functions (also called blending functions); that is,

$$\mathbf{B}^s(u) = \mathbf{U}\mathbf{N}^s = \begin{bmatrix} u^2 & u & 1 \end{bmatrix} \begin{bmatrix} 2 & -4 & 2 \\ -3 & 4 & -1 \\ 1 & 0 & 0 \end{bmatrix} \tag{2.15}$$

$$= \begin{bmatrix} 2u^2 - 3u + 1, & -4u^2 + 4u, & 2u^2 - u \end{bmatrix} = \begin{bmatrix} B_0^s(u), & B_1^s(u), & B_2^s(u) \end{bmatrix}$$

which are plotted in Figure 2.6a. Note that the superscript s denotes a spline curve.

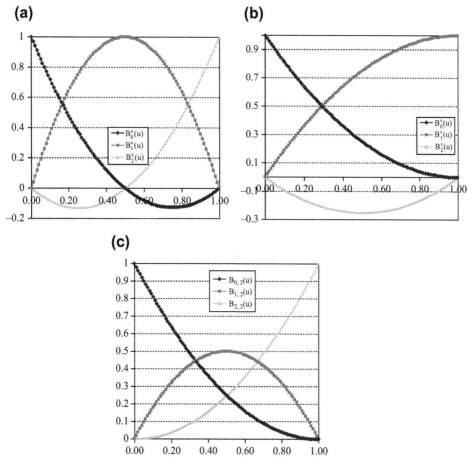

FIGURE 2.6 Basis Functions for Quadratic Curves. (a) Spline Curve. (b) Two End Points and a Vector. (c) Bézier Curve.

Therefore, Eqn (2.14) can be rewritten as

$$P(u) = B^sG^s = \begin{bmatrix} 2u^2 - 3u + 1, & -4u^2 + 4u, & 2u^2 - u \end{bmatrix} \begin{bmatrix} P_0 \\ P_1 \\ P_2 \end{bmatrix}$$

(2.16)

$$= B_0^s(u)P_0 + B_1^s(u)P_1 + B_2^s(u)P_2.$$

EXAMPLE 2.1

Given three points, $P_0 = [0,1]$, $P_1 = [1,2]$, and $P_2 = [2,0]$, derive the parametric equation and graph the spline curve formed by them.

Solutions

Using Eqn (2.16), we have

$$P(u) = B^sG^s = \begin{bmatrix} 2u^2 - 3u + 1, & -4u^2 + 4u, & 2u^2 - u \end{bmatrix} \begin{bmatrix} 0 & 1 \\ 1 & 2 \\ 2 & 0 \end{bmatrix}$$

$$= \begin{bmatrix} 2u, & -6u^2 + 5u + 1 \end{bmatrix}.$$

The spline curve is graphed in Matlab with script is shown below.

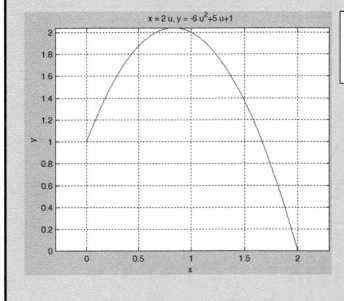

Matlab Script:

```
ezplot('2*u','-6*u^2+5*u+1',[0,1])
grid on
```

2.2.2.2 Two points and a vector

A quadratic curve can also be defined by its start and end points plus a vector, either at the start or the end point. For example, the curve shown in Figure 2.5b is defined by two points $\mathbf{P}_0$ and $\mathbf{P}_1$ at start and end of the curve and a tangent vector $\mathbf{P}_{1,u}$ at its end point. From Eqn (2.11), the tangent vector $\mathbf{P}_{1,u}$ is defined as

$$\mathbf{P}_{1,u} = \frac{\partial \mathbf{P}(u)}{\partial u}\bigg|_{u=1} = \frac{\partial (\mathbf{UA})}{\partial u}\bigg|_{u=1} = \frac{\partial \mathbf{U}}{\partial u}\bigg|_{u=1}\mathbf{A} = [2u \quad 1 \quad 0]|_{u=1}\,\mathbf{A} = [2 \quad 1 \quad 0]\mathbf{A} \qquad (2.17)$$

Following the same steps as before, we have

$$\begin{bmatrix} \mathbf{P}_0 \\ \mathbf{P}_1 \\ \mathbf{P}_{1,u} \end{bmatrix}_{3\times 3} = \begin{bmatrix} 0 & 0 & 1 \\ 1 & 1 & 1 \\ 2 & 1 & 0 \end{bmatrix}\mathbf{A}_{3\times 3} \qquad (2.18)$$

where $\mathbf{P}_0$, $\mathbf{P}_1$, and $\mathbf{P}_{1,u}$ are known. The matrix $\mathbf{A}$ can be obtained by

$$\mathbf{A} = \begin{bmatrix} 0 & 0 & 1 \\ 1 & 1 & 1 \\ 2 & 1 & 0 \end{bmatrix}^{-1} \begin{bmatrix} \mathbf{P}_0 \\ \mathbf{P}_1 \\ \mathbf{P}_{1,u} \end{bmatrix} = \begin{bmatrix} 1 & -1 & 1 \\ -2 & 2 & -1 \\ 1 & 0 & 0 \end{bmatrix} \begin{bmatrix} \mathbf{P}_0 \\ \mathbf{P}_1 \\ \mathbf{P}_{1,u} \end{bmatrix}. \qquad (2.19)$$

Hence from Eqn (2.11), we have

$$\mathbf{P}(u) = \mathbf{UA} = \begin{bmatrix} u^2 & u & 1 \end{bmatrix} \begin{bmatrix} 1 & -1 & 1 \\ -2 & 2 & -1 \\ 1 & 0 & 0 \end{bmatrix} \begin{bmatrix} \mathbf{P}_0 \\ \mathbf{P}_1 \\ \mathbf{P}_{1,u} \end{bmatrix} = \mathbf{UN}^v\mathbf{G}^v = \mathbf{B}^v\mathbf{G}^v \qquad (2.20)$$

where $\mathbf{N}^v$ is a constant 3×3 matrix for any given quadratic curve defined by the end points and a tangent vector at the end, and $\mathbf{B}^v$ is the 1×3 vector of the basis functions; that is,

$$\mathbf{B}^v(u) = \mathbf{UN}^v = \begin{bmatrix} u^2 & u & 1 \end{bmatrix} \begin{bmatrix} 1 & -1 & 1 \\ -2 & 2 & -1 \\ 1 & 0 & 0 \end{bmatrix}$$
$$= [u^2 - 2u + 1, \quad -u^2 + 2u, \quad u^2 - u] = \begin{bmatrix} B_0^v(u), & B_1^v(u), & B_2^v(u) \end{bmatrix} \qquad (2.21)$$

which are plotted in Figure 2.6b. Note that the superscript v denotes a curve with a tangent vector. Therefore, from Eqn (2.20), we have

$$\mathbf{P}(u) = \mathbf{B}^v\mathbf{G}^v = [u^2 - 2u + 1, \quad -u^2 + 2u, \quad u^2 - u] \begin{bmatrix} \mathbf{P}_0 \\ \mathbf{P}_1 \\ \mathbf{P}_2 \end{bmatrix} \qquad (2.22)$$
$$= B_0^v(u)\mathbf{P}_0 + B_1^v(u)\mathbf{P}_1 + B_2^v(u)\mathbf{P}_{1,u}.$$

Note that the geometric shape of the curve is controlled by the tangent vector in addition to its start and end points. The basis function associated with the tangent vector $B_2^v(u)$ is negative for $u \in [0,1]$, indicating that when the vector size increases, the curve is "pushed" backward, as to be shown in Example 2.2.

EXAMPLE 2.2

Given two points and a tangent vector at the end point, $\mathbf{P}_0 = [0,1]$, $\mathbf{P}_1 = [1,2]$, and $\mathbf{P}_{1,u} = [2,-7]$, derive the parametric equation and graph the curve formed by them. Also, graph the curve by changing the tangent vector to $\mathbf{P}_{1,u} = [3,-10.5]$.

Solutions

Using Eqn (2.22), we have

$$\mathbf{P}(u) = \mathbf{B}^v\mathbf{G}^v = \begin{bmatrix} u^2 - 2u + 1, & -u^2 + 2u, & u^2 - u \end{bmatrix} \begin{bmatrix} 0 & 1 \\ 2 & 0 \\ 2 & -7 \end{bmatrix}$$

$$= \begin{bmatrix} 2u, & -6u^2 + 5u + 1 \end{bmatrix}.$$

Using the tangent vector of a larger size, the curve (Curve 2) becomes

$$\mathbf{P}(u) = \mathbf{B}^v\mathbf{G}^v = \begin{bmatrix} u^2 - 2u + 1, & -u^2 + 2u, & u^2 - u \end{bmatrix} \begin{bmatrix} 0 & 1 \\ 2 & 0 \\ 3 & -10.5 \end{bmatrix}$$

$$= \begin{bmatrix} u^2 + u, & -9.5u^2 + 8.5u + 1 \end{bmatrix}.$$

The curves are graphed in Matlab with script shown below. It is clearly shown in the figure that when the vector size increases in Curve 2, the curve is "pushed" backward (in this case, upward and to the left).

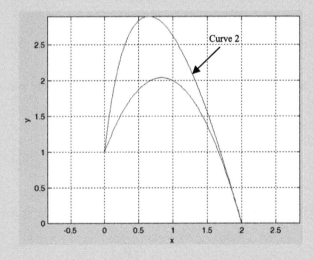

Matlab Script:

```
ezplot('2*u','-6*u^2+5*u+1',[0,1])
grid on
hold on
ezplot('u^2+u','-9.5*u^2+8.5*u+1',[0,1])
```

2.2.2.3 Bézier curve

The Bézier curve, originally developed by Pierre Bézier in the 1970s, has become one of the most commonly used curves for geometric modeling. As shown in Figure 2.6c, unlike a spline curve, a Bézier curve is defined by control points that do not necessarily stay on the curve. The control points form a control polygon (or characteristic polygon) that determines the shape of the curve. More specifically, in general, only the first and last control points stay on the curve; in fact, in this case they coincide with the start and end points of the curve, respectively. The curve is also tangent to the first and last line segments of the control polygon, which provides the designer with direct control of the geometric shape of the curve at the ends. In addition to controlling the tangent vectors of the curves at ends, changing the control point locations alters the shape of the curve, as illustrated in Figure 2.5c, in which the control point $\mathbf{P}_1$ is moved to a new location $\mathbf{P}_1{}'$. Figure 2.7 also illustrates this point, in which control point P2$'$ is relocated.

Mathematically, a Bézier curve is defined as

$$\mathbf{P}(u) = \sum_{i=0}^{n} \mathbf{P}_i B_{i,n}(u), \quad u \in [0, 1] \tag{2.23}$$

where $\mathbf{P}_i$ is the ith control point, n is the polynomial order of the curve, and $B_{i,n}(u)$ is the corresponding basis function, called the Bernstein polynomial, defined as

$$B_{i,n}(u) = C(n, i) u^i (1 - u)^{n-i}, \quad u \in [0, 1]. \tag{2.24}$$

$C(n,i)$ is the Binomial coefficient defined as

$$C(n, i) = \frac{n!}{i!(n - i)!} = \binom{n}{i}. \tag{2.25}$$

Note that we assume $u^0 = 1$, including when $u = 0$, and $0! = 1$, in Eqns (2.24) and (2.25), respectively.

For a quadratic Bézier curve, $n = 2$, and the curve is defined by three control points. From Eqn (2.24), the basic functions of a quadratic curve can be derived as follows:

$$\begin{aligned} B_{0,2}(u) &= (1 - u)^2 \\ B_{1,2}(u) &= 2u(1 - u) \\ B_{2,2}(u) &= u^2 \end{aligned} \tag{2.26}$$

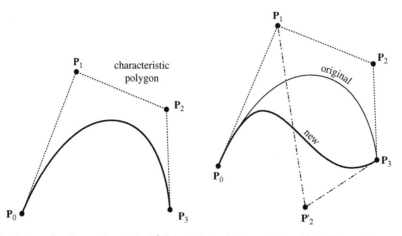

FIGURE 2.7 The Geometric Shape of a Cubic Bézier as Determined by Its Control Polygon.

which are plotted in Figure 2.6c. Note that the sum of all three basis functions is 1—that is, $\sum_{i=0}^{2} B_{i,2}(u) = 1$, implying that the basis functions form a partition of unity. The partition of unity is a very important property when utilizing Bernstein polynomials in geometric modeling. In particular, for any set of control points $\mathbf{P}_0, \mathbf{P}_1,\ldots, \mathbf{P}_n$, in a two- or three-dimensional space and for any u in [0,1], the expression of Eqn (2.23) is a convex combination of the set of points $\mathbf{P}_0, \mathbf{P}_1,\ldots, \mathbf{P}_n$. Note that a convex combination is a linear combination of vectors (e.g., in this case, control points in Bezier curve), where all coefficients are nonnegative and sum up to 1. In geometric modeling, a curve of convex combination implies the convex hull (or convex envelop) property, which ensures that the curve lies within the convex hull of the control points (that is, the control polygon).

It can be shown for a given n that the Bernstein polynomials $B_{i,n}(u)$ satisfy the following:

1. $0 \leq B_{i,n}(u) \leq 1$ for $u \in [0,1]$, and

2. $\displaystyle\sum_{i=0}^{n} B_{i,n}(u) = 1$

It is apparent that a Bézier curve, written in Eqn (2.23), is a convex combination of the control points; therefore the curve lies within the convex hull of the control points.

A quadratic Bézier curve shown in Eqn (2.23) can be explicitly written as

$$\mathbf{P}(u) = \sum_{i=0}^{2} \mathbf{P}_i B_{i,2}(u) = \mathbf{P}_0 B_{0,2}(u) + \mathbf{P}_1 B_{1,2}(u) + \mathbf{P}_2 B_{2,2}(u) \tag{2.27}$$

$$= \mathbf{P}_0(1-u)^2 + \mathbf{P}_1(2u(1-u)) + \mathbf{P}_2 u^2, \quad u \in [0,1].$$

Note that Eqn (2.27) can also be written in a matrix form, similar to those of Eqns (2.14) and (2.20), as

$$\mathbf{P}(u) = \begin{bmatrix} u^2 & u & 1 \end{bmatrix} \begin{bmatrix} 1 & -2 & 1 \\ -2 & 2 & 0 \\ 1 & 0 & 0 \end{bmatrix} \begin{bmatrix} \mathbf{P}_0 \\ \mathbf{P}_1 \\ \mathbf{P}_2 \end{bmatrix} = \mathbf{U}\mathbf{N}^B\mathbf{G}^B = \mathbf{B}^B\mathbf{G}^s \tag{2.28}$$

where $\mathbf{N}^B$ is a constant 3×3 matrix for any given quadratic Bézier curve, and $\mathbf{B}^B$ is the 1×3 vector of the basis functions; in this case, they are Bernstein polynomials.

EXAMPLE 2.3

Given three control points, $\mathbf{P}_0 = [0,1]$, $\mathbf{P}_1 = [1,3.5]$, and $\mathbf{P}_2 = [2,0]$, derive the parametric equation and graph the Bézier curve formed by them.

Solutions

Using Eqn (2.28), we have

$$\mathbf{P}(u) = \mathbf{B}^B\mathbf{G}^B = \begin{bmatrix} (1-u)^2, & 2u(1-u), & u^2 \end{bmatrix} \begin{bmatrix} 0 & 1 \\ 1 & 3.5 \\ 2 & 0 \end{bmatrix}$$

$$= [2u, \quad -6u^2 + 5u + 1].$$

EXAMPLE 2.3—CONT'D

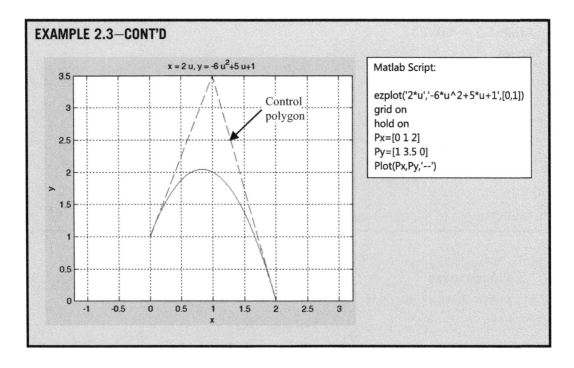

Matlab Script:

```
ezplot('2*u','-6*u^2+5*u+1',[0,1])
grid on
hold on
Px=[0 1 2]
Py=[1 3.5 0]
Plot(Px,Py,'--')
```

As shown in Examples 2.1–2.3, all three curves represented in their respective forms are identical, which implies that the same curve can be represented in different forms. This is because all forms are representing a quadratic curve, which is a second-order polynomial function. Mathematically, they are all identical; therefore, we have the following:

$$\mathbf{P}(u) = \mathbf{UA} = \mathbf{UN^sG^s} = \mathbf{UN^vG^v} = \mathbf{UN^BG^B}. \tag{2.29}$$

Equation (2.29) implies that curves can be transformed into various forms to meet different modeling requirements. For example, a quadratic spline curve can be converted to a two point and a vector form, as well as a Bézier curve as, respectively,

$$\mathbf{G^v} = \mathbf{N^{v^{-1}}N^sG^s} \tag{2.30a}$$

and

$$\mathbf{G^B} = \mathbf{N^{B^{-1}}N^sG^s}. \tag{2.30b}$$

The same is true for converting a two point and a vector or a Bézier curve to other forms.

EXAMPLE 2.4

Given a spline curve formed by three points, $\mathbf{P}_0 = [0,1]$, $\mathbf{P}_1 = [1,2]$, and $\mathbf{P}_2 = [2,0]$, convert the curve into two points with a vector form and then a Bézier curve.

Continued

EXAMPLE 2.4—CONT'D

Solutions

Using Eqns (2.30a) and (2.30b), we have

$$
\mathbf{G}^v = \mathbf{N}^{v^{-1}} \mathbf{N}^s \mathbf{G}^s =
\begin{bmatrix} 0 & 0 & 1 \\ 1 & 1 & 1 \\ 2 & 1 & 0 \end{bmatrix}
\begin{bmatrix} 2 & -4 & 2 \\ -3 & 4 & -1 \\ 1 & 0 & 0 \end{bmatrix}
\begin{bmatrix} 0 & 1 \\ 1 & 2 \\ 2 & 0 \end{bmatrix}
=
\begin{bmatrix} 0 & 1 \\ 2 & 0 \\ 2 & -7 \end{bmatrix}
$$

in which $\mathbf{G}^v$ contains the end points and tangent vector of the quadratic curve, and

$$
\mathbf{G}^B = \mathbf{N}^{B^{-1}} \mathbf{N}^s \mathbf{G}^s =
\begin{bmatrix} 1 & -2 & 1 \\ -2 & 2 & 0 \\ 1 & 0 & 0 \end{bmatrix}^{-1}
\begin{bmatrix} 2 & -4 & 2 \\ -3 & 4 & -1 \\ 1 & 0 & 0 \end{bmatrix}
\begin{bmatrix} 0 & 1 \\ 1 & 2 \\ 2 & 0 \end{bmatrix}
=
\begin{bmatrix} 0 & 1 \\ 1 & 3.5 \\ 2 & 0 \end{bmatrix}
$$

in which $\mathbf{G}^B$ contains the three control points of the quadratic Bézier curve.

2.2.3 Cubic curves

Similar to Eqn (2.11), a cubic curve can be written in the following parametric form:

$$
\mathbf{P}(u) = \begin{bmatrix} u^3 & u & 1 \end{bmatrix}_{1 \times 4}
\begin{bmatrix}
a_{3x} & a_{3y} & a_{3z} \\
a_{2x} & a_{2y} & a_{2z} \\
a_{1x} & a_{1y} & a_{1z} \\
a_{0x} & a_{0y} & a_{0z}
\end{bmatrix}_{4 \times 3}
= \mathbf{U}_{1 \times 4} \mathbf{A}_{4 \times 3}
\tag{2.31}
$$

where matrix $\mathbf{A}$ contains 12 (4×3) unknown coefficients.

2.2.3.1 Spline curve—four points

For a cubic spline curve, we assume the four distinct points are $\mathbf{P}_0 = \mathbf{P}(0)$, $\mathbf{P}_1 = \mathbf{P}(1/3)$, $\mathbf{P}_1 = \mathbf{P}(2/3)$, and $\mathbf{P}_3 = \mathbf{P}(1)$, as shown in Figure 2.8a. Note that $\mathbf{P}_1$ and $\mathbf{P}_2$ can be at locations other than $u = 1/3$ or 2/3.

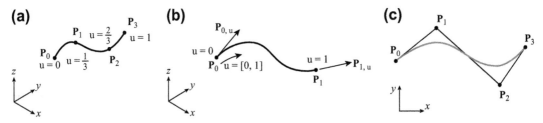

FIGURE 2.8 Cubic Curves Defined by (a) Four Distinct Points (Spline Curve), (b) End Points and End Vectors (Hermit Cubic Curve), and (c) Four Control Points (Bézier Curve).

Following the same idea as the quadratic curves, we plug $u = 0$, $u = 1/3$, $u = 2/3$, and $u = 1$ into Eqn (2.31) to yield

$$
\begin{bmatrix} \mathbf{P}(0) \\ \mathbf{P}(1/3) \\ \mathbf{P}(2/3) \\ \mathbf{P}(1) \end{bmatrix}_{4 \times 3} = \begin{bmatrix} \mathbf{P}_0 \\ \mathbf{P}_1 \\ \mathbf{P}_2 \\ \mathbf{P}_4 \end{bmatrix}_{4 \times 3} = \begin{bmatrix} 0 & 0 & 0 & 1 \\ 1/27 & 1/9 & 1/3 & 1 \\ 8/27 & 4/9 & 2/3 & 1 \\ 1 & 1 & 1 & 1 \end{bmatrix} \mathbf{A}_{4 \times 3}
\tag{2.32}
$$

where $\mathbf{P}_0$, $\mathbf{P}_1$, $\mathbf{P}_2$, and $\mathbf{P}_3$ must be known in order to define the curve. The matrix $\mathbf{A}$ can be obtained by

$$
\mathbf{A} = \begin{bmatrix} 0 & 0 & 0 & 1 \\ 1/27 & 1/9 & 1/3 & 1 \\ 8/27 & 4/9 & 2/3 & 1 \\ 1 & 1 & 1 & 1 \end{bmatrix}^{-1} \begin{bmatrix} \mathbf{P}_0 \\ \mathbf{P}_1 \\ \mathbf{P}_2 \\ \mathbf{P}_4 \end{bmatrix} = \begin{bmatrix} -9/2 & 27/2 & -27/2 & 9/2 \\ 9 & -45/2 & 18 & -9/2 \\ -11/2 & 9 & -9/2 & 1 \\ 1 & 0 & 0 & 0 \end{bmatrix} \begin{bmatrix} \mathbf{P}_0 \\ \mathbf{P}_1 \\ \mathbf{P}_2 \\ \mathbf{P}_4 \end{bmatrix}.
\tag{2.33}
$$

Hence, from Eqn (2.32), we have

$$
\mathbf{P}(u) = \mathbf{U}\mathbf{A} = \begin{bmatrix} u^3 & u^2 & u & 1 \end{bmatrix} \begin{bmatrix} -9/2 & 27/2 & -27/2 & 9/2 \\ 9 & -45/2 & 18 & -9/2 \\ -11/2 & 9 & -9/2 & 1 \\ 1 & 0 & 0 & 0 \end{bmatrix} \begin{bmatrix} \mathbf{P}_0 \\ \mathbf{P}_1 \\ \mathbf{P}_2 \\ \mathbf{P}_4 \end{bmatrix} = \mathbf{U}\mathbf{N}^s \mathbf{G}^s = \mathbf{B}^s \mathbf{G}^s
\tag{2.34}
$$

where $\mathbf{N}^s$ is a constant 4×4 matrix for any given cubic spline curve, and $\mathbf{B}^s$ is the 1×4 vector of the basis functions (also called blending functions); that is,

$$
\mathbf{B}^s(u) = \mathbf{U}\mathbf{N}^s = \begin{bmatrix} u^3 & u^2 & u & 1 \end{bmatrix} \begin{bmatrix} -9/2 & 27/2 & -27/2 & 9/2 \\ 9 & -45/2 & 18 & -9/2 \\ -11/2 & 9 & -9/2 & 1 \\ 1 & 0 & 0 & 0 \end{bmatrix}
$$

$$
= \begin{bmatrix} -9/2u^3 + 9u^2 - 11/2u + 1, & 27/2u^3 - 45/2u^2 + 9u,
\end{bmatrix}
$$

$$
-27/2u^3 + 18u^2 - 9/2u + 1, \quad 9/2u^3 + 9/2u^2 + u \Big]
$$

$$
= \begin{bmatrix} B_0^s(u), & B_1^s(u), & B_2^s(u), & B_3^s(u) \end{bmatrix}
\tag{2.35}
$$

which are plotted in Figure 2.9a. Therefore, from Eqn (2.34), we have

$$
\mathbf{P}(u) = \mathbf{B}^s \mathbf{G}^s = \begin{bmatrix} B_0^s(u), & B_1^s(u), & B_2^s(u), & B_3^s(u) \end{bmatrix} \begin{bmatrix} \mathbf{P}_0 \\ \mathbf{P}_1 \\ \mathbf{P}_2 \\ \mathbf{P}_4 \end{bmatrix}
\tag{2.36}
$$

$$
= B_0^s(u)\mathbf{P}_0 + B_1^s(u)\mathbf{P}_1 + B_2^s(u)\mathbf{P}_2 + B_3^s(u)\mathbf{P}_3.
$$

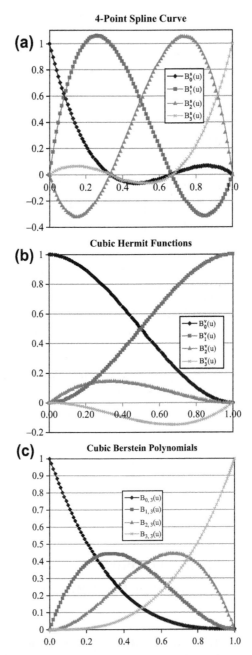

FIGURE 2.9 Basis Functions for Cubic Parametric Curves. (a) Four-point Spline Curve. (b) Hermit Cubic Curve. (c) Cubic Bernstein Polynomials.

EXAMPLE 2.5

Given three points, $\mathbf{P}_0 = [0,1]$, $\mathbf{P}_1 = [1,2]$, $\mathbf{P}_2 = [2,0]$, and $\mathbf{P}_3 = [4,-1]$, derive the parametric equation and graph the spline curve formed by them.

Solutions

Using Eqn (2.36), we have

$$\mathbf{P}(u) = \mathbf{B^sG^s} = \big[\, -9/2u^3 + 9u^2 - 11/2u + 1,\ \ 27/2u^3 - 45/2u^2 + 9u,$$

$$-27/2u^3 + 18u^2 - 9/2u + 1,\ \ 9/2u^3 + 9/2u^2 + u\big] \begin{bmatrix} 0 & 1 \\ 1 & 2 \\ 2 & 0 \\ 4 & -1 \end{bmatrix}$$

$$= \big[9/2u^3 - 9/2u^2 + 4u,\ \ 18u^3 - 63/2u^2 + 23/2u + 1\big].$$

The cubic spline curve is graphed in Matlab with script as shown below.

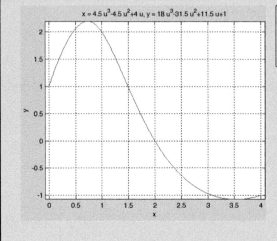

x = 4.5 u³-4.5 u²+4 u, y = 18 u³-31.5 u²+11.5 u+1

Matlab Script:

```
ezplot('4.5*u^3-4.5*u^2+4*u','18*u^3-31.5*u^2+11.5*u+1',
[0,1])
grid on
```

2.2.3.2 Hermit cubic curve (two end points and two end vectors)

Similar to the quadratic curve, a cubic curve can also be defined by its start and end points plus tangent vectors at the start and end points, as shown in Figure 2.8b, in which the tangent vectors $\mathbf{P}_{0,u}$ and $\mathbf{P}_{1,u}$ are at the start end points, respectively. This is called a Hermit cubic curve or curve of geometric format. Note that from Eqn (2.34), the tangent vectors $\mathbf{P}_{0,u}$ and $\mathbf{P}_{1,u}$ are defined as

$$\mathbf{P}_{0,u} = \frac{\partial \mathbf{P}(u)}{\partial u}\Big|_{u=0} = \frac{\partial (\mathbf{UA})}{\partial u}\Big|_{u=0} = \frac{\partial \mathbf{U}}{\partial u}\Big|_{u=0}\mathbf{A} = \begin{bmatrix} 3u^2 & 2u & 1 & 0 \end{bmatrix}\Big|_{u=0}\mathbf{A} = \begin{bmatrix} 0 & 0 & 1 & 0 \end{bmatrix}\mathbf{A}$$

$$\mathbf{P}_{1,u} = \frac{\partial \mathbf{P}(u)}{\partial u}\Big|_{u=1} = \frac{\partial (\mathbf{UA})}{\partial u}\Big|_{u=1} = \frac{\partial \mathbf{U}}{\partial u}\Big|_{u=1}\mathbf{A} = \begin{bmatrix} 3u^2 & 2u & 1 & 0 \end{bmatrix}\Big|_{u=1}\mathbf{A} = \begin{bmatrix} 3 & 2 & 1 & 0 \end{bmatrix}\mathbf{A}.$$

$$(2.37)$$

Following the same steps as before, we have

$$
\mathbf{A} = \begin{bmatrix} 0 & 0 & 0 & 0 \\ 1 & 1 & 1 & 1 \\ 0 & 0 & 1 & 0 \\ 3 & 2 & 1 & 0 \end{bmatrix}^{-1} \begin{bmatrix} \mathbf{P}_0 \\ \mathbf{P}_1 \\ \mathbf{P}_{0,u} \\ \mathbf{P}_{1,u} \end{bmatrix} = \begin{bmatrix} 2 & -2 & 1 & 1 \\ -3 & 3 & -2 & -1 \\ 0 & 0 & 1 & 0 \\ 1 & 0 & 0 & 0 \end{bmatrix} \begin{bmatrix} \mathbf{P}_0 \\ \mathbf{P}_1 \\ \mathbf{P}_{0,u} \\ \mathbf{P}_{1,u} \end{bmatrix}.
\tag{2.38}
$$

Hence from Eqn (2.32), we have

$$
\mathbf{P}(u) = \mathbf{U}\mathbf{A} = \begin{bmatrix} u^3 & u^2 & u & 1 \end{bmatrix} \begin{bmatrix} 2 & -2 & 1 & 1 \\ -3 & 3 & -2 & -1 \\ 0 & 0 & 1 & 0 \\ 1 & 0 & 0 & 0 \end{bmatrix} \begin{bmatrix} \mathbf{P}_0 \\ \mathbf{P}_1 \\ \mathbf{P}_{0,u} \\ \mathbf{P}_{1,u} \end{bmatrix} = \mathbf{U}\mathbf{N}^v\mathbf{G}^v = \mathbf{B}^v\mathbf{G}^v
\tag{2.39}
$$

where $\mathbf{N}^v$ is a constant 4×4 matrix for any given cubic curve defined by the end points and vectors, and $\mathbf{B}^v$ is the 1×4 vector of the basis functions called cubic Hermit functions; that is,

$$
\begin{aligned}
\mathbf{B}^v(u) = \mathbf{U}\mathbf{N}^v &= \begin{bmatrix} u^3 & u^2 & u & 1 \end{bmatrix} \begin{bmatrix} 2 & -2 & 1 & 1 \\ -3 & 3 & -2 & -1 \\ 0 & 0 & 1 & 0 \\ 1 & 0 & 0 & 0 \end{bmatrix} \\
&= \begin{bmatrix} 2u^3 - 3u^2 + 1, & -2u^3 + 3u^2 + 1, & u^3 - 2u^2 + u, & u^3 - u^2 \end{bmatrix} \\
&= \begin{bmatrix} B_0^v(u), & B_1^v(u), & B_2^v(u), & B_3^v(u) \end{bmatrix}
\end{aligned}
\tag{2.40}
$$

which are plotted in Figure 2.9b. Therefore, from Eqn (2.39), we have

$$
\begin{aligned}
\mathbf{P}(u) = \mathbf{B}^v\mathbf{G}^v &= \begin{bmatrix} 2u^3 - 3u^2 + 1, & -2u^3 + 3u^2 + 1, & u^3 - 2u^2 + u, & u^3 - u^2 \end{bmatrix} \begin{bmatrix} \mathbf{P}_0 \\ \mathbf{P}_1 \\ \mathbf{P}_{0,u} \\ \mathbf{P}_{1,u} \end{bmatrix} \\
&= B_0^v(u)\mathbf{P}_0 + B_1^v(u)\mathbf{P}_1 + B_2^v(u)\mathbf{P}_{0,u} + B_3^v(u)\mathbf{P}_{1,u}.
\end{aligned}
\tag{2.41}
$$

Similar to the quadratic curve, the geometric shape of the Hermit cubic curve is controlled by the tangent vectors in addition to its start and end points. The basis function $B_2^v(u)$, which is associated with the tangent vector at the start point of the curve, is positive for $u \in [0,1]$, indicating that when the size of the tangent vector $\mathbf{P}_{0,u}$ increases, the curve is "pulled" forward. On the other hand, the basis function $B_3^v(u)$ is negative for $u \in [0,1]$, indicating that when the size of the vector $\mathbf{P}_{1,u}$ increases, the curve is "pushed" backward similar to that of the quadratic curve.

EXAMPLE 2.6

Given two points and two vectors, $\mathbf{P}_0 = [0,0]$, $\mathbf{P}_1 = [1,1]$, $\mathbf{P}_{0,u} = [2,0]$, and $\mathbf{P}_{1,u} = [2,0]$, derive the parametric equation and graph the Hermit cubic curve formed by them. Graph the curve by changing the tangent vector from $\mathbf{P}_{0,u} = [2,0]$ to $\mathbf{P}_{0,u} = [4,0]$, and restore the original curve shape, and then change $\mathbf{P}_{1,u} = [2,0]$ to $\mathbf{P}_{1,u} = [4,0]$.

EXAMPLE 2.6—CONT'D

Solutions

Using Eqn (2.41), we have Curve 1:

$$\mathbf{P}(u) = \mathbf{B}^v\mathbf{G}^v = [2u^3 - 3u^2 + 1, \ -2u^3 + 3u^2 + 1, \ u^3 - 2u^2 + u, \ u^3 - u^2] \begin{bmatrix} 0 & 0 \\ 1 & 1 \\ 2 & 0 \\ 2 & 0 \end{bmatrix}$$

$$= [2u^3 - 3u^2 + 2u, \ -2u^3 + 3u^2].$$

Changing the tangent vector $\mathbf{P}_{0,u} = [2,0]$ to $\mathbf{P}_{0,u} = [4,0]$, the curve becomes Curve 2:

$$\mathbf{P}(u) = \mathbf{B}^v\mathbf{G}^v = [2u^3 - 3u^2 + 1, \ -2u^3 + 3u^2 + 1, \ u^3 - 2u^2 + u, \ u^3 - u^2] \begin{bmatrix} 0 & 0 \\ 1 & 1 \\ 4 & 0 \\ 2 & 0 \end{bmatrix}$$

$$= [4u^3 - 7u^2 + 4u, \ -2u^3 + 3u^2].$$

Go back to Curve 1 and change the tangent vector $\mathbf{P}_{1,u} = [2,0]$ to $\mathbf{P}_{1,u} = [4,0]$. The curve becomes Curve 3:

$$\mathbf{P}(u) = \mathbf{B}^v\mathbf{G}^v = [2u^3 - 3u^2 + 1, \ -2u^3 + 3u^2 + 1, \ u^3 - 2u^2 + u, \ u^3 - u^2] \begin{bmatrix} 0 & 0 \\ 1 & 1 \\ 2 & 0 \\ 4 & 0 \end{bmatrix}$$

$$= [4u^3 - 5u^2 + 2u, \ -2u^3 + 3u^2].$$

The curves are graphed in Matlab with the script shown below. Curve 2 clearly shows that when the vector size at the start point of the curve increases, the curve is "pulled" forward. On the other hand, Curve 3 shows that when the vector size increases at the end point of the curve, the curve is "pushed" backward.

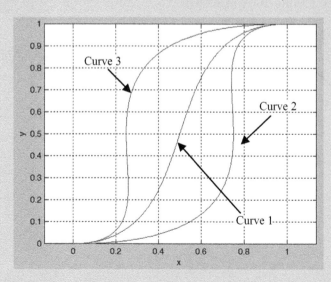

Matlab Script:

```
ezplot('2*u^3-3*u^2+2*u','-
2*u^3+3*u^2',[0,1])
grid on
hold on
ezplot('4*u^3-7*u^2+4*u','-
2*u^3+3*u^2',[0,1])
ezplot('4*u^3-5*u^2+2*u','-
2*u^3+3*u^2',[0,1])
```

2.2.3.3 Bézier curve

Following the discussion of the quadratic Bézier curve, a cubic Bézier curve consists of four control points. It can be derived as

$$\mathbf{P}(u) = \begin{bmatrix} u^3 & u^2 & u & 1 \end{bmatrix} \begin{bmatrix} -1 & 3 & -3 & 1 \\ 3 & -6 & 3 & 0 \\ -3 & 3 & 0 & 0 \\ 1 & 0 & 0 & 0 \end{bmatrix} \begin{bmatrix} \mathbf{P}_0 \\ \mathbf{P}_1 \\ \mathbf{P}_2 \\ \mathbf{P}_3 \end{bmatrix} = \mathbf{U}\mathbf{N}^B\mathbf{G}^B = \mathbf{B}^B\mathbf{G}^s \qquad (2.42)$$

where $\mathbf{N}^B$ is a constant 4×4 matrix for any given cubic Bézier curve, and $\mathbf{B}^B = [B_{0,3}(u), B_{1,3}(u), B_{2,3}(u), B_{3,3}(u)]$ is the 1×4 vector of the basis functions (Bernstein polynomials), as plotted in Figure 2.9c. Derivation of the basis functions is left as an exercise. One interesting point is that when a control point is added to the same location of existing one, the Bézier curve gets closer to the control polygon, as illustrated in Figure 2.10.

EXAMPLE 2.7

Given four control points, $\mathbf{P}_0 = [0,0]$, $\mathbf{P}_1 = [1,3]$, $\mathbf{P}_2 = [2,-2]$, and $\mathbf{P}_3 = [3,0]$, compute the parametric equation and graph the Bézier curve formed by them.

Solutions
Using Eqn (2.42), we have

$$\mathbf{P}(u) = \begin{bmatrix} u^3 & u^2 & u & 1 \end{bmatrix} \begin{bmatrix} -1 & 3 & -3 & 1 \\ 3 & -6 & 3 & 0 \\ -3 & 3 & 0 & 0 \\ 1 & 0 & 0 & 0 \end{bmatrix} \begin{bmatrix} 0 & 0 \\ 1 & 3 \\ 2 & -2 \\ 3 & 0 \end{bmatrix} = \begin{bmatrix} 3u, & 15u^3 - 24u^2 + 9u \end{bmatrix}.$$

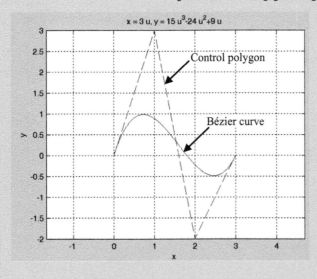

MATLAB Script:

```
ezplot('3*u','15*u^3-24*u^2+9*u',[0,1])
grid on
hold on
Px=[0 1 2 3]
Py=[0 3 -2 0]
plot(Px,Py,'--')
```

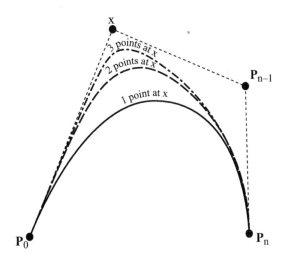

FIGURE 2.10 Increasing the "Pull" with Coincident Points at x.

2.2.4 **Continuities**

In geometric modeling, curves are joined with smooth transitions at the junction. As illustrated in Figure 2.11, when joining two curves, the end points must coincide; that is, $\mathbf{P}_1^A = \mathbf{P}_0^B$, in which $\mathbf{P}_1^A$ is the end point of curve A and $\mathbf{P}_0^B$ is the start point of curve B. This is so-called C^0-continuity. In order to maintain smoothness, the slope of the curves must be continuous across the junctions. For example, the tangent vectors of Hermit cubic curves shown in Figure 2.11a must be collinear (G^1-continuity) or identical (C^1-continuity) at the junction; that is, $\mathbf{P}_{1,u}^A = C\mathbf{P}_{0,u}^B$ ($C \neq 0$) or $\mathbf{P}_1^A = \mathbf{P}_0^B$, respectively. Also, for the Bézier curves shown in Figure 2.11b, the line segments of the respective control polygons must be either collinear (G^1-continuity) or identical (C^1-continuity); that is, $\mathbf{P}_2^A\mathbf{P}_3^A = C\mathbf{P}_0^B\mathbf{P}_1^B$ ($C \neq 0$) or $\mathbf{P}_2^A\mathbf{P}_3^A = \mathbf{P}_0^B\mathbf{P}_1^B$, respectively.

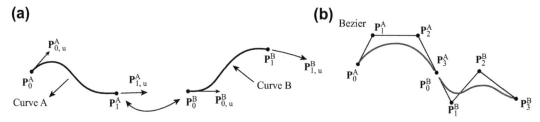

FIGURE 2.11 Curve Continuity. (a) Joining Two Hermit Cubic Curves. (b) Joining Two Bézier Curves.

2.2.5 B-spline curves

In general, the parametric curves discussed so far reveal several important characteristics. First, the polynomial order of the curve increases as more points are added. For example, three points form a quadratic spline curve, and four points form a cubic spline curve. Similarly, the polynomial order of a Bézier curve is determined by the total number of control points minus one. The issue with high-order polynomial curves is that as the polynomial order increases, there is a possibility that the curve may oscillate, which is undesirable for geometric modeling. Another characteristic of the curves we discussed so far is that when a point (or a tangent vector) is moved, the entire curve is affected. This is so-called global control, which may not be ideal when a designer only intends to alter the geometric shape of a part in a local area. Moreover, as shown in Section 2.4, conditions must be imposed at the curve junctions in order to maintain the desired geometric smoothness.

A better alternative that alleviates the above less-desirable characteristics is the B-spline curve. The power of B-spline curves is that designer can create with ease a very complex curve that is smoothly connected. The number of control points and the polynomial order are defined separately. In other words, adding a control point does not increase the polynomial order of the curve. In addition, when a control point is relocated, only a portion of the curve is affected, which is referred to as local control. One most important characteristic of the B-spline curve is that the entire curve is smooth. The derivative up to the $(p-1)^{th}$ order is continuous, where p is the polynomial order of the curve. That is, for a cubic B-spline curve, its curvature (involving a second-order derivative) is continuous throughout the curve.

2.2.5.1 Nonuniform B-spline curves

A B-spline curve, more specifically nonuniform B-spline curve, can be defined mathematically as

$$\mathbf{P}(u) = \sum_{i=0}^{n} \mathbf{P}_i N_{i,k}(u), \quad u \in [0, (n+1) - (k-1)] \tag{2.43}$$

where $n+1$ is the number of control points, $p = k-1$ is the polynomial order, and $N_{i,k}(u)$'s are the basis functions, which are defined recursively as

$$N_{i,k}(u) = \frac{(u-t_i)N_{i,k-1}(u)}{t_{i+k-1}-t_i} + \frac{(t_{i+k}-u)N_{i+1,k-1}(u)}{t_{i+k}-t_{i+1}} \tag{2.44}$$

and

$$N_{i,1}(u) = \begin{cases} 1, & t_i \le u \le t_{i+1} \\ 0, & \text{elsewhere.} \end{cases} \tag{2.45}$$

Note that t is called knots in Eqns (2.44) and (2.45), which is defined as

$$t_i = \begin{cases} 0, & i < k \\ i - k + 1, & k \le i \le n. \\ n - k + 2, & i > n \end{cases} \tag{2.46}$$

There are $n+k+1$ knots. Note that we defined $0/0 \equiv 0$ in Eqn (2.44).

We will use a quadratic B-spline curve example shown in Figure 2.12 to illustrate some of the important characteristics of the B-spline curves. In this example, six control points are given. They are $\mathbf{P}_0 = [1,0]$, $\mathbf{P}_1 = [0,1]$, $\mathbf{P}_2 = [0,2]$, $\mathbf{P}_3 = [1,4]$, $\mathbf{P}_4 = [1,6]$, and $\mathbf{P}_5 = [-3,8]$. Therefore, for this curve,

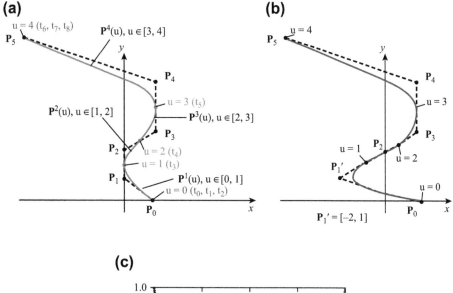

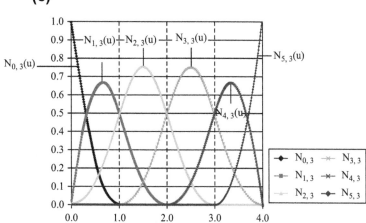

FIGURE 2.12 Quadratic B-spline Curve. (a) A Curve Defined by Six Control Points. (b) Geometric Shape of the Curve Altered Locally by Relocating a Control Point $P_1 = [0,1]$ to $P_1' = [-2,1]$. (c) Quadratic Basis Functions.

$n = 5, k = 3, u \in [0, (5+1)-(3-1) = 4]$, and we have $n + k + 1 = 5 + 3 + 1 = 9$ knots. These knots are defined, according to Eqn (2.46), as

$$t_{0,1,2} = 0$$
$$t_3 = 1$$
$$t_4 = 2 \tag{2.47}$$
$$t_5 = 3$$
$$t_{6,7,8} = 4.$$

Note that $t_{0,1,2} = 0$ and $t_{6,7,8} = 4$ are repeated knots, and $t_3 = 1$, $t_4 = 2$, and $t_5 = 3$ are nonrepeated.

The basis functions, according to Eqns (2.44) and (2.45), can be derived as follows (for details, see Appendix A):

$$N_{0,3}(u) = (1 - u)^2, \quad 0 \leq u \leq 1 \tag{2.48a}$$

$$N_{1,3}(u) = \begin{cases} \dfrac{1}{2}u(4 - 3u), & 0 \leq u \leq 1 \\[2mm] \dfrac{1}{2}(2 - u)^2, & 1 \leq u \leq 2 \end{cases} \tag{2.48b}$$

$$N_{2,3}(u) = \begin{cases} \dfrac{1}{2}u^2, & 1 \leq u \leq 2 \\[2mm] \dfrac{1}{2}(-2u^2 + 6u - 3), & 2 \leq u \leq 3 \\[2mm] \dfrac{1}{2}(3 - u)^2, & 3 \leq u \leq 4 \end{cases} \tag{2.48c}$$

$$N_{3,3}(u) = \begin{cases} \dfrac{1}{2}(u - 1)^2, & 1 \leq u \leq 2 \\[2mm] \dfrac{1}{2}(-2u^2 + 10u - 11), & 2 \leq u \leq 3 \\[2mm] \dfrac{1}{2}(4 - u)^2, & 3 \leq u \leq 4 \end{cases} \tag{2.48d}$$

$$N_{4,3}(u) = \begin{cases} \dfrac{1}{2}(u - 2)^2, & 2 \leq u \leq 3 \\[2mm] \dfrac{1}{2}(-3u^2 + 20u - 32), & 3 \leq u \leq 4 \end{cases} \tag{2.48e}$$

$$N_{5,3}(u) = (u - 3)^2, \quad 3 \leq u \leq 4. \tag{2.48f}$$

Hence from Eqn (2.43), the B-spline curve can be written as

$$
\begin{aligned}
\mathbf{P}(u) &= \sum_{i=0}^{5} \mathbf{P}_i N_{i,k}(u) \\
&= \mathbf{P}_0 N_{0,3}(u) + \mathbf{P}_1 N_{1,3}(u) + \mathbf{P}_2 N_{2,3}(u) + \mathbf{P}_3 N_{3,3}(u) + \mathbf{P}_4 N_{4,3}(u) + \mathbf{P}_5 N_{5,3}(u) \\
&= \begin{cases} (1 - u)^2 \mathbf{P}_0 + \dfrac{1}{2}u(4 - 3u)\mathbf{P}_1 + \dfrac{1}{2}u^2 \mathbf{P}_2, & 0 \leq u \leq 1 \\[2mm] \dfrac{1}{2}(2 - u)^2 \mathbf{P}_1 + \dfrac{1}{2}(-2u^2 + 6u - 3)\mathbf{P}_2 + \dfrac{1}{2}(u - 1)^2 \mathbf{P}_3, & 1 \leq u \leq 2 \\[2mm] \dfrac{1}{2}(3 - u)^2 \mathbf{P}_2 + \dfrac{1}{2}(-2u^2 + 10u - 11)\mathbf{P}_3 + \dfrac{1}{2}(u - 2)^2 \mathbf{P}_4, & 2 \leq u \leq 3 \\[2mm] \dfrac{1}{2}(4 - u)^2 \mathbf{P}_3 + \dfrac{1}{2}(-3u^2 + 20u - 32)\mathbf{P}_4 + \dfrac{1}{2}(u - 3)^2 \mathbf{P}_5, & 3 \leq u \leq 4 \end{cases}
\end{aligned} \tag{2.49}
$$

There are several important observations:

1. As illustrated in Eqn (2.49), there are actually four piecewise quadratic B-spline curve segments joined with C^1-continuity at the nonrepeated knots, that is, at $t_3 = 1$, $t_4 = 2$, and $t_5 = 3$. You may want to verify the statement regarding the C^1-continuity by taking derivatives of the curve segment equations and plugging in respective u values at the junctions.

2. The B-spline curve starts and ends at the first and last control points, respectively, and is tangent to the first and the last line segments of the control polygon, respectively, similar to that of a Bézier curve.

3. The knot $t_3 = 1$ (where $u = 1$) is the midpoint of the line segment $\mathbf{P}_1\mathbf{P}_2$ of the control polygon, which is also the junction of the first and second B-spline curve segments, $\mathbf{P}^1(u)$ and $\mathbf{P}^2(u)$, respectively. The same is true for other nonrepeated knots. For a quadratic B-spline curve, its curve segments touch the midpoint of their respective line segments of the control polygon, and are tangent to the line segments at the contact points. Note that this is only true for quadratic curves. Increasing the polynomial order of a B-spline curve results in "pulling" the curve away from its control polygon, as illustrated in Figure 2.13 with a quadratic and a cubic B-spline curves.

4. As revealed in Eqn (2.49), the four curve segments are controlled by their respective control polygons. For example, curve segment 1, $\mathbf{P}^1(u)$ with $u \in [0,1]$, is controlled by polygon $\mathbf{P}_0\mathbf{P}_1\mathbf{P}_2$; curve segment 2, $\mathbf{P}^2(u)$ with $u \in [1,2]$, is controlled by polygon $\mathbf{P}_1\mathbf{P}_2\mathbf{P}_3$; and so on. Therefore, when a control point is relocated, for example, moving control point $\mathbf{P}_1 = [0,1]$ to $\mathbf{P}_1' = [-2,1]$, as shown in Figure 2.12b, instead of the entire curve, only curve segments 1 and 2 are affected. This local-control characteristic is desirable for fine-tuning the local geometric shape for part design.

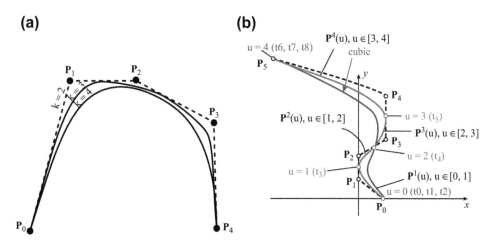

FIGURE 2.13 B-spline Curves Being "Pushed" Away from the Control Polygon When Increasing Their Polynomial Order. (a) k = 2 Linear, k = 3 Quadratic, and k = 4 Cubic Curves. (b) Quadratic and Cubic B-spline Curves Defined by the Same Set of Control Points.

5. The B-spline curve and Bézier curve are identical when $(n + 1) - (k - 1) = n - k + 2 = 1$, implying there is only one curve segment in the B-spline curve. In this case, $n = k - 1$; that is, the n and k are coupled and the polynomial order of the curve $(k - 1)$ equals n, which is the number of control points minus 1. As discussed before, adding control points to a Bézier curve increases its polynomial order. However, adding control points to a B-spline curve increases number of curve segments (and hence the parameter u domain) while keeping the same polynomial order. This is one of the most desirable characteristics of the B-spline curve in geometric modeling.

6. As illustrated in Figure 2.12c, the six basis functions are symmetric in pairs. For example, $N_{0,3}(u)$ and $N_{5,3}(u)$, and $N_{1,3}(u)$ and $N_{4,3}(u)$ are symmetric. $N_{3,3}(u)$ and $N_{4,3}(u)$ are not only symmetric, but their geometric shape is in fact identical. The basis functions $N_{3,3}(u)$ and $N_{4,3}(u)$ are referred to as uniform, which are employed to construct uniform B-spline curves to be discussed next. The remaining four basis functions are nonuniform. Uniform or nonuniform basis functions stem from the spans of the knots. As shown in Eqn (2.44), the basis functions are defined recursively, with the lowest-order functions, step functions, defined in Eqn (2.45). All basis functions are strongly influenced by knots. As illustrated in Eqn (2.47), three knots repeat at 0 and another three repeat at 4. The spans between consecutive knots are nonuniform; that is, some are 0 (between the repeated knots) and some are 1. According to Eqn (2.45), basis functions $N_{0,1} = N_{1,1} = N_{6,1} = N_{7,1} = 0$ due to the zero-span between repeated knots. As shown in Figure 2.14, such zero-span affects basis functions all the way to $N_{0,3}(u)$, $N_{1,3}(u)$, $N_{4,3}(u)$, and $N_{5,3}(u)$, which are called nonuniform. B-spline curves constructed by using the basis functions, including nonuniform ones, are called nonuniform B-spline curves. One important characteristic of a nonuniform B-spline curve is that the curve starts and ends at the first and last control points, and it is tangent at the respective end control points to the control polygon.

7. Similar to the Bernstein polynomials, the sum of all basis functions is 1; that is, $\sum_{i=0}^{n} N_{i,k}(u) = 1$, implying that the basis functions form a partition of unity. In addition, $0 \le N_{i,k}(u) \le 1$; therefore, the convex hull property prevails, which ensures that the curve lies within the convex hull of the control points, just like a Bézier curve.

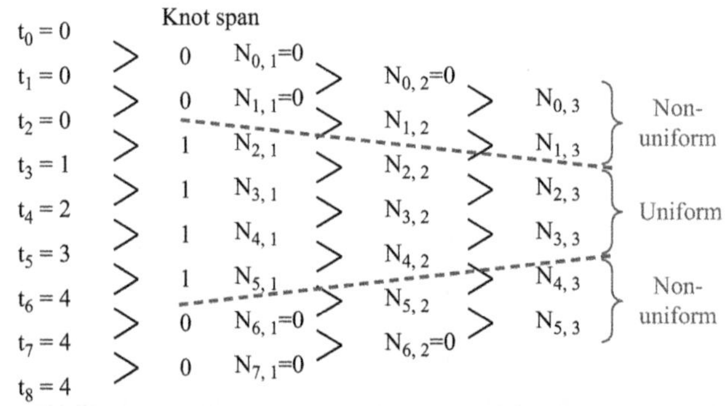

FIGURE 2.14 Uniform and Nonuniform Basis Functions Stem from the Uniform and Nonuniform Spans between Neighboring Knots.

2.2.5.2 Uniform B-spline curves

A uniform B-spline curve is constructed by only using the uniform basis functions. As discussed in the earlier example, the basis functions $N_{2,3}(u)$ and $N_{3,3}(u)$, shown in Eqns (2.48c) and (2.48d), respectively, are uniform. A B-spline curve constructed only by using such functions is called a uniform B-spline curve. By closely examining Eqns (2.48c) and (2.48d), the uniform basis functions can be generalized for quadratic curves as the following:

$$N_{i,3}(u) = \begin{cases} \dfrac{1}{2}(-u+i)^2 \\[2mm] \dfrac{1}{2}[(u-i+2)(-u+i)+(-u+i+1)(u-i+1)], \quad u \in [i-1,i]. \\[2mm] \dfrac{1}{2}(u-i+1)^2 \end{cases} \quad (2.50)$$

A B-spline curve segment can then be constructed as

$$P^i(u) = \frac{1}{2}(-u+i)^2 P_{i-1} + \frac{1}{2}[(u-i+2)(-u+i)+(-u+i+1)(u-i+1)]P_i$$
$$+ \frac{1}{2}(u-i+1)^2 P_{i+1}, \quad u \in [i-1,i]. \quad (2.51)$$

By replacing u with $u+i-1$ in Eqn (2.51) to eliminate the i in the parentheses on the right hand side of Eqn (2.51), we have

$$P^i(u) = \frac{1}{2}(1-u)^2 P_{i-1} + \frac{1}{2}(-2u^2+2u+1)P_i + \frac{1}{2}u^2 P_{i+1}, \quad u \in [0,1], \; i \in [1, n-1] \quad (2.52)$$

where $n+1$ is the total number of control points. Note that a set of $n+1$ control points defines $n-1$ quadratic uniform B-spline curve segments by using Eqn (2.52).

It is apparent that Eqn (2.52) is more desirable for constructing uniform B-spline curves because the index i is removed in the basis functions and the u domain is converted back to [0,1]. Another advantage of Eqn (2.52) is that it can be written in a matrix form, similar to those discussed before; that is,

$$P^i(u) = \begin{bmatrix} u^2 & u & 1 \end{bmatrix} \frac{1}{2} \begin{bmatrix} 1 & -2 & 1 \\ -2 & 2 & 0 \\ 1 & 1 & 0 \end{bmatrix} \begin{bmatrix} P_{i-1} \\ P_i \\ P_{i+1} \end{bmatrix} = U_{1\times3} M^3_{3\times3} \begin{bmatrix} P_{i-1} \\ P_i \\ P_{i+1} \end{bmatrix}, \quad u \in [0,1], \; i \in [1, n-1]$$
$$(2.53)$$

where M^3 is a constant 3×3 matrix. Note that a 4×4 M^4 matrix can be derived by following the same steps (left as an exercise) for a cubic uniform B-spline curve:

$$P^i(u) = \begin{bmatrix} u^3 & u^2 & u & 1 \end{bmatrix}_{1\times4} M^4_{4\times4} \begin{bmatrix} P_{i-1} \\ P_i \\ P_{i+1} \\ P_{i+2} \end{bmatrix} = U_{1\times4} M^4_{4\times4} \begin{bmatrix} P_{i-1} \\ P_i \\ P_{i+1} \\ P_{i+2} \end{bmatrix}, \quad u \in [0,1], \; i \in [1, n-2]$$
$$(2.54)$$

where

$$\mathbf{M}_{4\times4}^4 = \frac{1}{6}\begin{bmatrix} -1 & 3 & -3 & 1 \\ 3 & -6 & 3 & 0 \\ -3 & 0 & 3 & 0 \\ 1 & 4 & 1 & 0 \end{bmatrix}. \tag{2.55}$$

EXAMPLE 2.8

Use the same six control points as before, $\mathbf{P}_0 = [1,0]$, $\mathbf{P}_1 = [0,1]$, $\mathbf{P}_2 = [0,2]$, $\mathbf{P}_3 = [1,4]$, $\mathbf{P}_4 = [1,6]$, and $\mathbf{P}_5 = [-3,8]$ to construct a quadratic uniform B-spline curve like the one shown below. Note that curve segment $\mathbf{P}^1(u)$ does not start from the control point $\mathbf{P}_0$ and curve segment $\mathbf{P}^4(u)$ does not end at the last control point $\mathbf{P}_5$. All neighboring curve segments join at their respective nonrepeated knots and are tangent to the control polygon at these knots.

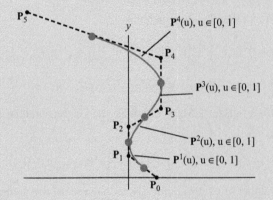

Solutions

Using Eqn (2.53), the following equations describe the respective four curve segments:

$$\mathbf{P}^1(u) = \begin{bmatrix} u^2 & u & 1 \end{bmatrix}\frac{1}{2}\begin{bmatrix} 1 & -2 & 1 \\ -2 & 2 & 0 \\ 1 & 1 & 0 \end{bmatrix}\begin{bmatrix} \mathbf{P}_0 \\ \mathbf{P}_1 \\ \mathbf{P}_2 \end{bmatrix} = \begin{bmatrix} u^2 & u & 1 \end{bmatrix}\frac{1}{2}\begin{bmatrix} 1 & -2 & 1 \\ -2 & 2 & 0 \\ 1 & 1 & 0 \end{bmatrix}\begin{bmatrix} 1 & 0 \\ 0 & 1 \\ 0 & 2 \end{bmatrix} = \begin{bmatrix} \frac{1}{2}u^2 - u + \frac{1}{2}, & u + \frac{1}{2} \end{bmatrix}$$

$$\mathbf{P}^2(u) = \begin{bmatrix} u^2 & u & 1 \end{bmatrix}\frac{1}{2}\begin{bmatrix} 1 & -2 & 1 \\ -2 & 2 & 0 \\ 1 & 1 & 0 \end{bmatrix}\begin{bmatrix} \mathbf{P}_1 \\ \mathbf{P}_2 \\ \mathbf{P}_3 \end{bmatrix} = \begin{bmatrix} u^2 & u & 1 \end{bmatrix}\frac{1}{2}\begin{bmatrix} 1 & -2 & 1 \\ -2 & 2 & 0 \\ 1 & 1 & 0 \end{bmatrix}\begin{bmatrix} 0 & 1 \\ 0 & 2 \\ 1 & 4 \end{bmatrix} = \begin{bmatrix} \frac{1}{2}u^2, & \frac{1}{2}u^2 + u + \frac{3}{2} \end{bmatrix}$$

EXAMPLE 2.8—CONT'D

$$\mathbf{P}^3(u) = \begin{bmatrix} u^2 & u & 1 \end{bmatrix} \frac{1}{2} \begin{bmatrix} 1 & -2 & 1 \\ -2 & 2 & 0 \\ 1 & 1 & 0 \end{bmatrix} \begin{bmatrix} \mathbf{P}_2 \\ \mathbf{P}_3 \\ \mathbf{P}_4 \end{bmatrix} = \begin{bmatrix} u^2 & u & 1 \end{bmatrix} \frac{1}{2} \begin{bmatrix} 1 & -2 & 1 \\ -2 & 2 & 0 \\ 1 & 1 & 0 \end{bmatrix} \begin{bmatrix} 0 & 2 \\ 1 & 4 \\ 1 & 6 \end{bmatrix} = \begin{bmatrix} -\frac{1}{2}u^2 + u + \frac{1}{2}, & 2u + 3 \end{bmatrix}$$

$$\mathbf{P}^4(u) = \begin{bmatrix} u^2 & u & 1 \end{bmatrix} \frac{1}{2} \begin{bmatrix} 1 & -2 & 1 \\ -2 & 2 & 0 \\ 1 & 1 & 0 \end{bmatrix} \begin{bmatrix} \mathbf{P}_3 \\ \mathbf{P}_4 \\ \mathbf{P}_5 \end{bmatrix} = \begin{bmatrix} u^2 & u & 1 \end{bmatrix} \frac{1}{2} \begin{bmatrix} 1 & -2 & 1 \\ -2 & 2 & 0 \\ 1 & 1 & 0 \end{bmatrix} \begin{bmatrix} 1 & 4 \\ 1 & 6 \\ -3 & 8 \end{bmatrix} = \begin{bmatrix} -2u^2 + 1, & 2u + 5 \end{bmatrix}$$

The curves are graphed in Matlab with the script shown below.

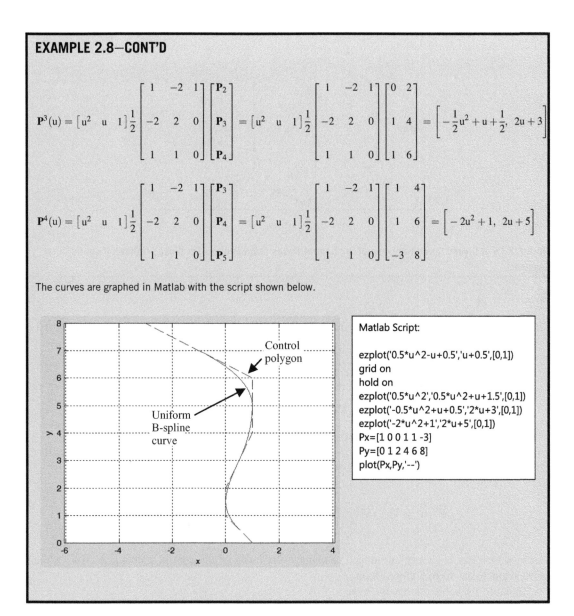

Matlab Script:

```
ezplot('0.5*u^2-u+0.5','u+0.5',[0,1])
grid on
hold on
ezplot('0.5*u^2','0.5*u^2+u+1.5',[0,1])
ezplot('-0.5*u^2+u+0.5','2*u+3',[0,1])
ezplot('-2*u^2+1','2*u+5',[0,1])
Px=[1 0 0 1 1 -3]
Py=[0 1 2 4 6 8]
plot(Px,Py,'--')
```

2.2.5.3 Closed uniform B-spline curves

The curve shown above is an open B-spline curve, in which the start and end control points do not coincide. Uniform B-spline curves are well suited for modeling part geometry of a smooth closed profile. In this case, its control polygon must be closed, which can be achieved by simply aligning the first and the last control points. For example, the six control points shown in Figure 2.15 form a closed

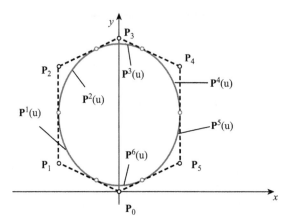

FIGURE 2.15 A Closed Control Polygon of Six Control Points that Encloses Six Closed Uniform B-spline Curve Segments.

control polygon by connecting control points $\mathbf{P}_5$ back to $\mathbf{P}_0$. A quadratic uniform B-spline curve can be constructed by using Eqn (2.53) as

$$\mathbf{P}^i(u) = \mathbf{U}_{1\times3}\mathbf{M}^3_{3\times3}\begin{bmatrix} \mathbf{P}_{(i-1)\mathrm{mod}(n+1)} \\ \mathbf{P}_{(i)\mathrm{mod}(n+1)} \\ \mathbf{P}_{(i+1)\mathrm{mod}(n+1)} \end{bmatrix}, \quad u \in [0, 1], \quad i \in [1, n+1] \tag{2.56}$$

in which "mod" is the remaining operator. For example, if $i = 6$ and $n = 5$, then $(i-1)\mathrm{mod}(n+1) = 5 \bmod 6 = 5$; $(i) \bmod(n+1) = 6 \bmod 6 = 0$; and $(i+1)\mathrm{mod}(n+1) = 7 \bmod 6 = 1$. Therefore, from Eqn (2.56), the sixth curve segment shown in Figure 2.15 can be found as

$$\mathbf{P}^6(u) = \mathbf{U}_{1\times3}\mathbf{M}^3_{3\times3}\begin{bmatrix} \mathbf{P}_{(6-1)\mathrm{mod}(5+1)} \\ \mathbf{P}_{(6)\mathrm{mod}(5+1)} \\ \mathbf{P}_{(6+1)\mathrm{mod}(5+1)} \end{bmatrix} = \mathbf{U}_{1\times3}\mathbf{M}^3_{3\times3}\begin{bmatrix} \mathbf{P}_5 \\ \mathbf{P}_0 \\ \mathbf{P}_1 \end{bmatrix}.$$

The mod operator is simply introduced to manage the index of the control points as well as adding curve segments to form a closed loop.

Similarly, for cubic curves, we have

$$\mathbf{P}^i(u) = \mathbf{U}_{1\times4}\mathbf{M}^4_{4\times4}\begin{bmatrix} \mathbf{P}_{(i-1)\mathrm{mod}(n+1)} \\ \mathbf{P}_{(i)\mathrm{mod}(n+1)} \\ \mathbf{P}_{(i+1)\mathrm{mod}(n+1)} \\ \mathbf{P}_{(i+2)\mathrm{mod}(n+1)} \end{bmatrix}, \quad u \in [0, 1], \quad i \in [1, n+1]. \tag{2.57}$$

The following example illustrates the characteristics of the closed uniform B-spline curves in more detail, using both quadratic and cubic curves.

EXAMPLE 2.9

Use the four control points, $\mathbf{P}_0 = [1,0]$, $\mathbf{P}_1 = [2,1]$, $\mathbf{P}_2 = [1,2]$, and $\mathbf{P}_3 = [0,1]$, which form a closed control polygon, to construct both a quadratic and a cubic uniform B-spline curve, similar to those shown below.

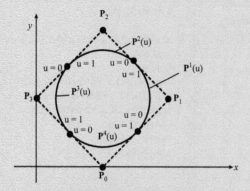

The closed quadratic uniform B-spline curve

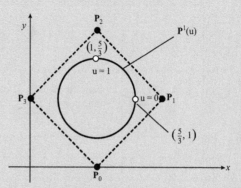

The closed cubic uniform B-spline curve

Using Eqn (2.56) for a quadratic curve, we have

$$\mathbf{P}^1(u) = \begin{bmatrix} u^2 & u & 1 \end{bmatrix} \frac{1}{2} \begin{bmatrix} 1 & -2 & 1 \\ -2 & 2 & 0 \\ 1 & 1 & 0 \end{bmatrix} \begin{bmatrix} \mathbf{P}_0 \\ \mathbf{P}_1 \\ \mathbf{P}_2 \end{bmatrix} = \begin{bmatrix} u^2 & u & 1 \end{bmatrix} \frac{1}{2} \begin{bmatrix} 1 & -2 & -1 \\ -2 & 2 & 0 \\ 1 & 1 & 0 \end{bmatrix} \begin{bmatrix} 1 & 0 \\ 2 & 1 \\ 1 & 2 \end{bmatrix} = \begin{bmatrix} -u^2 + u + \frac{3}{2}, u + \frac{1}{2} \end{bmatrix}$$

$$\mathbf{P}^2(u) = \begin{bmatrix} u^2 & u & 1 \end{bmatrix} \frac{1}{2} \begin{bmatrix} 1 & -2 & 1 \\ -2 & 2 & 0 \\ 1 & 1 & 0 \end{bmatrix} \begin{bmatrix} \mathbf{P}_1 \\ \mathbf{P}_2 \\ \mathbf{P}_3 \end{bmatrix} = \begin{bmatrix} u^2 & u & 1 \end{bmatrix} \frac{1}{2} \begin{bmatrix} 1 & -2 & 1 \\ -2 & 2 & 0 \\ 1 & 1 & 0 \end{bmatrix} \begin{bmatrix} 2 & 1 \\ 1 & 2 \\ 0 & 1 \end{bmatrix} = \begin{bmatrix} -u + \frac{3}{2}, -u^2 + u + \frac{3}{2} \end{bmatrix}$$

$$\mathbf{P}^3(u) = \begin{bmatrix} u^2 & u & 1 \end{bmatrix} \frac{1}{2} \begin{bmatrix} 1 & -2 & 1 \\ -2 & 2 & 0 \\ 1 & 1 & 0 \end{bmatrix} \begin{bmatrix} \mathbf{P}_2 \\ \mathbf{P}_3 \\ \mathbf{P}_0 \end{bmatrix} = \begin{bmatrix} u^2 & u & 1 \end{bmatrix} \frac{1}{2} \begin{bmatrix} 1 & -2 & 1 \\ -2 & 2 & 0 \\ 1 & 1 & 0 \end{bmatrix} \begin{bmatrix} 1 & 2 \\ 0 & 1 \\ 1 & 0 \end{bmatrix} = \begin{bmatrix} u^2 - u + \frac{1}{2}, -u + \frac{3}{2} \end{bmatrix}$$

$$\mathbf{P}^4(u) = \begin{bmatrix} u^2 & u & 1 \end{bmatrix} \frac{1}{2} \begin{bmatrix} 1 & -2 & 1 \\ -2 & 2 & 0 \\ 1 & 1 & 0 \end{bmatrix} \begin{bmatrix} \mathbf{P}_3 \\ \mathbf{P}_0 \\ \mathbf{P}_1 \end{bmatrix} = \begin{bmatrix} u^2 & u & 1 \end{bmatrix} \frac{1}{2} \begin{bmatrix} 1 & -2 & 1 \\ -2 & 2 & 0 \\ 1 & 1 & 0 \end{bmatrix} \begin{bmatrix} 0 & 1 \\ 1 & 0 \\ 2 & 1 \end{bmatrix} = \begin{bmatrix} u + \frac{1}{2}, u^2 - u + \frac{1}{2} \end{bmatrix}$$

Continued

EXAMPLE 2.9—CONT'D

The curve is graphed in Matlab with the script shown below.

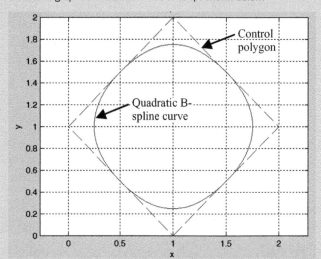

Matlab Script:

```
ezplot('-u^2+u+1.5','u+0.5',[0,1])
grid on
hold on
ezplot('-u+1.5','-u^2+u+1.5',[0,1])
ezplot('u^2-u+0.5','-u+1.5',[0,1])
ezplot('u+0.5','u^2-u+0.5',[0,1])
Px=[1 2 1 0 1]
Py=[0 1 2 1 0]
plot(Px,Py,'--')
```

Now, we use Eqn (2.57) for a cubic curve.

$$\mathbf{P}^1(u) = \begin{bmatrix} u^3 & u^2 & u & 1 \end{bmatrix} \mathbf{M}^4 \begin{bmatrix} \mathbf{P}_0 \\ \mathbf{P}_1 \\ \mathbf{P}_2 \\ \mathbf{P}_4 \end{bmatrix} = \begin{bmatrix} u^3 & u^2 & u & 1 \end{bmatrix} \frac{1}{6} \begin{bmatrix} -1 & 3 & -3 & 1 \\ 3 & -6 & 3 & 0 \\ -3 & 0 & 3 & 0 \\ 1 & 4 & 1 & 0 \end{bmatrix} \begin{bmatrix} 1 & 0 \\ 2 & 1 \\ 1 & 2 \\ 0 & 1 \end{bmatrix} = \begin{bmatrix} \frac{1}{3}u^3 - u^2 + \frac{5}{3}, & -\frac{1}{3}u^3 + u + 1 \end{bmatrix}$$

$$\mathbf{P}^2(u) = \begin{bmatrix} u^3 & u^2 & u & 1 \end{bmatrix} \mathbf{M}^4 \begin{bmatrix} \mathbf{P}_1 \\ \mathbf{P}_2 \\ \mathbf{P}_3 \\ \mathbf{P}_0 \end{bmatrix} = \begin{bmatrix} u^3 & u^2 & u & 1 \end{bmatrix} \frac{1}{6} \begin{bmatrix} -1 & 3 & -3 & 1 \\ 3 & -6 & 3 & 0 \\ -3 & 0 & 3 & 0 \\ 1 & 4 & 1 & 0 \end{bmatrix} \begin{bmatrix} 2 & 1 \\ 1 & 2 \\ 0 & 1 \\ 1 & 0 \end{bmatrix} = \begin{bmatrix} \frac{1}{3}u^3 - u + 1, & \frac{1}{3}u^3 - u^2 + \frac{5}{3} \end{bmatrix}$$

$$\mathbf{P}^3(u) = \begin{bmatrix} u^3 & u^2 & u & 1 \end{bmatrix} \mathbf{M}^4 \begin{bmatrix} \mathbf{P}_2 \\ \mathbf{P}_3 \\ \mathbf{P}_0 \\ \mathbf{P}_1 \end{bmatrix} = \begin{bmatrix} u^3 & u^2 & u & 1 \end{bmatrix} \frac{1}{6} \begin{bmatrix} -1 & 3 & -3 & 1 \\ 3 & -6 & 3 & 0 \\ -3 & 0 & 3 & 0 \\ 1 & 4 & 1 & 0 \end{bmatrix} \begin{bmatrix} 1 & 2 \\ 0 & 1 \\ 1 & 0 \\ 2 & 1 \end{bmatrix} = \begin{bmatrix} -\frac{1}{3}u^3 + u^2 + \frac{1}{3}, & \frac{1}{3}u^3 - u + 1 \end{bmatrix}$$

$$\mathbf{P}^4(u) = \begin{bmatrix} u^3 & u^2 & u & 1 \end{bmatrix} \mathbf{M}^4 \begin{bmatrix} \mathbf{P}_3 \\ \mathbf{P}_0 \\ \mathbf{P}_1 \\ \mathbf{P}_2 \end{bmatrix} = \begin{bmatrix} u^3 & u^2 & u & 1 \end{bmatrix} \frac{1}{6} \begin{bmatrix} -1 & 3 & -3 & 1 \\ 3 & -6 & 3 & 0 \\ -3 & 0 & 3 & 0 \\ 1 & 4 & 1 & 0 \end{bmatrix} \begin{bmatrix} 0 & 1 \\ 1 & 0 \\ 2 & 1 \\ 1 & 2 \end{bmatrix} = \begin{bmatrix} -\frac{1}{3}u^3 + u + 1, & -\frac{1}{3}u^3 + u^2 + \frac{1}{3} \end{bmatrix}$$

EXAMPLE 2.9—CONT'D

The curve is graphed in Matlab with the script shown below.

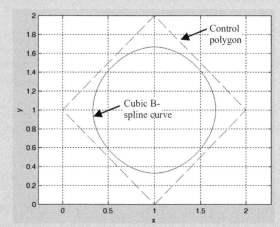

Matlab Script:

```
ezplot('1/3*u^3-u^2+5/3','-1/3*u^3+u+1',[0,1])
grid on
hold on
ezplot('1/3*u^3-u+1','1/3*u^3-u^2+5/3',[0,1])
ezplot('-1/3*u^3+u^2+1/3','1/3*u^3-u+1',[0,1])
ezplot('-1/3*u^3+u+1','-
1/3*u^3+u^2+1/3',[0,1])
Px=[1 2 1 0 1]
Py=[0 1 2 1 0]
plot(Px,Py,'--')
```

As shown in this example, unlike the quadratic B-spline curve, the cubic curve does not contact the control polygon, as also indicated in Figure 2.13. The cubic B-spline is composed of four curve segments. How smooth is the cubic curve? Is the continuity C^1 or C^2 at the junctions of the curve segments? What are the Cartesian coordinates of the junction points of curve segments? For example, for curve segment 1, what are the locations of its start and end points? This is left as an exercise. The cubic curve looks much like a circle: Is it a true circle?

2.2.6 NURB curves

Can a B-spline curve represent a true circle? The answer is no. The polynomial order of a B-spline curve is finite. For a cubic curve, its polynomial order is 3; for a true circle, its polynomial order is infinite. Why infinite? As discussed in Section 2.2.2.1, a true circle on a plane can be represented in a polar coordinate system as

$$x = a + r\cos\theta, \text{ and } y = b + r\sin\theta, \quad \theta \in [0, 2\pi]. \tag{2.58}$$

A Taylor series expansion for the trigonometry functions above, such as $\sin\theta$, is

$$\sin\theta = \theta - \frac{\theta^3}{3!} + \frac{\theta^5}{5!} - \frac{\theta^7}{7!} + \dots = \sum_{n=1}^{\infty} \frac{(-1)^{n-1}\theta^{2n-1}}{(2n-1)!} \tag{2.59}$$

which is a function of an infinite polynomial order.

A parametric curve that is capable of representing geometric entities, such as a circle or any other conic curves, is NURB, which is one of the most versatile and general curves employed for geometric modeling.

Mathematically, a NURB curve is defined as

$$\mathbf{P}(u) = \frac{\sum_{i=0}^{n} h_i \mathbf{P}_i N_{i,k}(u)}{\sum_{i=0}^{n} h_i N_{i,k}(u)}, \quad u \in [0, (n+1) - (k-1)] \tag{2.60}$$

where $N_{i,k}(u)$'s are the basis functions of the B-spline curve (discussed previously), $\mathbf{P}_i$ is the ith control point, h_i is the weight associated with the control point $\mathbf{P}_i$, and $n + 1$ is the total number of control points.

Note that when all the weights are in unity (i.e., $h_i = 1$), the NURB curve is no longer rational; in fact, it degenerates to a B-spline curve because the sum of the B-spline basis functions in the denominator of Eqn (2.60) is 1; i.e., $\sum_{i=0}^{n} h_i N_{i,k}(u) = \sum_{i=0}^{n} N_{i,k}(u) = 1$.

In addition, the weights h_i play a significant role in determining the geometric shape of the NURB curve. For example, for a quadratic NURB of $n = 2$ and $k = 3$ shown in Figure 2.16a, we set $h_0 = h_2 = 1$, and vary h_1. When $h_1 = 0$, the curve becomes a straight line connecting $\mathbf{P}_0$ and $\mathbf{P}_2$. When h_1 is increased with a positive value, the curve is "pulled" closer to the control polygon. A negative h_1 value is "pushing" the curve to the opposite of the control polygon. Note that the convex hull property does not hold if $h_1 < 0$.

Now, is it possible to find a value for the weight h_1 that allows the quadratic NURB curve enclosed by the control polygon to analytically represent a 90° circular arc of radius 1? The answer is yes (e.g., see Figure 2.16b). Let's take a look at the NURB curve in Example 2.10.

EXAMPLE 2.10

Use the three control points, $\mathbf{P}_0 = [0,1]$, $\mathbf{P}_1 = [1,1]$, and $\mathbf{P}_2 = [1,0]$ shown in Figure 2.16b, which form a control polygon to construct a quadratic NURB curve that represents a 90-degree circular arc of radius 1 analytically.

Solutions

Any conic (including circles) can be parameterized in terms of rational quadratic functions. Hence, an arc of a conic has a NURB representation (Piegl and Tiller, 1987). It is shown mathematically in Appendix B that when using a quadratic NURB curve enclosed by a control polygon $\mathbf{P}_0\mathbf{P}_1\mathbf{P}_2$ to represent a circular arc, the weight h_1 is determined as $h_1 = \sin a$, where a is the angle of $\mathbf{P}_0\mathbf{P}_1\mathbf{0}$ shown in Figure 2.16b. Therefore, for this example, the angle a is 45°; therefore, $h_1 = \sin a = \frac{1}{\sqrt{2}}$, and the NURB curve can be written using Eqn (2.60) as

$$\mathbf{P}(u) = \frac{\sum_{i=0}^{2} h_i \mathbf{P}_i N_{i,3}(u)}{\sum_{i=0}^{2} h_i N_{i,3}(u)} = \frac{h_0 \mathbf{P}_0 N_{0,3}(u) + h_1 \mathbf{P}_1 N_{1,3}(u) + h_2 \mathbf{P}_2 N_{2,3}(u)}{h_0 N_{0,3}(u) + h_1 N_{1,3}(u) + h_2 N_{2,3}(u)}$$

$$= \frac{(1)[0\ 1](1-u)^2 + \dfrac{1}{\sqrt{2}}[1\ 1]2u(1-u) + (1)[1\ 0]u^2}{(1)(1-u)^2 + \dfrac{1}{\sqrt{2}}2u(1-u) + (1)u^2}.$$

EXAMPLE 2.10—CONT'D

And in component form, we have

$$P_x(u) = \frac{\sqrt{2}u(1-u) + u^2}{(1-u)^2 + \sqrt{2}u(1-u) + u^2}$$

$$P_y(u) = \frac{(1-u)^2 + \sqrt{2}u(1-u)}{(1-u)^2 + \sqrt{2}u(1-u) + u^2}.$$

The NURB curve and a true circular arc are graphed in Matlab with the script shown below.

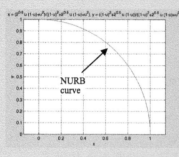
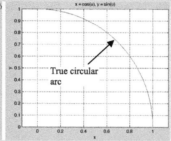

Matlab Script:

```
ezplot('(2^0.5*u*(1-u)+u^2)/((1-
u)^2+2^0.5*u*(1-u)+u^2)', '((1-
u)^2+2^0.5*u*(1-u))/((1-
u)^2+2^0.5*u*(1-u)+u^2)',[0,1])
grid on
hold on
Px=[0 1 1]
Py=[1 1 0]
plot(Px,Py,'--')
ezplot('cos(u)', 'sin(u)',[0,3.14159/2])
```

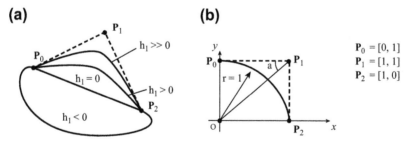

FIGURE 2.16 Quadratic NURB Curve. (a) Effect of the Weight h_1 to the Geometric Shape of the Curve. (b) Representing a 90-degree circular arc.

2.3 Parametric surfaces

A parametric surface is a surface in the Euclidean space R^3, which is defined by parametric equations with two parameters (u,w). Parametric representation is probably the most general way to specify a surface. The curvature and arc length of curves on the surface, surface area, differential geometric invariants such as the first and second fundamental forms, Gaussian, mean, and principal curvatures can all be computed from a given parameterization. Due to their generality, parametric surfaces are widely adopted in geometric modeling for support of product design and manufacturing, among many other applications.

In this section, we discuss several parametric surfaces that are commonly found in geometric modeling. In the next section (Section 2.4), we focus on CAD-generated surfaces represented in parametric form.

2.3.1 Parametric representation

Similar to parametric curves, a parametric surface in space can be written in the following parametric form:

$$\mathbf{S}(u, w) = \left[S_x(u, w), S_y(u, w), S_z(u, w)\right]_{1\times3}, \quad (u, w) \in [0, 1] \times [0, 1] \tag{2.61}$$

where u and w are the parametric coordinates of the surface. Usually, these parametric coordinates range between 0 and 1.

2.3.1.1 Bicubic surface patch

A bicubic surface patch can be defined in terms of cubic polynomials as

$$
\begin{aligned}
\mathbf{S}(u, w) &= \sum_{i=0}^{3} \sum_{j=0}^{3} \mathbf{a}_{ij} u^i w^j \\
&= \mathbf{a}_{33} u^3 w^3 + \mathbf{a}_{32} u^3 w^2 + \mathbf{a}_{31} u^3 w + \mathbf{a}_{30} u^3 + \mathbf{a}_{23} u^2 w^3 + \mathbf{a}_{22} u^2 w^2 + \mathbf{a}_{21} u^2 w \\
&\quad + \mathbf{a}_{20} u^2 + \mathbf{a}_{13} u w^3 + \mathbf{a}_{12} u w^2 + \mathbf{a}_{11} u w + \mathbf{a}_{10} u + \mathbf{a}_{03} w^3 + \mathbf{a}_{02} w^2 + \mathbf{a}_{01} w \\
&\quad + \mathbf{a}_{00}, \quad (u, w) \in [0, 1] \times [0, 1]
\end{aligned}
\tag{2.62}
$$

where $\mathbf{a}_{ij}$ is a 1×3 vector; that is, $\mathbf{a}_{ij} = [a_{ijx}, a_{ijy}, a_{ijz}]$. Hence, for example, the x-component of a parametric surface is

$$
\begin{aligned}
S_x(u, w) &= \sum_{i=0}^{3} \sum_{j=0}^{3} a_{ijx} u^i w^j \\
&= a_{33x} u^3 w^3 + a_{32x} u^3 w^2 + a_{31x} u^3 w + a_{30x} u^3 \\
&\quad + a_{23x} u^2 w^3 + a_{22x} u^2 w^2 + a_{21x} u^2 w + a_{20x} u^2 \\
&\quad + a_{13x} u w^3 + a_{12x} u w^2 + a_{11x} u w + a_{10x} u \\
&\quad + a_{03x} w^3 + a_{02x} w^2 + a_{01x} w + a_{00x}
\end{aligned}
\tag{2.63a}
$$

$$
= \begin{bmatrix} u^3 & u^2 & u & 1 \end{bmatrix}
\begin{bmatrix}
a_{33x} & a_{32x} & a_{31x} & a_{30x} \\
a_{23x} & a_{22x} & a_{21x} & a_{20x} \\
a_{13x} & a_{12x} & a_{11x} & a_{10x} \\
a_{03x} & a_{02x} & a_{01x} & a_{00x}
\end{bmatrix}
\begin{bmatrix} w^3 \\ w^2 \\ w \\ 1 \end{bmatrix}
$$

$$
= \mathbf{U}_{1\times4} \mathbf{A}_{x_{4\times4}} \mathbf{W}^T_{4\times1}, \quad (u, w) \in [0, 1] \times [0, 1]
$$

where $\mathbf{A_x}$ is a 4×4 matrix of 16 coefficients, which are to be determined. Similarly,

$$S_y(u, w) = \begin{bmatrix} u^3 & u^2 & u & 1 \end{bmatrix} \begin{bmatrix} a_{33y} & a_{32y} & a_{31y} & a_{30y} \\ a_{23y} & a_{22y} & a_{21y} & a_{20y} \\ a_{13y} & a_{12y} & a_{11y} & a_{10y} \\ a_{03y} & a_{02y} & a_{01y} & a_{00y} \end{bmatrix} \begin{bmatrix} w^3 \\ w^2 \\ w \\ 1 \end{bmatrix} \tag{2.63b}$$

$$= \mathbf{U}_{1\times4}\mathbf{A}_{y_{4\times4}}\mathbf{W}_{4\times1}^T, \quad (u, w) \in [0, 1] \times [0, 1]$$

and

$$S_z(u, w) = \begin{bmatrix} u^3 & u^2 & u & 1 \end{bmatrix} \begin{bmatrix} a_{33z} & a_{32z} & a_{31z} & a_{30z} \\ a_{23z} & a_{22z} & a_{21z} & a_{20z} \\ a_{13z} & a_{12z} & a_{11z} & a_{10z} \\ a_{03z} & a_{02z} & a_{01z} & a_{00z} \end{bmatrix} \begin{bmatrix} w^3 \\ w^2 \\ w \\ 1 \end{bmatrix} \tag{2.63c}$$

$$= \mathbf{U}_{1\times4}\mathbf{A}_{z_{4\times4}}\mathbf{W}_{4\times1}^T, \quad (u, w) \in [0, 1] \times [0, 1].$$

Similarly to Eqn (2.63), Eqn (2.62) can be written in a matrix form as

$$\mathbf{S}(u, w) = \begin{bmatrix} u^3 & u^2 & u & 1 \end{bmatrix} \begin{bmatrix} \mathbf{a}_{33} & \mathbf{a}_{32} & \mathbf{a}_{31} & \mathbf{a}_{30} \\ \mathbf{a}_{23} & \mathbf{a}_{22} & \mathbf{a}_{21} & \mathbf{a}_{20} \\ \mathbf{a}_{13} & \mathbf{a}_{12} & \mathbf{a}_{11} & \mathbf{a}_{10} \\ \mathbf{a}_{03} & \mathbf{a}_{02} & \mathbf{a}_{01} & \mathbf{a}_{00} \end{bmatrix} \begin{bmatrix} w^3 \\ w^2 \\ w \\ 1 \end{bmatrix} \tag{2.64}$$

$$= \mathbf{U}_{1\times4}\mathbf{A}_{4\times4\times3}\mathbf{W}_{4\times1}^T, \quad (u, w) \in [0, 1] \times [0, 1]$$

where $\mathbf{A}$ is a 4×4×3 matrix of 48 coefficients (or a tensor of order 2), which are to be determined. Note that in Eqn (2.64), the sizes of vectors and matrix do not match; thus, the multiplications cannot be actually carried out. We simply use the equation to describe the parametric surface in a more compact form. When performing multiplications, the x-, y-, and z-components of the surface equations in Eqn (2.61) must be carried out separately (e.g., like that of Eqn (2.63a) for the x-component of the surface).

2.3.1.2 16-Point format

A bicubic surface patch can be created by 16 distinct points arranged in a 4×4 matrix form, as shown in Figure 2.17a. Similar to the cubic spline curve, these points are assumed at the 0, 1/3, 2/3, and 1 locations of the parametric coordinates u and w; hence, the surface equation can be written as

$$\mathbf{S}(u, w) = \mathbf{U}\mathbf{N}^s\mathbf{G}^s\mathbf{N}^{s^T}\mathbf{W}^T, \quad (u, w) \in [0, 1] \times [0, 1] \tag{2.65}$$

where

$$\mathbf{N}^s = \begin{bmatrix} -9/2 & 27/2 & -27/2 & 9/2 \\ 9 & -45/2 & 18 & -9/2 \\ -11/2 & 9 & -9/2 & 1 \\ 1 & 0 & 0 & 0 \end{bmatrix} \tag{2.66}$$

which is identical to that of the cubic spline curve, and

$$\mathbf{G}^s = \begin{bmatrix} \mathbf{P}_{00} & \mathbf{P}_{01} & \mathbf{P}_{02} & \mathbf{P}_{03} \\ \mathbf{P}_{10} & \mathbf{P}_{11} & \mathbf{P}_{12} & \mathbf{P}_{13} \\ \mathbf{P}_{20} & \mathbf{P}_{21} & \mathbf{P}_{22} & \mathbf{P}_{23} \\ \mathbf{P}_{30} & \mathbf{P}_{31} & \mathbf{P}_{32} & \mathbf{P}_{33} \end{bmatrix}_{4\times4\times3} = \begin{bmatrix} \mathbf{P}(0,0) & \mathbf{P}\left(0,\frac{1}{3}\right) & \mathbf{P}\left(0,\frac{2}{3}\right) & \mathbf{P}(0,1) \\ \mathbf{P}\left(\frac{1}{3},0\right) & \mathbf{P}\left(\frac{1}{3},\frac{1}{3}\right) & \mathbf{P}\left(\frac{1}{3},\frac{2}{3}\right) & \mathbf{P}\left(\frac{1}{3},1\right) \\ \mathbf{P}\left(\frac{2}{3},0\right) & \mathbf{P}\left(\frac{2}{3},\frac{1}{3}\right) & \mathbf{P}\left(\frac{2}{3},\frac{2}{3}\right) & \mathbf{P}\left(\frac{2}{3},1\right) \\ \mathbf{P}(1,0) & \mathbf{P}\left(1,\frac{1}{3}\right) & \mathbf{P}\left(1,\frac{2}{3}\right) & \mathbf{P}(1,1) \end{bmatrix}_{4\times4\times3}$$

(2.67)

which is defined by the Cartesian coordinates of the 16 points.

2.3.1.3 Coons patch

A Coons patch (named after Steven Anson Coons, 1912–1979) is a bicubic parametric surface formed by four corner points, eight tangent vectors (two vectors in the u and w directions, respectively, at each of the four corners), and four twister vectors at the respective four corner points, as shown in Figure 2.17b.

Mathematically, a Coons patch is defined as

$$\mathbf{S}(u,w) = \mathbf{U}\,\mathbf{N}^v\,\mathbf{G}^v\mathbf{N}^{v^{\mathrm{T}}}\mathbf{W}^{\mathrm{T}}, \quad (u,w) \in [0,1] \times [0,1]$$

(2.68)

where

$$\mathbf{N}^v = \begin{bmatrix} 2 & -2 & 1 & 1 \\ -3 & 3 & -2 & -1 \\ 0 & 0 & 1 & 0 \\ 1 & 0 & 0 & 0 \end{bmatrix}$$

(2.69)

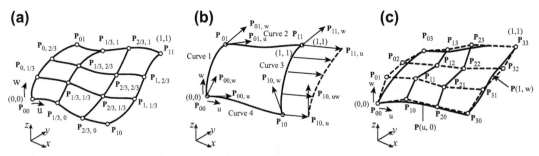

FIGURE 2.17 Bicubic Surface Patches as Defined by (a) 16 Distinct Points; (b) Corner Points, Tangents Vectors, and Twist Vectors (Coons Patch); and (c) 16 Control Points (Bézier Surface Patch).

which is identical to that of the Hermit cubic curve (two point and two vector format), and

$$
\mathbf{G}^v = \begin{bmatrix}
\mathbf{P}_{00} & \mathbf{P}_{01} & \mathbf{P}_{00,w} & \mathbf{P}_{01,w} \\
\mathbf{P}_{10} & \mathbf{P}_{11} & \mathbf{P}_{10,w} & \mathbf{P}_{11,w} \\
\mathbf{P}_{00,u} & \mathbf{P}_{01,u} & \mathbf{P}_{00,uw} & \mathbf{P}_{01,uw} \\
\mathbf{P}_{10,u} & \mathbf{P}_{11,u} & \mathbf{P}_{10,uw} & \mathbf{P}_{11,uw}
\end{bmatrix}_{4\times4\times3}
\tag{2.70}
$$

where $\mathbf{P}_{00} = \mathbf{S}(0,0)$, $\mathbf{P}_{01} = \mathbf{S}(0,1)$, $\mathbf{P}_{10} = \mathbf{S}(1,0)$, and $\mathbf{P}_{11} = \mathbf{S}(1,1)$ are the four corner points; $\mathbf{P}_{00,u} = \partial\mathbf{S}/\partial u|_{u=w=0}$, $\mathbf{P}_{01,u}$, $\mathbf{P}_{10,u}$, and $\mathbf{P}_{11,u}$ are the tangent vectors in the u direction at the four corner points; $\mathbf{P}_{00,w} = \partial\mathbf{S}/\partial w|_{u=w=0}$, $\mathbf{P}_{01,w}$, $\mathbf{P}_{10,w}$, and $\mathbf{P}_{11,w}$ are the tangent vectors in the w direction at the four corner points; and $\mathbf{P}_{00,uw} = \partial^2\mathbf{S}/\partial u\partial w|_{u=w=0}$, $\mathbf{P}_{01,uw}$, $\mathbf{P}_{10,uw}$, and $\mathbf{P}_{11,uw}$ are the twister vectors at the four corner points.

Note that a twister vector represents changes of tangent vector in u (or w) direction at a corner point along a boundary curve in the w (or u) direction. For example, $\mathbf{P}_{10,uw} = \partial/\partial w(\partial\mathbf{S}/\partial u)|_{u=1,\,w=0} = \partial\mathbf{P}_{10,u}/\partial w$ is the derivative of the tangent vector along the u direction at $\mathbf{P}_{10}$ with respect to w (i.e., along boundary curve 3 shown in Figure 2.17b). Geometrically, this twister vector represents the changes of the tangent vector $\mathbf{P}_{10,u}$ along boundary curve 3, as shown in Figure 2.17b. The same twister vector can also be interpreted as $\mathbf{P}_{10,uw} = \partial/\partial u(\partial\mathbf{S}/\partial w)|_{u=1,\,w=0} = \partial\mathbf{P}_{10,w}/\partial u$, which is the derivative of the tangent vector along the w direction at $\mathbf{P}_{10}$ with respect to u, representing the changes of the tangent vector $\mathbf{P}_{10,w}$ along boundary curve 4. Also, the first two rows of the matrix $\mathbf{G}^v$ are boundary curves 1 and 3, respective; and columns 1 and 2 are boundary curves 4 and 2, respectively.

C^0-continuity of composite Coons patches can be imposed by joining their neighboring boundary edges. For example, to ensure C^0-continuity across the two patches A and B, the patches depicted in Figure 2.18a must have

$$
\mathbf{S}^A(1,w) = \mathbf{S}^B(0,w), \text{ or } \mathbf{P}^A_{10} = \mathbf{P}^B_{00}, \mathbf{P}^A_{11} = \mathbf{P}^B_{00}, \mathbf{P}^A_{10,w} = \mathbf{P}^B_{00,w}, \text{ and } \mathbf{P}^A_{11,w} = \mathbf{P}^B_{00,w}.
\tag{2.71}
$$

For G^1-continuity, the tangent vectors across the joining boundary of the surfaces must be collinear; that is,

$$
\mathbf{S}^A_{,u}(1,w) = C\mathbf{S}^B_{,u}(0,w), \text{ or } \mathbf{P}^A_{10,u} = C\mathbf{P}^B_{00,u}, \mathbf{P}^A_{11,u} = C\mathbf{P}^B_{00,u}, \mathbf{P}^A_{10,uw} = C\mathbf{P}^B_{00,uw}, \text{ and } \mathbf{P}^A_{11,uw}
$$
$$
= C\mathbf{P}^B_{00,uw}, C \neq 0.
\tag{2.72}
$$

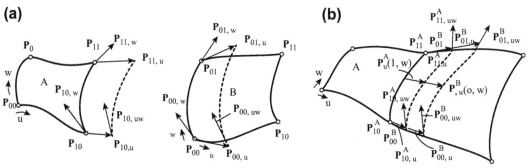

FIGURE 2.18 Continuity of Composite Coons Patches. (a) C^0-continuity. (b) G^1- or C^1-continuity.

For C^1-continuity, the constant $C = 1$.

The Coons patch is popular for support of geometric modeling. One key characteristic of the patch is that its geometric shape can be controlled or adjusted by not only altering the corner points, but also the tangent vectors and twister vectors. By imposing required properties on these vectors, we can designate a Coons patch to represent specific surfaces. For example, a cylindrical surface shown in Figure 2.19 can be represented by a Coons patch, which is illustrated next.

In CAD, a cylindrical surface is created when we extrude a sketch profile in the depth (or extrusion) direction. In geometric modeling, a cylindrical surface can be thought of as sweeping a straight line along a path curve, as shown in Figure 2.19a, in which the path curve $P(u)$ is assumed as a cubic curve, and the straight line is along the w direction, defined by a vector $\mathbf{r}$. A Coons patch that represents this cylindrical surface is shown in Figure 2.19b, in which boundary curve 4 is the path curve and boundary curve 1 is the straight line.

The matrix $\mathbf{G}^v$ that defines the cylindrical surface is written in Eqn (2.73), in which the first two rows defines the straight boundary edges 1 and 3, respectively; and the first two columns are boundary curves 4 and 2, respectively. In fact, the first column of matrix $\mathbf{G}^v$ is the path curve $P(u)$, and the first row is the straight line that sweeps along the path curve. Note that all twister vectors are 0 because tangent vectors are not varying along any of the boundary edges.

$$\mathbf{G}^v = \begin{bmatrix} \mathbf{P}_0 & \mathbf{P}_2 & \mathbf{P}_2 - \mathbf{P}_0 & \mathbf{P}_2 - \mathbf{P}_0 \\ \mathbf{P}_1 & \mathbf{P}_1 + \mathbf{P}_2 - \mathbf{P}_0 & \mathbf{P}_2 - \mathbf{P}_0 & \mathbf{P}_2 - \mathbf{P}_0 \\ \mathbf{P}_{0,u} & \mathbf{P}_{0,u} & 0 & 0 \\ \mathbf{P}_{1,u} & \mathbf{P}_{1,u} & 0 & 0 \end{bmatrix}_{4 \times 4 \times 3} \tag{2.73}$$

2.3.1.4 Bézier surface

Mathematically, a Bézier surface (or patch) is defined as

$$\mathbf{S}(u, w) = \sum_{i=0}^{n} \sum_{j=0}^{m} \mathbf{P}_{ij} B_{i,n}(u) B_{j,m}(w), \quad (u, w) \in [0, 1] \times [0, 1] \tag{2.74}$$

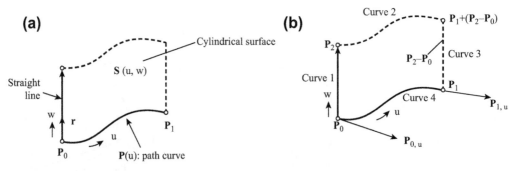

FIGURE 2.19 Representing a Cylindrical Surface Using a Coons Patch. (a) Cylindrical Surface. (b) Coons Patch Representing the Cylindrical Surface.

where $B_{i,n}(u)$ and $B_{j,m}(w)$ are Bernstein polynomials of order n and m, respectively, in u and w; and P_{ij} is the control point of the ith row at the jth location in the $(n+1) \times (m+1)$ control point matrix. Note that n does not have to be equal to m, implying that the polynomial orders of the Bézier surface along the u and w direction do not have to be identical.

As a special case, a bicubic Bézier surface is defined by 16 control points arranged in a 4×4 matrix form that forms a control polyhedron, as shown in Figure 2.17c. From Eqn (2.74), a bicubic Bézier surface is then defined as

$$S(u,w) = \sum_{i=0}^{3} \sum_{j=0}^{3} P_{ij} B_{i,3}(u) B_{j,3}(w), \quad (u,w) \in [0,1] \times [0,1] \tag{2.75}$$

which can also be written in a matrix (or tensor) form as

$$S(u,w) = U\, N^B\, G^B N^{B^T} W^T, \quad (u,w) \in [0,1] \times [0,1] \tag{2.76}$$

where

$$N^B = \begin{bmatrix} -1 & 3 & -3 & 1 \\ 3 & -6 & 3 & 0 \\ -3 & 3 & 0 & 0 \\ 1 & 0 & 0 & 0 \end{bmatrix} \tag{2.77}$$

which is identical to that of the cubic Bézier curve, and

$$G^B = \begin{bmatrix} P_{00} & P_{01} & P_{02} & P_{03} \\ P_{10} & P_{11} & P_{12} & P_{13} \\ P_{20} & P_{21} & P_{22} & P_{23} \\ P_{30} & P_{31} & P_{32} & P_{33} \end{bmatrix}_{4 \times 4 \times 3} \tag{2.78}$$

which consists of the 16 control points arranged in a 4×4 matrix form.

2.3.2 B-spline surface

Similar to a Bézier surface, a B-spline surface is defined by basis functions and the control polyhedron as

$$S(u,w) = \sum_{i=0}^{n} \sum_{j=0}^{m} P_{ij} N_{i,k}(u) N_{j,\ell}(w), \quad (u,w) \in [0, n-k+2] \times [0, m-\ell+2] \tag{2.79}$$

where $N_{i,k}(u)$ and $N_{j,\ell}(w)$ are the same basis functions as those of the B-spline curves and P_{ij} is the control point of the ith row at the jth location in the $(n+1) \times (m+1)$ matrix. In Eqn (2.79), the polynomial orders of the basis functions $N_{i,k}(u)$ and $N_{j,\ell}(w)$ are $k-1$ and $\ell-1$, respectively. Note that k does not have to be equal to ℓ, implying that the polynomial orders of the B-spline surface along the u and w directions do not have to be identical.

Depending on the choice of the basis functions (e.g., uniform or nonuniform and polynomial orders), numerous types of surfaces can be adequately modeled using B-spline surfaces, as illustrated in Figure 2.20. The three surfaces in Figure 2.20a are open–open (i.e., open in both u and w directions) and the two in Figure 2.20b are open-close (open in the w direction and closed in u direction). The surface on the left in Figure 2.20a employs nonuniform basis functions in the u direction, and the surface in the middle employs uniform basis functions. In both surfaces, a straight line is assumed in the w direction. The surface on the right assumes uniform basis functions in both the u and w directions. Both are quadratic. Both surfaces in Figure 2.20b assume a quadratic B-spline curve in the u direction. The surface on the left assumes a straight line along the w direction, and the one on the right employs nonuniform quadratic basis functions.

Similar to the B-spline surfaces, a NURB surface can be defined as

$$S(u, w) = \frac{\sum_{i=0}^{n} \sum_{j=0}^{m} h_{ij} \mathbf{P}_{ij} N_{i,k}(u) M_{j,\ell}(w)}{\sum_{i=0}^{n} \sum_{j=0}^{m} h_{ij} N_{i,k}(u) M_{j,\ell}(w)}, \quad (u, w) \in [0, n - k + 2] \times [0, m - \ell + 2]. \qquad (2.80)$$

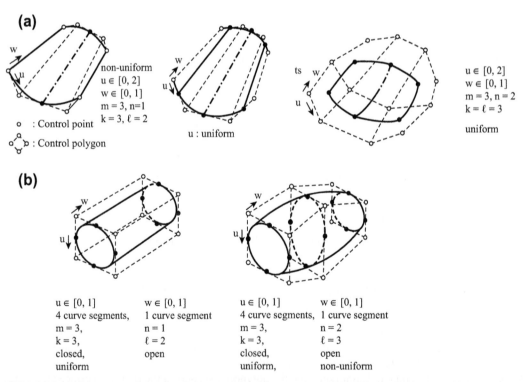

FIGURE 2.20 Various Types of B-spline Surfaces. (a) Open–open (Left: Nonuniform; Middle: Uniform in the u Direction; Right: Uniform in both u and w Directions). (b) Open-close: Open in the w Direction and Closed in the u Direction (Left: Linear in w Direction, Right: Quadratic in w Direction).

The major difference between a NURB and a B-spline surface is that the NURB surface is able to represent regular surfaces, such as a sphere, ellipsoid, and so on, just like that of NURB curves being able to represent Conic curves, such as circle or ellipse.

2.4 CAD-generated surfaces

With the knowledge of basic geometric modeling discussed in Sections 2.2 and 2.3, we are moving one step further to discuss surfaces generated by CAD. In CAD, we sketch an open profile and protrude it for a surface or protrude a closed profile for a solid feature. The protrusion capabilities commonly available in CAD include extrusion, blend (or loft), revolve, and sweep, as illustrated in Figure 2.21.

From a geometric modeling perspective, extruding a profile curve generates a cylindrical surface (Figure 2.21a). Sweeping a profile along a path curve leads to a sweep surface (Figure 2.21b). Revolving a sketch profile along an axis produces a surface of revolution (or revolved surface), as shown in Figure 2.21c. Lofting two parallel sketch profiles without guide curves yields a ruled surface. Lofting more than two parallel sketch profiles (or two profiles with guide curves shown in Figure 2.21d) creates a loft surface.

In this section, we discuss mathematic representations for the parametric surfaces generated by the four types of protrusion discussed.

2.4.1 Cylindrical surfaces

As discussed earlier, in geometric modeling, a cylindrical surface can be considered as sweeping a straight line along a path curve $\mathbf{P}(u)$, as shown in Figure 2.22a. Mathematically, such a surface can be written in a parametric form as

$$\mathbf{S}(u, w) = \mathbf{P}(u) + w\mathbf{r}, \quad (u, w) \in [0, 1] \times [0, 1] \tag{2.81}$$

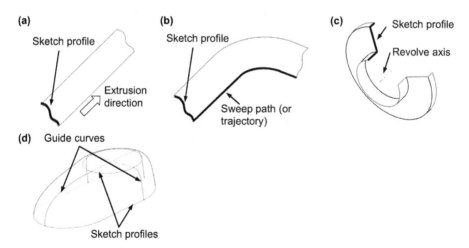

FIGURE 2.21 Protrusion of a Profile for Surface or Solid. (a) Extrusion. (b) Sweep. (c) Revolve. (d) Loft (or blend).

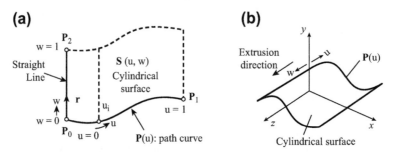

FIGURE 2.22 Cylindrical Surface. (a) Sweeping Straight Line along a Path Curve P(u). (b) Extruding Curve P(u) along a Straight Line Perpendicular to the Sketch Curve P(u).

in which u and w are the parametric coordinates of the surface, and $\mathbf{r}$ is the vector of the straight line. Note that the vector $\mathbf{r}$ can also be written as $\mathbf{r} = \mathbf{P}_2 - \mathbf{P}_0$, where $\mathbf{P}_0$ and $\mathbf{P}_2$ are the start and end point of the straight line, respectively. Note that the curve $\mathbf{P}(u)$ and the straight line are in space in general. Certainly, a same cylindrical surface can be generated by extruding the curve $\mathbf{P}(u)$ along the straight line. If curve $\mathbf{P}(u)$ is part of a sketch profile in CAD, as shown in Figure 2.22b, when the sketch is extruded, a cylindrical surface is generated, representing the boundary geometry of a solid feature. In this case, the straight line is always perpendicular to the sketch plane where the curve $\mathbf{P}(u)$ resides. The same equation in 2.81 represents the cylindrical surface.

The following example illustrates more details in constructing the mathematical representation for a cylindrical surface. We also include a Matlab script to graph the surface. Note that instead of using a Matlab surface graph function (e.g., surface(x,y,z)) to plot the surface, we use plot3(x,y,z) to plot points and line segments that show a surface with mesh. We hope such graphs offer you more insights in understanding the mathematic representation of parametric surfaces.

EXAMPLE 2.11

Find the parametric equation of the cylindrical surface generated by extruding a cubic spline curve on the x–y plane along the z-direction for 5 units, as shown below. Note that the four points that form the cubic spline curve are given as $\mathbf{P}_0 = [0,0,0]$, $\mathbf{P}_1 = [1,2.5,0]$, $\mathbf{P}_2 = [2,1,0]$, and $\mathbf{P}_3 = [4,2,0]$.

EXAMPLE 2.11—CONT'D

Solutions

Using Eqn (2.34), the parametric equation for the curve $P(u)$ can be written as

$$P(u) = U N^s G^s = \begin{bmatrix} u^3 & u^2 & u & 1 \end{bmatrix} \begin{bmatrix} -9/2 & 27/2 & -27/2 & 9/2 \\ 9 & -45/2 & 18 & -9/2 \\ -11/2 & 9 & -9/2 & 1 \\ 1 & 0 & 0 & 0 \end{bmatrix} \begin{bmatrix} 0 & 0 & 0 \\ 1 & 2.5 & 0 \\ 2 & 1 & 0 \\ 4 & 2 & 0 \end{bmatrix}$$

$$= [4.5u^3 - 4.5u^2 + 4u, 29.5u^3 - 47.25u^2 + 20u, 0], \quad u \in [0, 1].$$

The extrusion vector r is
$r = [0, 0, 5]$.

Therefore, from Eqn (2.81), the parametric equation of the cylindrical surface is

$$S(u, w) = P(u) + rw = \left[4.5u^3 - 4.5u^2 + 4u, 29.5u^3 - 47.25u^2 + 20u, 5w\right], (u, w) \in [0, 1] \times [0, 1].$$

The surface is graphed in Matlab with the script shown below.

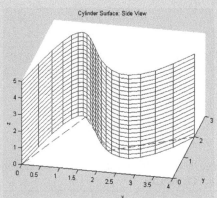

Cylinder Surface: Side View

```
Matlab Script:

N=21;                              % Create u value from 0 to 1 with increment
0.05 [1/(N-1)]
u=linspace(0,1,N);                 % Define vector u of N entries, with numbers
from 0 to 1
w= linspace(0,1,N);                % Evenly spaced. Define vector w of N entries
Px=[0 1 2 4];                      % Control points (and next two lines)
Py=[0 2.5 1 2];
Pz=[0 0 0 0];
r=5;                               % Define extrusion magnitude r=5
for i=1:1:N                        % Start a loop for fixing w value
x=4.5*u.^3-4.5*u.^2+4*u;
y=29.5*u.^3-47.25*u.^2+20*u;
z=(w(i)*r)*ones(1,N);              % Define a vector of N entries with identical
value 1
figure(2);                         % Define a figure pointer for plot
plot3(x,y,z);                      % Make a 3D plot
hold on;                           % Hold the plot for additional data, curves, etc.
end;                               % End the loop
for i=1:1:N                        % Start a loop for fixing u value
x=(4.5*u(i)^3-4.5*u(i)^2+4*u(i))*ones(1,N);
y=(29.25*u(i)^3-47.25*u(i)^2+20*u(i))*ones(1,N);
z=(w*r);
figure(2);
plot3(x,y,z);
hold on;
end;
plot3(Px,Py,Pz,'--');
xlabel('x'),ylabel('y'),zlabel('z');   % Add labels
title('Cylinder Surface: Front View');% Add title
view(-55,-25);                         %View(azimuth, elevation)
title('Cylinder Surface: Back View');
view(55,-25);
title('Cylinder Surface: Side View');
hold off
```

2.4.2 **Ruled surfaces**

A ruled surface is defined by two path curves on the opposite sides of the surface, in which the trace of a straight line with its start and end points pass through the respective path curves with the same parametric value generates a ruled surface, as illustrated in Figure 2.23a. The simplest of all ruled surfaces are plane, cone, and cylindrical surfaces. In addition, a surface with boundaries formed by four straight lines that are not coplanar is not a flat surface but a ruled surface, as shown in Figure 2.23b. In this case, both path curves $\mathbf{P}(u)$ and $\mathbf{Q}(u)$ are straight lines, which are not necessarily co-planar.

Given two distinct curves $\mathbf{P}(u)$ and $\mathbf{Q}(u)$ as shown in Figure 2.24a, a ruled surface is constructed by joining two points of the same u value (e.g., u^*) on the curves $\mathbf{P}(u)$ and $\mathbf{Q}(u)$, respectively, with a straight line; and sweeping the straight line along the two path curves with the same u value. Because the line connecting the two points of the same u value (such as u^*) respectively on curves $\mathbf{P}(u)$ and $\mathbf{Q}(u)$ is a straight line, this straight line can be written as

$$\mathbf{S}(u^*, w) = (1 - w)\,\mathbf{P}(u^*) + w\mathbf{Q}(u^*), \quad w \in [0, 1]. \tag{2.82a}$$

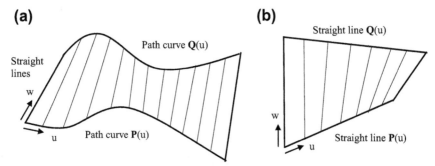

FIGURE 2.23 Ruled Surfaces. (a) General Ruled Surface Formed by Two Path Curves P(u) and Q(u). (b) Ruled Surface Formed by Two Straight Lines P(u) and Q(u) that are not Co-planar.

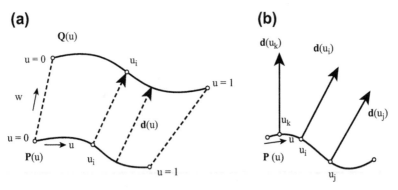

FIGURE 2.24 Ruled Surfaces. (a) General Ruled Surface Formed by Two Path Curves P(u) and Q(u). (b) the same ruled surface generated by sweeping a non constant vector d(u) = Q(u) − P(u) along the path curve P(u).

Because Eqn (2.82a) is true for any u in [0,1], it can be generalized as a ruled surface by setting u ∈ [0,1] (i.e., removing the superscript * for the parameter u in Eqn (2.82a)) as

$$\mathbf{S}(u, w) = (1 - w)\,\mathbf{P}(u) + w\mathbf{Q}(u), \quad (u, w) \in [0, 1] \times [0, 1] \tag{2.82b}$$

which can also be rewritten as

$$\mathbf{S}(u, w) = \mathbf{P}(u) + w(\mathbf{Q}(u) - \mathbf{P}(u)) = \mathbf{P}(u) + w\,\mathbf{d}(u), \quad (u, w) \in [0, 1] \times [0, 1] \tag{2.82c}$$

which indicates that the same ruled surface can be generated by sweeping a nonconstant vector $\mathbf{d}(u) = \mathbf{Q}(u) - \mathbf{P}(u)$ along the path curve $\mathbf{P}(u)$, as shown in Figure 2.24b.

The following example shows more details in constructing parametric equations for a ruled surface. Again, a Matlab script is included to graph the surface.

EXAMPLE 2.12

Find the parametric equation of the ruled surface generated by two path curves of cubic spline curves. Curve $\mathbf{P}(u)$ is the same as that of Example 2.11 and resides on the x–y plane. Curve $\mathbf{Q}(u)$ is resides on a plane that is parallel to x–y plane and offset 5 units along the z-direction, as shown below. Note that the four points that form the cubic spline curve $\mathbf{Q}(u)$ are given as $\mathbf{Q}_0 = [0,2,5]$, $\mathbf{Q}_1 = [1,1,5]$, $\mathbf{Q}_2 = [2,2.5,5]$, and $\mathbf{Q}_3 = [4,1,5]$.

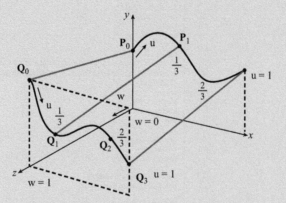

Solutions

Using Eqn (2.34), the parametric equation for curve $\mathbf{Q}(u)$ can be written as

$$\mathbf{Q}(u) = \mathbf{U}^s\mathbf{N}^s\mathbf{G}^s = \begin{bmatrix} u^3 & u^2 & u & 1 \end{bmatrix} \begin{bmatrix} -9/2 & 27/2 & -27/2 & 9/2 \\ 9 & -45/2 & 18 & -9/2 \\ -11/2 & 9 & -9/2 & 1 \\ 1 & 0 & 0 & 0 \end{bmatrix} \begin{bmatrix} 0 & 2 & 5 \\ 1 & 1 & 5 \\ 2 & 2.5 & 5 \\ 4 & 1 & 5 \end{bmatrix}$$

$$= [4.5u^3 - 4.5u^2 + 4u, 24.75u^3 + 36u^2 + 12.25u + 2.5, 0], \quad u \in [0, 1].$$

Continued

EXAMPLE 2.12—CONT'D

Therefore, from Eqn (2.82b), the parametric equation of the ruled surface is

$$\mathbf{S}(u,w) = (1-w)\mathbf{P}(u) + w\,\mathbf{Q}(u) = (1-w)[4.5u^3 - 4.5u^2 + 4u, 29.5u^3 - 47.25u^2 + 20u, 0]$$
$$+w[4.5u^3 - 4.5u^2 + 4u, 24.75u^3 + 36u^2 + 12.25u + 2.5,\ 0], \quad (u,w) \in [0,1] \times [0,1].$$

The surface is graphed in Matlab with the script shown below.

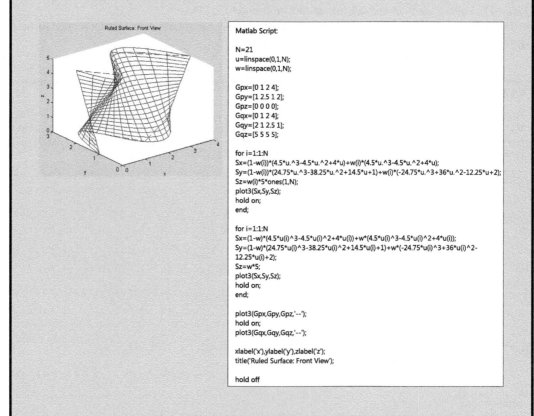

Matlab Script:

```
N=21
u=linspace(0,1,N);
w=linspace(0,1,N);

Gpx=[0 1 2 4];
Gpy=[1 2.5 1 2];
Gpz=[0 0 0 0];
Gqx=[0 1 2 4];
Gqy=[2 1 2.5 1];
Gqz=[5 5 5 5];

for i=1:1:N
Sx=(1-w(i))*(4.5*u.^3-4.5*u.^2+4*u)+w(i)*(4.5*u.^3-4.5*u.^2+4*u);
Sy=(1-w(i))*(24.75*u.^3-38.25*u.^2+14.5*u+1)+w(i)*(-24.75*u.^3+36*u.^2-12.25*u+2);
Sz=w(i)*5*ones(1,N);
plot3(Sx,Sy,Sz);
hold on;
end;

for i=1:1:N
Sx=(1-w)*(4.5*u(i)^3-4.5*u(i)^2+4*u(i))+w*(4.5*u(i)^3-4.5*u(i)^2+4*u(i));
Sy=(1-w)*(24.75*u(i)^3-38.25*u(i)^2+14.5*u(i)+1)+w*(-24.75*u(i)^3+36*u(i)^2-
12.25*u(i)+2);
Sz=w*5;
plot3(Sx,Sy,Sz);
hold on;
end;

plot3(Gpx,Gpy,Gpz,'--');
hold on;
plot3(Gqx,Gqy,Gqz,'--');

xlabel('x'),ylabel('y'),zlabel('z');
title('Ruled Surface: Front View');

hold off
```

2.4.3 Loft (or blend) surfaces

In CAD, when we loft a solid or surface feature using more than two sketch profiles, we generate a loft (or blend) surface, instead of a ruled surface. For example, a loft surface can be constructed by lofting three curves $\mathbf{P}(u)$, $\mathbf{Q}(u)$, and $\mathbf{R}(u)$ on three respective parallel sketch planes along the w direction, as shown in Figure 2.25a. If we assume that the curve $\mathbf{Q}(u)$ is located at $w = 1/2$ of the loft surface, then any curve along the w direction that is formed by a fixed u value at the three respective curves; for

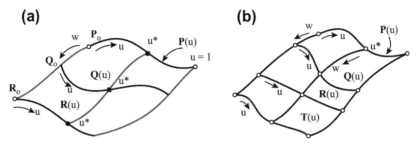

FIGURE 2.25 Loft Surfaces. (a) Quadratic along the w Direction by Lofting Three Curves. (b) Cubic along the w Direction by Lofting Four Curves.

example, u^*, shown in Figure 2.25, is a quadratic spline curve formed by $P(u^*)$, $Q(u^*)$, and $R(u^*)$, just like that of Eqn (2.14); i.e.,

$$C(w) = \begin{bmatrix} w^2 & w & 1 \end{bmatrix} \begin{bmatrix} 2 & -4 & 2 \\ -3 & 4 & -1 \\ 1 & 0 & 0 \end{bmatrix} \begin{bmatrix} P(u^*) \\ Q(u^*) \\ R(u^*) \end{bmatrix} = W_{1\times3}N^S_{3\times3} \begin{bmatrix} P(u^*) \\ Q(u^*) \\ R(u^*) \end{bmatrix}, \quad w \in [0,1]. \quad (2.83)$$

Note that Eqn (2.83) is true for all u values in [0,1]; therefore, the parametric equation of the loft surface can be written as

$$S(u,w) = \begin{bmatrix} w^2 & w & 1 \end{bmatrix} \begin{bmatrix} 2 & -4 & 2 \\ -3 & 4 & -1 \\ 1 & 0 & 0 \end{bmatrix} \begin{bmatrix} P(u) \\ Q(u) \\ R(u) \end{bmatrix} = W_{1\times3}N^S_{3\times3} \begin{bmatrix} P(u) \\ Q(u) \\ R(u) \end{bmatrix}, \quad (u,w) \in [0,1] \times [0,1].$$

$$(2.84)$$

Following the same fashion, a surface that lofts from four curves shown in Figure 2.25b can be written as follows, assuming that the four curves are located along the w direction of the loft surface at $w = 0$, 1/3, 2/3, and 1, respectively:

$$S(u,w) = \begin{bmatrix} w^3 & w^2 & w & 1 \end{bmatrix} \begin{bmatrix} -9/2 & 27/2 & -27/2 & 9/2 \\ 9 & -45/2 & 18 & -9/2 \\ -11/2 & 9 & -9/2 & 1 \\ 1 & 0 & 0 & 0 \end{bmatrix} \begin{bmatrix} P(u) \\ Q(u) \\ R(u) \\ T(u) \end{bmatrix} \quad (2.85)$$

$$= W_{1\times4}N^S_{4\times4} \begin{bmatrix} P(u) \\ Q(u) \\ R(u) \\ T(u) \end{bmatrix}, \quad (u,w) \in [0,1] \times [0,1].$$

The following example shows more detail in constructing parametric equations for a loft surface with three curves.

EXAMPLE 2.13

Find the parametric equation of the loft surface generated by lofting three curves on respective parallel planes (parallel to the x–y plane) with uniformed space of 5 units, as shown below. The cubic spline curves $P(u)$ and $Q(u)$ are the same as that of Example 2.12. Curve $R(u)$ is a straight line defined by $R_0 = [0,1,10]$ and $R_1 = [4,1,10]$.

Solutions
Using Eqn (2.10), the parametric equation for the straight line $R(u)$ can be written as

$$R(u) = (1-u)R_0 + uR_0 = (1-u)[0,1,10] + u[4,1,10] = [4u,1,10].$$

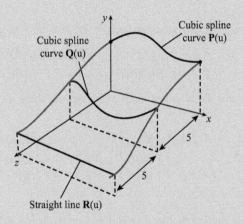

Therefore, from Eqn (2.84), the parametric equation of the loft surface is

$$S(u,w) = \begin{bmatrix} w^2 & w & 1 \end{bmatrix} \begin{bmatrix} 2 & -4 & 2 \\ -3 & 4 & -1 \\ 1 & 0 & 0 \end{bmatrix} \begin{bmatrix} 4.5u^3 - 4.5u^2 + 4u, & 29.5u^3 - 47.25u^2 + 20u, & 0 \\ 4.5u^3 - 4.5u^2 + 4u, & 24.75u^3 + 36u^2 + 12.25u + 2.5, & 0 \\ 4u, & 0, & 10 \end{bmatrix}$$

$$= \begin{bmatrix} 2w^2 - w + 1, & -4w^2 + 4w, & 2w^2 - w \end{bmatrix} \begin{bmatrix} 4.5u^3 - 4.5u^2 + 4u, & 29.5u^3 - 47.25u^2 + 20u, & 0 \\ 4.5u^3 - 4.5u^2 + 4u, & 24.75u^3 + 36u^2 + 12.25u + 2.5, & 0 \\ 4u, & 0, & 10 \end{bmatrix}, \quad (u,w) \in [0,1] \times [0,1].$$

The surface is graphed in Matlab with the script shown below on the next page.

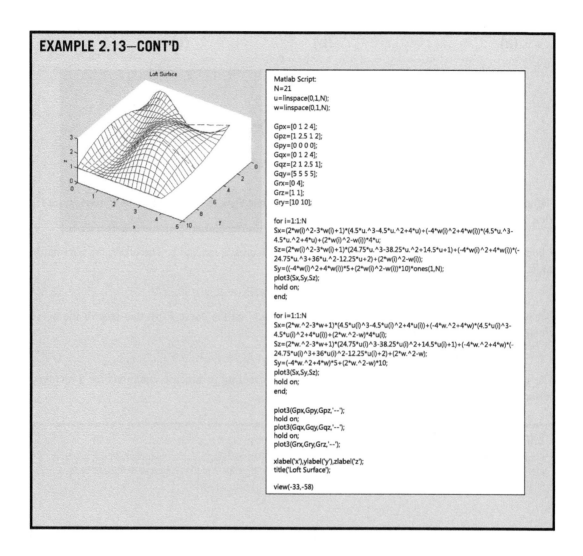

EXAMPLE 2.13—CONT'D

Loft Surface

```
Matlab Script:
N=21
u=linspace(0,1,N);
w=linspace(0,1,N);

Gpx=[0 1 2 4];
Gpz=[1 2.5 1 2];
Gpy=[0 0 0 0];
Gqx=[0 1 2 4];
Gqz=[2 1 2.5 1];
Gqy=[5 5 5 5];
Grx=[0 4];
Grz=[1 1];
Gry=[10 10];

for i=1:1:N
Sx=(2*w(i)^2-3*w(i)+1)*(4.5*u.^3-4.5*u.^2+4*u)+(-4*w(i)^2+4*w(i))*(4.5*u.^3-
4.5*u.^2+4*u)+(2*w(i)^2-w(i))*4*u;
Sz=(2*w(i)^2-3*w(i)+1)*(24.75*u.^3-38.25*u.^2+14.5*u+1)+(-4*w(i)^2+4*w(i))*(-
24.75*u.^3+36*u.^2-12.25*u+2)+(2*w(i)^2-w(i));
Sy=((-4*w(i)^2+4*w(i))*5+(2*w(i)^2-w(i))*10)*ones(1,N);
plot3(Sx,Sy,Sz);
hold on;
end;

for i=1:1:N
Sx=(2*w.^2-3*w+1)*(4.5*u(i)^3-4.5*u(i)^2+4*u(i))+(-4*w.^2+4*w)*(4.5*u(i)^3-
4.5*u(i)^2+4*u(i))+(2*w.^2-w)*4*u(i);
Sz=(2*w.^2-3*w+1)*(24.75*u(i)^3-38.25*u(i)^2+14.5*u(i)+1)+(-4*w.^2+4*w)*(-
24.75*u(i)^3+36*u(i)^2-12.25*u(i)+2)+(2*w.^2-w);
Sy=(-4*w.^2+4*w)*5+(2*w.^2-w)*10;
plot3(Sx,Sy,Sz);
hold on;
end;

plot3(Gpx,Gpy,Gpz,'--');
hold on;
plot3(Gqx,Gqy,Gqz,'--');
hold on;
plot3(Grx,Gry,Grz,'--');

xlabel('x'),ylabel('y'),zlabel('z');
title('Loft Surface');

view(-33,-58)
```

2.4.4 **Revolved surfaces**

In CAD, when we sketch a profile and revolve it along an axis, the trace of the profile forms a revolved surface or surface of revolution. How do we represent the revolved surface in a parametric form? Consider the curve $\mathbf{P}(u)$ on the x–z plane, shown in Figure 2.26a. If we revolve the curve along the z-axis counter clockwise for an angle of $\pi/2$ and pick just one point on the curve (e.g., $\mathbf{P}(u^*)$), to follow its trace, we will see that the trace of the point is a quarter circle (see Figure 2.26b in iso-view and viewed from the top, shown in Figure 2.26c) with center point O and radius $P_x(u^*)$. The quarter circle is located on a plane that is parallel with x–y plane, but is elevated at a height $P_z(u^*)$, which is the

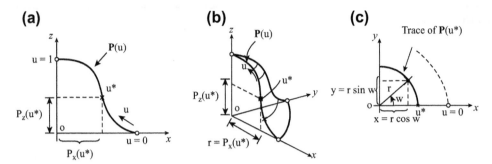

FIGURE 2.26 Surface of Revolution. (a) Sketch Profile P(u) in Front View. (b) Revolved Surface in Iso-view. (c) Top View of the Trace on the Curve P(u˙).

z-component of the curve $\mathbf{P}(u)$ at $u = u^*$, as shown in Figures 2.26a and b. Therefore, the parametric equation of the quarter circle can be written as

$$\mathbf{P}(w) = [P_X(u^*)\cos w, P_X(u^*)\sin w, P_Z(u^*)], w \in [0, \pi/2] \qquad (2.86)$$

Because the above equation is true for any point $u \in [0,1]$ on the curve $\mathbf{P}(u)$, the trace of the curve creates a revolved surface written as

$$\mathbf{S}(u, w) = [P_X(u)\cos w, P_X(u)\sin w, P_Z(u)], u \in [0, 1], w \in [0, \pi/2]. \qquad (2.87)$$

The following two examples show more details in constructing parametric equations for a revolved surface.

EXAMPLE 2.14

Find the parametric equation of the revolved surface generated by revolving a cubic spline curve P(u), which is identical to that of Example 2.11 on the x–z plane shown below with respect to the z-axis counterclockwise for a π/2 angle.

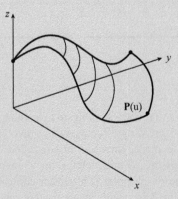

EXAMPLE 2.14—CONT'D

Solutions

Using Eqn (2.87), the parametric equation for the revolved surface can be written as

$$\mathbf{S}(u,w) = [P_x(u)\cos w, P_x(u)\sin w, P_z(u)]$$

$$= [(4.5u^3 - 4.5u^2 + 4u)\cos w, (4.5u^3 - 4.5u^2 + 4u)\sin w, 29.5u^3 - 47.25u^2 + 20u], u \in [0, 1],$$

$$w \in \left[0, \pi/2\right].$$

The surface is graphed in Matlab with the script shown below.

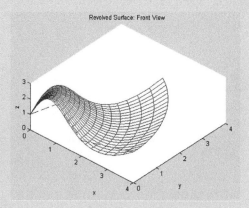

```
Matlab Script:

N=21
u=linspace(0,1,N);
w=linspace(0,1,N);

Gpx=[0 1 2 4];
Gpy=[0 0 0 0];
Gpz=[1 2.5 1 2];

for i=1:1:N
Sx=(4.5*u.^3-4.5*u.^2+4*u)*cos(w(i));
Sy=(4.5*u.^3-4.5*u.^2+4*u)*sin(w(i));
Sz=24.75*u.^3-38.25*u.^2+14.5*u+1;
plot3(Sx,Sy,Sz);
hold on;
end;

for i=1:1:N
Sx=(4.5*u(i)^3-4.5*u(i)^2+4*u(i))*cos(w);
Sy=(4.5*u(i)^3-4.5*u(i)^2+4*u(i))*sin(w);
Sz=(24.75*u(i)^3-38.25*u(i)^2+14.5*u(i)+1)*ones(1,N);
plot3(Sx,Sy,Sz);
hold on;
end;

plot3(Gpx,Gpy,Gpz,'--');
hold on;

xlabel('x'),ylabel('y'),zlabel('z');
title('Revolved Surface: Front View');

view(42,60);

hold off;
```

EXAMPLE 2.15

Find the parametric equation of the revolved surface generated by revolving the quarter circle of radius 1 on the x–z plane shown below with respect to the z axis for a $\pi/2$ angle.

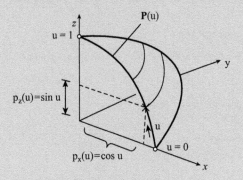

Solutions

Using Eqn (2.87), the parametric equation for the revolved surface can be written as

$$S(u, w) = [P_x(u) \cos w, P_x(u) \sin w, P_z(u)]$$
$$= [\cos u \cos w, \cos u \sin w, \sin u], u \in [0, 1], w \in [0, \pi/2].$$

The surface is graphed in Matlab with the script shown below. Note that the circular arc **P**(u) can also be represented in a NURB form, such as by using equation of Example 2.10. This is left as an exercise.

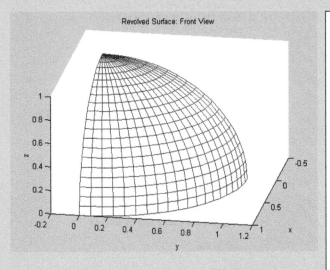

Matlab Script:

```
N=21
u=linspace(0,1.570796327,N);
w=linspace(0,1.570796327,N);

for i=1:1:N
Sx=cos(u)*cos(w(i));
Sy=cos(u)*sin(w(i));
Sz=sin(u);
plot3(Sx,Sy,Sz);
hold on;
end;

for i=1:1:N
Sx=cos(u(i))*cos(w);
Sy=cos(u(i))*sin(w);
Sz=sin(u(i))*ones(1,N);
plot3(Sx,Sy,Sz);
hold on;
end;

xlabel('x'),ylabel('y'),zlabel('z');
title('Revolved Surface: Front View');

view(100,30);

hold off
```

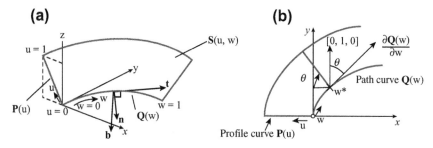

FIGURE 2.27 Sweep Surface Generated by Sweeping Profile Curve P(u) along Path Curve Q(w). (a) The Frenet Frame (t, b, n). (b) Top View of the Sweep Surface.

2.4.5 Sweep surfaces

The trace of moving a profile curve $\mathbf{P}(u)$ along a path (or trajectory) curve $\mathbf{Q}(w)$ is a sweep surface $\mathbf{S}(u,w)$. For example, moving the straight line $\mathbf{P}(u)$ along the path curve $\mathbf{Q}(w)$ shown in Figure 2.27 generates a sweep surface $\mathbf{S}(u,w)$.

In CAD, we create a profile on a sketch plane and a path curve on a plane that is perpendicular with the profile sketch, and then we sweep the profile along the path to create a sweep solid feature or a sweep surface. If the path curve is a straight line, the sweep surface generated is nothing but a cylindrical surface. If the path curve is a circular arc, the resulting sweep surface is a surface of revolution. Therefore, both cylindrical and revolved surfaces can be considered as special cases of sweep surface. For sweeping a curve $\mathbf{P}(u)$ along a straight line, the orientation of the curve $\mathbf{P}(u)$ is not changing. While sweeping a curve $\mathbf{P}(u)$ along a circular arc, the curve orientation is constantly changing in order to ensure that the curve $\mathbf{P}(u)$ is always perpendicular to the path curve $\mathbf{Q}(w)$. If the path curve $\mathbf{Q}(w)$ is a general parametric curve, how can we orient the path curve $\mathbf{P}(u)$ properly so that it is always perpendicular to the path curve? This is an important characteristic of a sweep surface and requires our attention. The trick is attached a smoothly-varying coordinate system, a so-called Frenet frame, at any given location along the path curve $\mathbf{Q}(w)$.

A Frenet frame is defined by three independent direction vectors for a spatial curve. The vectors are (i) a normalized tangent vector, $\mathbf{t}(w)$, defined as

$$\mathbf{t}(w) = \text{normalized}\left(\frac{\partial \mathbf{Q}(w)}{\partial w}\right) = \frac{\mathbf{Q}_{,w}(w)}{\|\mathbf{Q}_{,w}(w)\|};$$
(2.88a)

(ii) a normalized binormal vector, defined as

$$\mathbf{b}(w) = \text{normalized}\left(\mathbf{Q}_{,w}(w) \times \mathbf{Q}_{,ww}(w)\right) = \frac{\mathbf{Q}_{,w}(w) \times \mathbf{Q}_{,ww}(w)}{\|\mathbf{Q}_{,w}(w) \times \mathbf{Q}_{,ww}(w)\|};$$
(2.88b)

and (iii) a normalized normal vector, defined as

$$\mathbf{n}(w) = \text{normalized}(\mathbf{b}(w) \times \mathbf{t}(w)) = \frac{\mathbf{b}(w) \times \mathbf{t}(w)}{\|\mathbf{b}(w) \times \mathbf{t}(w)\|}.$$
(2.88c)

The Frenet coordinate system (or frame) $(\mathbf{t}, \mathbf{b}, \mathbf{n})$ varies smoothly, as we move along the path curve $\mathbf{Q}(w)$, as long as the curve is second-order differentiable; that is, $\mathbf{Q}_{,ww}(w)$ exists for all w. Using the smoothly varying Frenet frame, the trace of the profile curve $\mathbf{P}(u)$ at any given w value along the path curve (hence the sweep surface) can be determined by placing $\mathbf{P}(u)$ on the normal plane (spanned by vectors $\mathbf{b}$ and $\mathbf{n}$), placing the start point of $\mathbf{P}(u)$ on the path curve $\mathbf{Q}(w)$, aligning $P_x(u)$ with vector $\mathbf{n}$, and aligning $P_z(u)$ with vector $\mathbf{b}$.

Mathematically, the parametric equation for a sweep surface, generated by sweeping a planar profile $\mathbf{P}(u)$ along a path curve $\mathbf{Q}(w)$ in space, can be defined as

$$\mathbf{S}(u, w) = \mathbf{Q}(w) + \mathbf{R}(w)\,(s(w)\,\mathbf{P}(u)), \quad u \in [0, 1], w \in [0, 1] \tag{2.89}$$

where $\mathbf{R}(w)$ is a rotation matrix that rotates the profile curve $\mathbf{P}(u)$ so that its x- and z-components align with vectors $\mathbf{n}$ and $\mathbf{b}$, respectively; and $s(w)$ is a scale factor that scales the profile curve.

Note that in most sweep features in CAD, the scale factor is set to unity, and the path curve is usually a planar curve. For example, in Figure 2.27a, the path curve $\mathbf{Q}(w)$ is sketched on the x–y plane. In this subsection, we assume that the path curve is placed on a plane that is perpendicular to that of the profile curve in order to simplify the mathematical equations of the sweep surface. Viewing from the top, as shown in Figure 2.27b, it is apparent that in order to keep the profile curve $\mathbf{P}(u)$ perpendicular with the path curve $\mathbf{Q}(w)$ at any given w value (e.g., w^* in Figure 2.27b), the profile curve $\mathbf{P}(u)$ must rotate an θ angle clockwise along the z-axis. The θ angle can be calculated as

$$\theta(w) = \cos^{-1}\big([0 \quad 1 \quad 0] \cdot \mathbf{t}^T(w)\big). \tag{2.90}$$

The rotation matrix is then obtained as

$$\mathbf{T}(\theta) = \begin{bmatrix} \cos\theta & \sin\theta & 0 \\ -\sin\theta & \cos\theta & 0 \\ 0 & 1 & 1 \end{bmatrix}. \tag{2.91}$$

At the given value $w = w^*$, the profile curve $\mathbf{P}(u)$ is rotated clockwise with an θ angle clockwise along the z-axis, and then moved to the location of $\mathbf{Q}(w^*)$; that is,

$$\mathbf{P}'(u) = \left[\mathbf{T}(\theta)\,\mathbf{P}(u)^T\right]^T + \mathbf{Q}(w^*), \quad u \in [0, 1] \tag{2.92}$$

which is true for $w \in [0,1]$. Hence, the parametric equation of the sweep surface can be written as

$$\mathbf{S}(u, w) = \left[\mathbf{T}(\theta)\,\mathbf{P}(u)^T\right]^T + \mathbf{Q}(w), \quad u \in [0, 1], w \in [0, 1]. \tag{2.93}$$

The following example shows more details in constructing parametric equations for a simple sweep surface.

EXAMPLE 2.16

Find the parametric equation of the sweep surface generated by sweeping a straight line along a cubic Bézier curve shown below on the next page. The straight line is formed by connecting $\mathbf{P}_0 = [0,0,0]$ and $\mathbf{P}_1 = [-5,0,5]$; and the four control points of the cubic Bézier curve are $\mathbf{Q}_0 = [0,0,0]$, $\mathbf{Q}_1 = [0,5,0]$, $\mathbf{Q}_2 = [7.5,5,0]$, and $\mathbf{Q}_3 = [7.5,0,0]$.

EXAMPLE 2.16—CONT'D

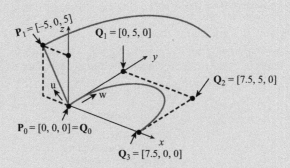

Solutions
From Eqns (2.10) and (2.42), the parametric equations of the straight line and the cubic Bézier curve can be written, respectively, as

$$\mathbf{P}(u) = (1-u)[0,0,0] + u[-5,0,5] = [-5u, 0, 5u]$$

and

$$\mathbf{Q}(w) = \begin{bmatrix} w^3 & w^2 & w & 1 \end{bmatrix} \begin{bmatrix} -1 & 3 & -3 & 1 \\ 3 & -6 & 3 & 0 \\ -3 & 3 & 0 & 0 \\ 1 & 0 & 0 & 0 \end{bmatrix} \begin{bmatrix} \mathbf{Q}_0 \\ \mathbf{Q}_1 \\ \mathbf{Q}_2 \\ \mathbf{Q}_3 \end{bmatrix} = \begin{bmatrix} w^3 & w^2 & w & 1 \end{bmatrix} \begin{bmatrix} -1 & 3 & -3 & 1 \\ 3 & -6 & 3 & 0 \\ -3 & 3 & 0 & 0 \\ 1 & 0 & 0 & 0 \end{bmatrix} \begin{bmatrix} 0 & 0 & 0 \\ 0 & 5 & 0 \\ 7.5 & 5 & 0 \\ 7.5 & 0 & 0 \end{bmatrix}$$

$$= \begin{bmatrix} -15w^3 + 22.5w^2, & -15w^2 + 15w, & 0 \end{bmatrix}.$$

The normalized tangent vector of the path curve $\mathbf{Q}(w)$ is

$$\mathbf{t}(w) = \frac{\mathbf{Q}_{,w}(w)}{\|\mathbf{Q}_{,w}(w)\|} = \frac{[-45w^2 + 45w, 30w + 15, 0]}{\sqrt{(-45w^2 + 45w)^2 + (30w + 15)^2 + 0^2}}$$

and the rotation angle θ can be obtained as

$$\theta = \cos^{-1}([0 \quad 1 \quad 0] \cdot \mathbf{t}^T) = \cos^{-1} \left([0 \quad 1 \quad 0] \cdot \begin{bmatrix} \dfrac{-45w^2 + 45w}{\sqrt{(-45w^2 + 45w)^2 + (30w + 15)^2 + 0^2}} \\ \dfrac{-30w + 15}{\sqrt{(-45w^2 + 45w)^2 + (-30w + 15)^2 + 0^2}} \\ 0 \end{bmatrix} \right)$$

$$= \cos^{-1} \left(\frac{-30w + 15}{\sqrt{(-45w^2 + 45w)^2 + (-30w + 15)^2 + 0^2}} \right).$$

Now, rotating the curve $\mathbf{P}(u)$ an θ angle clockwise along the z-axis, we have

$$\mathbf{T}(\theta)\mathbf{P}(u)^T = \begin{bmatrix} \cos\theta & \sin\theta & 0 \\ -\sin\theta & \cos\theta & 0 \\ 0 & 1 & 1 \end{bmatrix} \begin{bmatrix} -5u \\ 0 \\ 5u \end{bmatrix} = \begin{bmatrix} -5u\cos\theta \\ 5u\sin\theta \\ 5u \end{bmatrix}.$$

Continued

EXAMPLE 2.16—CONT'D

Hence, from Eqn (2.93), the sweep surface can be obtained as

$$\mathbf{S}(u,w) = \left[\mathbf{T}(\theta)\,\mathbf{P}(u)^T\right]^T + \mathbf{Q}(w)$$
$$= \left[-5u\cos\theta, 5u\sin\theta, 5u\right] + \left[-15w^3 + 22.5w^2, -15w^2 + 15, 0\right]$$
$$= \left[-5u\cos\theta - 15w^3 + 22.5w^2, 5u\sin\theta - 15w^2 + 15, 5u\right], u \in [0,1], w \in [0,1].$$

The surface is graphed in Matlab with the script shown below.

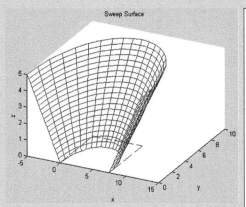

Sweep Surface

```
Matlab Script:

N=21
u=linspace(0,1,N);
w=linspace(0,1,N);

Px=[0 0 7.5 7.5];
Py=[0 5 5 0];
Pz=[0 0 0 0];

n=[0 1 0];

for i=1:1:N
tmag=((-45*w(i)^2+45*w(i))^2+(-30*w(i)+15)^2)^0.5;
t=[(-45*w(i)^2+45*w(i))/tmag (-30*w(i)+15)/tmag 0];
theta=acos(n*t');
Sx=-5*u.*cos(theta)-15*w(i)^3+22.5*w(i)^2;
Sy=5*u.*sin(theta)-15*w(i)^2+15*w(i);
Sz=5*u;
plot3(Sx,Sy,Sz);
hold on;
end;

tmag=((-45*w.^2+45*w).^2+(-30*w+15).^2).^0.5;
for i=1:1:N
t=[(-45*w(i)^2+45*w(i))./tmag(i) (-30*w(i)+15)./tmag(i) 0];
theta(i)=acos(n*t');
end;

for i=1:1:N
Sx=-5*u(i)*cos(theta)-15*w.^3+22.5*w.^2;
Sy=5*u(i)*sin(theta)-15*w.^2+15*w;
Sz=5*u(i)*ones(1,N);
plot3(Sx,Sy,Sz);
hold on;
end;

plot3(Px,Py,Pz,'--');
hold on;

xlabel('x'),ylabel('y'),zlabel('z');
title('Sweep Surface');

view(27,36);

hold off;
```

2.5 **Geometric transformations**

In geometric modeling, geometric entities, such as curves and surfaces, need to be constantly transformed for numerous purposes. The Euclidean transformations are the most commonly used transformations. A Euclidean transformation is a translation, a rotation, or a mirror. Euclidean transformations preserve length and angle measure. Moreover, the shape of a geometric entity will not change. That is, lines transform to lines, planes transform to planes, circles transform to circles, and ellipsoids transform to ellipsoids. Only the position and orientation of the object will change.

Another transformation, called affine transformation, is a generalization of Euclidean transformation. Under affine transformations, lines transform to lines; however, circles may become ellipses. Length and angle are not preserved. Essentially, an affine transformation is any transformation that preserves collinearity (i.e., all points lying on a line initially still lie on a line after transformation) and ratios of distances (e.g., the midpoint of a line segment remains the midpoint after transformation). Although an affine transformation preserves proportions on lines, it does not necessarily preserve angles or lengths. Geometric contraction, expansion, dilation, reflection, rotation, shear, similarity transformations, spiral similarities, and translation are all affine transformations, as are their combinations.

In this subsection, we discuss only the most basic transformations, including scaling, translation, and rotation. Note that by combining a number of transformations, a more sophisticated transformation, such as mirror or rotating along an arbitrarily axis, can be carried out.

Transformation of a parametric curve or surface can be accomplished by transforming its characteristic points, such as control points of Bézier or B-spline curves or surfaces, as well as tangent vectors (e.g., Hermit cubic curves), and twister vectors (e.g., the Coons patch). Mathematically, applying an affine (or Euclidean) transformation to a geometric entity, such as a B-spline curve $\mathbf{P}(u)$, can be expressed as

$$\mathbf{P}'(u) = \mathbf{T}\,\mathbf{P}(u) = \mathbf{T}\left(\sum_{i=0}^{n}\mathbf{P}_i N_{i,k}(u)\right) = \sum_{i=0}^{n}(\mathbf{T}\mathbf{P}_i)N_{i,k}(u) = \sum_{i=0}^{n}\mathbf{P}_i' N_{i,k}(u) \qquad (2.94)$$

where $\mathbf{T}$ is the affine transformation matrix, $\mathbf{P}_i'$ are the transformed characteristic points (in this case, control points), and the curve $\mathbf{P}'(u)$ is the transformed B-spline curve. In this section, we assume all curves and points are in column vector form.

Affine transformation is powerful and uniform mathematically. It is ideal for support of geometric transformations. To understand affine transformations, we need to first discuss homogeneous coordinates.

2.5.1 **Homogeneous coordinates**

Every point (x,y) in a 2D Cartesian plane has a corresponding set of homogeneous coordinates (hx, hy, h) in the 3D projective space (also called the homogeneous space). When $h = 1$, (hx, hy, h) becomes $(x, y, 1)$, projecting (hx, hy, h) point to the $h = 1$ plane, as illustrated in Figure 2.28a. Therefore, representing planar curves on a 2D Cartesian plane is a special case of the more general homogeneous coordinates.

Also, every point in the 3D Cartesian space (x, y, z) has a corresponding set of homogeneous coordinates (hx, hy, hz, h) in the four-dimensional (4D) projective space (again, called the homogeneous

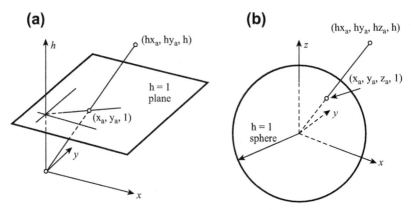

FIGURE 2.28 Homogeneous Coordinates. (a) Two-dimensional. (b) Three-dimensional.

space). As illustrated in Figure 2.28b, when h = 1, (hx, hy, hz, h) becomes (x, y, z, 1), which projects the point (hx, hy, hz, h) to the h = 1 sphere. Again, representing spatial curves and surfaces in 3D Cartesian space (h = 1) is a special case of the more general homogeneous coordinates.

Note that geometric transformations can be handled more effectively in the homogeneous coordinates than ordinary Cartesian coordinates, which is illustrated in the following example.

Let $\mathbf{P}_i$, i = 0, n, be (n + 1) control points of a B-spline curve in the 4D homogeneous space with the same h; that is,

$$\mathbf{P}_i = [\, hx_i\ hy_i\ hz_i\ h\,]^T_{1\times4}. \tag{2.95}$$

Again, all points, such as $\mathbf{P}_i$, are represented in column vector form.

In homogeneous coordinates, the transformation matrix for an affine transformation can be defined by a 4×4 matrix as

$$\mathbf{T} = \begin{bmatrix} A & B & C & M \\ D & E & F & N \\ G & H & I & O \\ J & K & L & S \end{bmatrix} \tag{2.96}$$

in which the 3×3 matrix $\begin{bmatrix} A & B & C \\ D & E & F \\ G & H & I \end{bmatrix}$ defines the scaling and rotation transformations, the 3×1 column vector $[M\ N\ O]^T$ determines the geometric translation, and the scalar $[S]$ specifies the uniform global scaling. Note that the 1×3 row vector $[J\ K\ L]$ is usually set to $[0\ 0\ 0]$.

With this transformation matrix defined in Eqn (2.95), an affine transformation of points $\mathbf{P}_i$, i = 0, n, can be obtained as

$$\mathbf{P}'_i = \mathbf{T}\,\mathbf{P}_i, \text{ for } i = 0, n \tag{2.97}$$

where

$$\mathbf{P}'_i = \begin{bmatrix} x'_i \; y'_i \; z'_i \; 1 \end{bmatrix}^T \text{is the point } \mathbf{P}_i = \begin{bmatrix} x_i \; y_i \; z_i \; 1 \end{bmatrix}^T \text{after transformation.}$$

2.5.2 Scaling

The transformation matrix for scaling a geometric entity is defined as

$$\mathbf{T}_s = \begin{bmatrix} A & 0 & 0 & 0 \\ 0 & E & 0 & 0 \\ 0 & 0 & I & 0 \\ 0 & 0 & 0 & 1 \end{bmatrix} \tag{2.98}$$

where A, E, and I are the scaling factors for x-, y-, and z-coordinates, respectively, as illustrated in Figure 2.29a. The rectangle of size a×b defined by four corner points $\mathbf{P}_0$ $\mathbf{P}_1$ $\mathbf{P}_2$ $\mathbf{P}_3$ is scaled to be of size Aa×Eb, defined by the transformed corners points $\mathbf{P}_0'$ $\mathbf{P}_1'$ $\mathbf{P}_2'$ $\mathbf{P}_3'$, as shown below.

$$\begin{bmatrix} \mathbf{P}'_0 & \mathbf{P}'_1 & \mathbf{P}'_2 & \mathbf{P}'_3 \end{bmatrix}_s = \mathbf{T}_s \begin{bmatrix} \mathbf{P}_0 & \mathbf{P}_1 & \mathbf{P}_2 & \mathbf{P}_3 \end{bmatrix} = \begin{bmatrix} A & 0 & 0 & 0 \\ 0 & E & 0 & 0 \\ 0 & 0 & I & 0 \\ 0 & 0 & 0 & 1 \end{bmatrix} \begin{bmatrix} P_{0x} & P_{1x} & P_{2x} & P_{3x} \\ P_{0y} & P_{1y} & P_{2y} & P_{3y} \\ P_{0z} & P_{1z} & P_{2z} & P_{3z} \\ 1 & 1 & 1 & 1 \end{bmatrix}$$

$$= \begin{bmatrix} A & 0 & 0 & 0 \\ 0 & E & 0 & 0 \\ 0 & 0 & I & 0 \\ 0 & 0 & 0 & 1 \end{bmatrix} \begin{bmatrix} 0 & a & a & 0 \\ 0 & 0 & b & b \\ 0 & 0 & 0 & 0 \\ 1 & 1 & 1 & 1 \end{bmatrix} = \begin{bmatrix} 0 & Aa & Aa & 0 \\ 0 & 0 & Eb & Eb \\ 0 & 0 & 0 & 0 \\ 1 & 1 & 1 & 1 \end{bmatrix} \tag{2.99}$$

Note that the four points $\mathbf{P}_0$, $\mathbf{P}_1$, $\mathbf{P}_2$, and $\mathbf{P}_3$ can be control points of a parametric curve, such as a cubic Bézier curve. In this case, the same procedure shown above applies, as illustrated in Figure 2.29b.

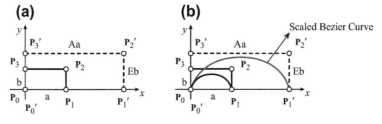

FIGURE 2.29 Scaling Transformations. (a) Scaling a Rectangle. (b) Scaling a Cubic Bézier Curve.

The 4×4 transformation matrix for a uniform global scaling can be defined as

$$\mathbf{T}_s^g = \begin{bmatrix} 1 & 0 & 0 & 0 \\ 0 & 1 & 0 & 0 \\ 0 & 0 & 1 & 0 \\ 0 & 0 & 0 & S \end{bmatrix} \qquad (2.100)$$

where S is the scale factor. If we scale the rectangle shown in Figure 2.29a with a scale factor S, the rectangle of size a×b defined by four corner points $\mathbf{P}_0\,\mathbf{P}_1\,\mathbf{P}_2\,\mathbf{P}_3$ becomes a/S×b/S, as shown below.

$$\begin{bmatrix} \mathbf{P}_0' & \mathbf{P}_1' & \mathbf{P}_2' & \mathbf{P}_3' \end{bmatrix}_s = \mathbf{T}_s^g \begin{bmatrix} \mathbf{P}_0 & \mathbf{P}_1 & \mathbf{P}_2 & \mathbf{P}_3 \end{bmatrix} = \begin{bmatrix} 1 & 0 & 0 & 0 \\ 0 & 1 & 0 & 0 \\ 0 & 0 & 1 & 0 \\ 0 & 0 & 0 & S \end{bmatrix} \begin{bmatrix} P_{0x} & P_{1x} & P_{2x} & P_{3x} \\ P_{0y} & P_{1y} & P_{2y} & P_{3y} \\ P_{0z} & P_{1z} & P_{2z} & P_{3z} \\ 1 & 1 & 1 & 1 \end{bmatrix}$$

$$\qquad (2.101)$$

$$= \begin{bmatrix} 1 & 0 & 0 & 0 \\ 0 & 1 & 0 & 0 \\ 0 & 0 & 1 & 0 \\ 0 & 0 & 0 & S \end{bmatrix} \begin{bmatrix} 0 & a & a & 0 \\ 0 & 0 & b & b \\ 0 & 0 & 0 & 0 \\ 1 & 1 & 1 & 1 \end{bmatrix} = \begin{bmatrix} 0 & a & a & 0 \\ 0 & 0 & b & b \\ 0 & 0 & 0 & 0 \\ S & S & S & S \end{bmatrix}$$

Note that the matrix $\begin{bmatrix} 0 & a & a & 0 \\ 0 & 0 & b & b \\ 0 & 0 & 0 & 0 \\ S & S & S & S \end{bmatrix}$ is in the 4D homogeneous coordinates, which can be brought

back to the Cartesian coordinates by dividing the entries by S; that is,

$$\begin{bmatrix} \mathbf{P}_0' & \mathbf{P}_1' & \mathbf{P}_2' & \mathbf{P}_3' \end{bmatrix}_s = \begin{bmatrix} 0 & \dfrac{a}{S} & \dfrac{a}{S} & 0 \\ 0 & 0 & \dfrac{b}{S} & \dfrac{b}{S} \\ 0 & 0 & 0 & 0 \\ 1 & 1 & 1 & 1 \end{bmatrix}. \qquad (2.102)$$

2.5.3 Translation

The transformation matrix for translating a geometric entity is defined as

$$\mathbf{T}_t = \begin{bmatrix} 1 & 0 & 0 & M \\ 0 & 1 & 0 & N \\ 0 & 0 & 1 & O \\ 0 & 0 & 0 & 1 \end{bmatrix} \qquad (2.103)$$

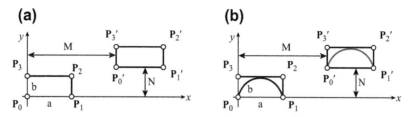

FIGURE 2.30 Geometry Translation. (a) Translating a Rectangle. (b) Translating a Cubic Bézier Curve.

where M, N, and O are the translation factors for the x-, y-, and z-coordinates, respectively, as illustrated in Figure 2.30a. The rectangle of size $a \times b$ defined by four corner points $\mathbf{P_0} \, \mathbf{P_1} \, \mathbf{P_2} \, \mathbf{P_3}$ is translated to a new location, defined by the transformed corner points $\mathbf{P_0'} \, \mathbf{P_1'} \, \mathbf{P_2'} \, \mathbf{P_3'}$, as shown below.

$$
\begin{bmatrix} \mathbf{P_0'} & \mathbf{P_1'} & \mathbf{P_2'} & \mathbf{P_3'} \end{bmatrix}_t = \mathbf{T}_t \begin{bmatrix} \mathbf{P_0} & \mathbf{P_1} & \mathbf{P_2} & \mathbf{P_3} \end{bmatrix} = \begin{bmatrix} 1 & 0 & 0 & M \\ 0 & 1 & 0 & N \\ 0 & 0 & 1 & O \\ 0 & 0 & 0 & 1 \end{bmatrix} \begin{bmatrix} P_{0x} & P_{1x} & P_{2x} & P_{3x} \\ P_{0y} & P_{1y} & P_{2y} & P_{3y} \\ P_{0z} & P_{1z} & P_{2z} & P_{3z} \\ 1 & 1 & 1 & 1 \end{bmatrix}
$$

$$
= \begin{bmatrix} 1 & 0 & 0 & M \\ 0 & 1 & 0 & N \\ 0 & 0 & 1 & O \\ 0 & 0 & 0 & 1 \end{bmatrix} \begin{bmatrix} 0 & a & a & 0 \\ 0 & 0 & b & b \\ 0 & 0 & 0 & 0 \\ 1 & 1 & 1 & 1 \end{bmatrix} = \begin{bmatrix} M & a+M & a+M & M \\ N & N & b+N & b+N \\ 0 & 0 & 0 & 0 \\ 1 & 1 & 1 & 1 \end{bmatrix}
$$

(2.104)

Note that, like before, the four points $\mathbf{P_0}$, $\mathbf{P_1}$, $\mathbf{P_2}$, and $\mathbf{P_3}$ can be control points of a parametric curve, such as a cubic Bézier curve. In this case, the same procedure shown above applies, as illustrated in Figure 2.30b.

2.5.4 Rotations

The transformation matrix for rotating a geometric entity on the x–y plane, such as a point $\mathbf{P}$ shown in Figure 2.31a, along the z-axis at a positive angle θ can be written as

$$
\mathbf{T}_{rz} = \begin{bmatrix} \cos\theta & -\sin\theta & 0 & 0 \\ \sin\theta & \cos\theta & 0 & 0 \\ 0 & 0 & 1 & 0 \\ 0 & 0 & 0 & 1 \end{bmatrix}.
$$

(2.105a)

Similarly, the matrices for rotating along the y- and x-axes, shown in Figures 2.31b and c, respectively, can be written as

$$
\mathbf{T}_{ry} = \begin{bmatrix} \cos\theta & -\sin\theta & 0 & 0 \\ 0 & 1 & 0 & 0 \\ \sin\theta & \cos\theta & 0 & 0 \\ 0 & 0 & 0 & 1 \end{bmatrix}
$$

(2.105b)

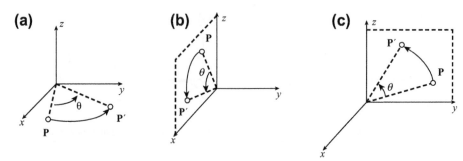

FIGURE 2.31 Rotation Transformations. (a) Rotating a Point along the z-axis. (b) Rotating a Point along the y-axis. (c) Rotating a Point along the x-axis.

and

$$\mathbf{T}_{rx} = \begin{bmatrix} 1 & 0 & 0 & 0 \\ 0 & \cos\theta & -\sin\theta & 0 \\ 0 & \sin\theta & \cos\theta & 0 \\ 0 & 0 & 0 & 1 \end{bmatrix}. \tag{2.105c}$$

Is the order of rotation transformations interchangeable? For example, is $\mathbf{T}_{rx}(\alpha)\mathbf{T}_{ry}(\beta) = \mathbf{T}_{ry}(\beta)\mathbf{T}_{rx}(\alpha)$? The answer is generally no, unless the rotation angles α and β are infinitesimally small. Another important property worth mentioning is that these rotation transformation matrices shown in Eqns (2.105a–c) are orthogonal; that is,

$$\mathbf{A}^T\mathbf{A} = \mathbf{I} \tag{2.106a}$$

where matrix $\mathbf{A}$ is the rotation part of the transformation matrix; that is,

$$\mathbf{T}_r = \begin{bmatrix} & & & 0 \\ & \mathbf{A} & & 0 \\ & & & 0 \\ 0 & 0 & 0 & 1 \end{bmatrix}.$$

In addition, the determinant of the matrix $\mathbf{A}$ is 1; that is,

$$|\mathbf{A}| = 1 \tag{2.106b}$$

2.5.5 Composite transformations

On many occasions, a geometric transformation is accomplished by multiple transformations. For example, rotating a rectangle shown in the figure of Example 2.17 at a point other than the origin of the Cartesian coordinate system requires first translating the entity to a location where the rotating point coincides with the origin of the coordinate system. After rotating the entity with respect to the origin, the rotated entity must be translated back to its original location. Such a transformation is called a composite transformation. The 4×4 transformation matrix $\mathbf{T}_c$ for a composite transformation consists of multiplications of individual transform matrices in a prescribed order. In general, the order is not interchangeable.

EXAMPLE 2.17

Find the composite transformation matrix that rotates the rectangle shown below at the point $\mathbf{S} = [3,2]^T$, a 30° angle counterclockwise. Note that the corner points of the rectangle are $\mathbf{P}_1 = [1,1]^T$, $\mathbf{P}_2 = [2,1]^T$, $\mathbf{P}_3 = [2,3]^T$ and $\mathbf{P}_4 = [1,3]^T$.

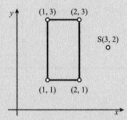

Solutions

There are three individual transformations involved. They are translating $(-3,-2)$, rotating a 30° angle along the z-axis, and translating $(3,2)$, as shown below. Note that because the transformation takes place on the x–y plane, we omit entities relevant to z-component in the transformation matrices.

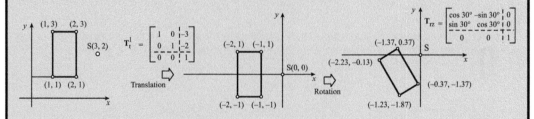

The individual transformation matrices are defined as

$$\mathbf{T}_t^1 = \begin{bmatrix} 1 & 0 & -3 \\ 0 & 1 & -2 \\ 0 & 0 & 1 \end{bmatrix} \quad \mathbf{T}_{rz} = \begin{bmatrix} \cos 30° & -\sin 30° & 0 \\ \sin 30° & \cos 30° & 0 \\ 0 & 0 & 1 \end{bmatrix} \quad \mathbf{T}_t^2 = \begin{bmatrix} 1 & 0 & 3 \\ 0 & 1 & 2 \\ 0 & 0 & 1 \end{bmatrix}.$$

Continued

EXAMPLE 2.17—CONT'D

Therefore, the composite transformation matrix can be calculated as

$$\mathbf{T_c} = \mathbf{T_t^2 T_{rz} T_t^1} = \begin{bmatrix} 1 & 0 & 3 \\ 0 & 1 & 2 \\ 0 & 0 & 1 \end{bmatrix} \begin{bmatrix} \cos 30° & -\sin 30° & 0 \\ \sin 30° & \cos 30° & 0 \\ 0 & 0 & 1 \end{bmatrix} \begin{bmatrix} 1 & 0 & -3 \\ 0 & 1 & -2 \\ 0 & 0 & 1 \end{bmatrix} = \begin{bmatrix} 0.866 & -0.5 & 1.40 \\ 0.5 & 0.866 & -1.23 \\ 0 & 0 & 1 \end{bmatrix}.$$

Hence, the transformed rectangle is defined by the four transformed corner points as

$$\mathbf{P'} = \mathbf{T_c P} = \begin{bmatrix} 0.866 & -0.5 & 1.40 \\ 0.5 & 0.866 & -1.23 \\ 0 & 0 & 1 \end{bmatrix} \begin{bmatrix} 1 & 2 & 2 & 1 \\ 1 & 1 & 3 & 3 \\ 1 & 1 & 1 & 1 \end{bmatrix} = \begin{bmatrix} 1.77 & 2.63 & 1.63 & 0.77 \\ 0.13 & 0.63 & 2.37 & 1.87 \\ 1 & 1 & 1 & 1 \end{bmatrix}.$$

EXAMPLE 2.18

Mirror the isosceles triangle shown below along a 45°-axis. Note that the corner points of the triangle are $\mathbf{P_1} = [1,1]^T$, $\mathbf{P_2} = [1,3]^T$, and $\mathbf{P_3} = [3,3]^T$.

Solutions

This mirror transformation can be accomplished by first rotating the triangle 45° clockwise along the z-axis so that its hypotenuse aligns with the x-axis. The triangle is then rotated along the x-axis by a 180° angle. Then, it is rotated 45° counter clockwise along the z-axis.

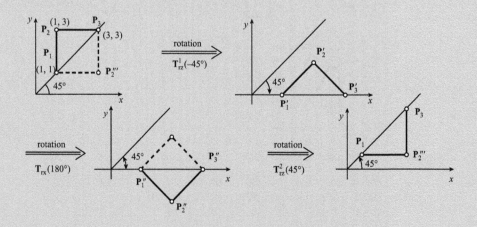

EXAMPLE 2.18—CONT'D

The composite transformation matrix, consisting of three individual rotation matrices, can be found as

$$\mathbf{T_c} = \mathbf{T_{rz}^2}\mathbf{T_{rx}}\mathbf{T_{rz}^1} = \begin{bmatrix} \cos(45°) & -\sin(45°) & 0 & 0 \\ \sin(45°) & \cos(45°) & 0 & 0 \\ 0 & 0 & 1 & 0 \\ 0 & 0 & 0 & 1 \end{bmatrix} \begin{bmatrix} 1 & 0 & 0 & 0 \\ 0 & \cos(180°) & -\sin(180°) & 0 \\ 0 & \sin(180°) & \cos(180°) & 0 \\ 0 & 0 & 0 & 1 \end{bmatrix}$$

$$\times \begin{bmatrix} \cos(-45°) & -\sin(-45°) & 0 & 0 \\ \sin(-45°) & \cos(-45°) & 0 & 0 \\ 0 & 0 & 1 & 0 \\ 0 & 0 & 0 & 1 \end{bmatrix}$$

$$= \begin{bmatrix} \frac{1}{\sqrt{2}} & -\frac{1}{\sqrt{2}} & 0 & 0 \\ \frac{1}{\sqrt{2}} & \frac{1}{\sqrt{2}} & 0 & 0 \\ 0 & 0 & 1 & 0 \\ 0 & 0 & 0 & 1 \end{bmatrix} \begin{bmatrix} 1 & 0 & 0 & 0 \\ 0 & -1 & 0 & 0 \\ 0 & 0 & -1 & 0 \\ 0 & 0 & 0 & 1 \end{bmatrix} \begin{bmatrix} \frac{1}{\sqrt{2}} & \frac{1}{\sqrt{2}} & 0 & 0 \\ -\frac{1}{\sqrt{2}} & \frac{1}{\sqrt{2}} & 0 & 0 \\ 0 & 0 & 1 & 0 \\ 0 & 0 & 0 & 1 \end{bmatrix} = \begin{bmatrix} 0 & 1 & 0 & 0 \\ 1 & 0 & 0 & 0 \\ 0 & 0 & -1 & 0 \\ 0 & 0 & 0 & 1 \end{bmatrix}.$$

Hence, the transformed triangle is defined by the three transformed corner points as

$$\mathbf{P'} = \mathbf{T_c}\mathbf{P} = \begin{bmatrix} 0 & 1 & 0 & 0 \\ 1 & 0 & 0 & 0 \\ 0 & 0 & -1 & 0 \\ 0 & 0 & 0 & 1 \end{bmatrix} \begin{bmatrix} 1 & 1 & 3 \\ 1 & 3 & 3 \\ 0 & 0 & 0 \\ 1 & 1 & 1 \end{bmatrix} = \begin{bmatrix} 1 & 3 & 3 \\ 1 & 1 & 3 \\ 0 & 0 & 0 \\ 1 & 1 & 1 \end{bmatrix}.$$

Is this the only way to perform the mirror transformation? The answer is no. You may try another composite transformation to mirror the triangle.

2.6 Case studies

Two case studies are included in this section. They are the curve fitting and surface skinning techniques, and applications of the techniques to engineering applications. We include four examples to demonstrate the modeling technique, including integration of topology and shape optimization, human middle ear, human tooth, and reverse engineering of an airplane tubing.

2.6.1 Curve fitting and surface skinning

In many engineering applications, discrete points extracted from a physical object often serve as a starting point for geometric model construction. One example is tracing the histological sections of a biological object. Through tracing outlines of the sections, discrete points are obtained and are employed to construct B-spline curves that represent the exterior contours of the components using a curve fitting technique. The surface skinning technique is then employed to quilt the B-spline curves for smooth boundary surfaces of the object using B-spline surfaces.

2.6.1.1 Curve fitting

The curve fitting technique employs the least square fitting for discrete points measured on a preselected section of an object. The best fitting curve can be obtained by minimizing the sum of the distance between the curve and the geometric points. Mathematically, the distance sum f is defined as

$$f = \sum_{j=0}^{r} \|\mathbf{P}_j - \mathbf{x}(u_j)\|^2 \tag{2.107}$$

where $\mathbf{P}_j$ is the position vector of the jth discrete point, and $r + 1$ is the total number of points captured in the section contour; $\|\cdot\|$ is the norm of the vector $\cdot$, $\mathbf{x}(u)$ is the fitting B-spline curve, and $\mathbf{x}(u_j) = [x_1(u_j), x_2(u_j), x_3(u_j)]$ is the position vector of the fitting B-spline curve at u_j, where u is the parametric coordinate of the curve. The u_j in Eqn (2.107) is defined by the length ratio of the polygon formed by the geometric points $\mathbf{P}_j$, as illustrated in Figure 2.32a. Mathematically, the values of u_j can be calculated by

$$u_0 = 0, \quad u_j = (r+1)\sum_{k=0}^{j-1}\left|\mathbf{P}_{(k+1)\bmod(r+1)} - \mathbf{P}_k\right| \bigg/ \sum_{k=0}^{r}\left|\mathbf{P}_{(k+1)\bmod(r+1)} - \mathbf{P}_k\right|, \quad (j = 1, r). \tag{2.108}$$

The B-spline curve is defined as

$$\mathbf{x}(u) = \sum_{i=0}^{n} \mathbf{B}_i \, N_{i,k}(u) \tag{2.109}$$

(a)   **(b)**

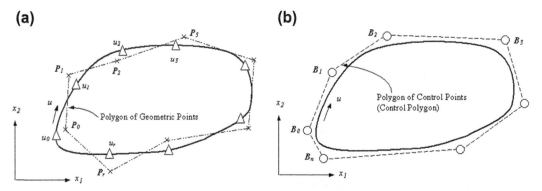

FIGURE 2.32 B-spline Curve Fitting. (a) Curve Fitting for Geometric Points $\mathbf{P}_j$. (b) B-spline Curve with Control Points $\mathbf{B}_i$.

where $\mathbf{B}_i$ is the ith control point shown in Figure 2.32b, $n+1$ is the number of control points, and $N_{i,k}(u)$ is the basis function of the B-spline curve, defined recursively as

$$N_{i,k}(u) = \frac{(u - t_i)N_{i,k-1}(u)}{t_{i+k-1} - t_i} + \frac{(t_{i+k} - u)N_{i+1,k-1}(u)}{t_{i+k} - t_{i+1}} \quad \text{and} \quad \begin{cases} N_{i,1}(u) = 1, & \text{if } t_i \leq u \leq t_{i+1} \\ N_{i,1}(u) = 0, & \text{otherwise} \end{cases}$$

(2.110)

where $[t_i, t_{i+1}]$ is a knot span formed by the two consecutive knots t_i and t_{i+1}, and $k - 1$ is the polynomial order of the basis functions.

To minimize f, the derivatives of f with respect to the $n + 1$ control points are set to zero. For simplicity, considering only the ℓth control point, one has

$$\frac{df}{d\mathbf{B}_\ell} = \sum_{j=0}^{r} \left\| -2\mathbf{P}_j \sum_{i=0}^{n} N_{i,k}(u_j) + 2\sum_{i=0}^{n} N_{i,k}(u_j) \left(\sum_{i=0}^{n} N_{i,k}(u_j)\mathbf{B}_\ell \right) \right\| = 0.$$

(2.111)

For $\ell = 0, n$, the above expression can be rewritten in a matrix form as

$$\mathbf{N}^T\mathbf{N}\mathbf{B} = \mathbf{N}^T\mathbf{P}$$

(2.112)

where $\mathbf{N} \in \mathbf{R}^{(r+1)\times(n+1)}$, $\mathbf{B} = \mathbf{R}^{(n+1)\times 3}$, $\mathbf{P} = \mathbf{R}^{(r+1)\times 3}$, and

$$\mathbf{N} = \begin{bmatrix} N_{0,k}(u_0) & N_{1,k}(u_0) & \cdots & N_{n,k}(u_0) \\ N_{0,k}(u_1) & N_{1,k}(u_1) & \ddots & N_{n,k}(u_1) \\ \vdots & \ddots & \ddots & \vdots \\ N_{0,k}(u_r) & N_{1,k}(u_r) & \cdots & N_{n,k}(u_r) \end{bmatrix}_{(r+1)\times(n+1)}$$

(2.113)

Note that $\mathbf{N}^T\mathbf{N}$ is invertible if $N_{i,k}(u_j) \neq 0$. This is true if and only if $t_{i-k+1} < u_j < t_{i+1}$, for $i = 0,n$, and $j = 0,r$. This implies that there must exist at least one u_j in at least one knot span so that $N_{i,k}(u_j) \neq 0$ for all basis functions. This requirement can be achieved by adjusting the knot values of the basis functions. The curve fitting error can be controlled by adjusting the polynomial order and the number of control points. The output of the curve fitting is a set of control points and basis functions that describe the smoothed section contour.

2.6.1.2 Surface skinning

The fitting B-spline curves discussed above are then "quilted" across sections to form an open B-spline surface, as shown in Figure 2.33, using the surface skinning technique. Note that in this process, the number of control points of the B-spline curves must be kept identical across sections. In addition, the polynomial order of the basis functions and knot values of the B-spline curves must be identical in all sections. The control points are connected to their corresponding points across sections, as shown in Figure 2.33a, to form a control polyhedron. The enclosed B-spline surface is then constructed, as shown in Figure 2.33b, by

$$x(u, w) = \sum_{i=0}^{n} \sum_{j=0}^{m} \mathbf{B}_{ij}N_{i,k}(u)\, M_{j,\ell}(w),$$

(2.114)

where $n + 1$ and $m + 1$ are the numbers of control points in the u- and w-parametric directions, respectively; and $k - 1$ and $\ell - 1$ are the polynomial orders of the basis functions $N_{i,k}(u)$ and $M_{j,\ell}(w)$,

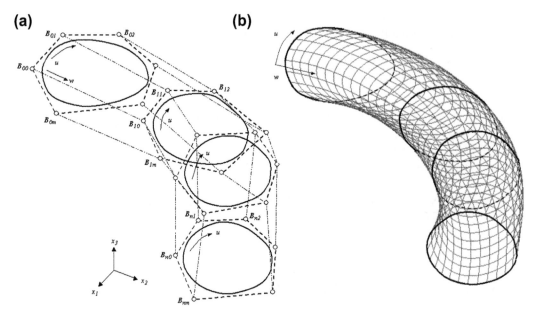

FIGURE 2.33 B-spline Surface Skinning. (a) Control Polyhedron and Section Curves. (b) B-spline Surface Enclosed by the Control Polyhedron.

respectively. Note that the B-spline surface constructed is C^2-continuous in both u- and w-parametric directions, if cubic basis functions are assumed. The control points and basis functions of the B-spline surface can be imported into CAD software to support solid modeling.

2.6.2 Engineering applications

The first application is the integration of topology and shape optimization. Topology optimization (Bendsoe and Sigmund, 2003) has drawn significant attention in the recent development of structural optimization. This method has been proven effective in determining the initial geometric shape for structural designs. The main drawback of the method, however, is that the topology optimization always leads to a nonsmooth structural geometry, while most of the engineering applications require a smooth geometric shape, especially for manufacturing. On the other hand, shape optimization (Chang and Choi, 1992) starts with a smooth geometric model that can be manufactured much easier. However, the optimal shape is confined to the topology of the initial structural geometry. No additional holes can be created during the shape optimization process. It is desirable to combine topology and shape optimizations to support structural design effectively by taking advantage of both methods. The curve fitting and surface skinning technique discussed in Section 2.6.1 is ideal to support integration of topology optimization and shape optimization.

(a)   **(b)**   **(c)**

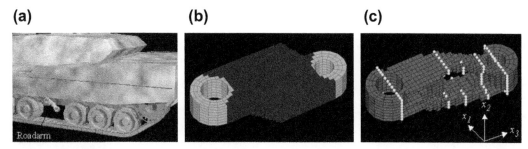

FIGURE 2.34 The Tracked Vehicle Roadarm Example. (a) Physical Model. (b) Initial Finite Element Model. (c) Topologically Optimized Model.

To demonstrate the technique, a tracked vehicle roadarm shown in Figure 2.34a is optimized using topology optimization from initial shape shown in Figure 2.34b to that of Figure 2.34c (Tang and Chang, 2001).

The optimal design is unsmooth and cannot be manufactured. Geometric points of five representative sections (Step 2 in Figure 2.35) of the roadarm are selected and fitted with B-spline curves (Steps 3a and 3b in Figure 2.35). Following the surface skinning method, an outer polyhedron formed by the 6×5 control points and the enclosed B-spline surface are created (Step 4a). Similarly, an inner B-spline surface (4×3 control points) that represents the hole in the roadarm is created (Step 4b). These B-spline surfaces are imported into SolidWorks for solid model construction. In SolidWorks, the outer and inner solid models are created by filling up the cavities enclosed by the outer and inner B-spline surfaces, respectively. The final solid model is obtained by subtracting the inner solid from the outer one (Step 5) and uniting the subtracted solid model with two end half cylinders, as shown in Figure 2.35.

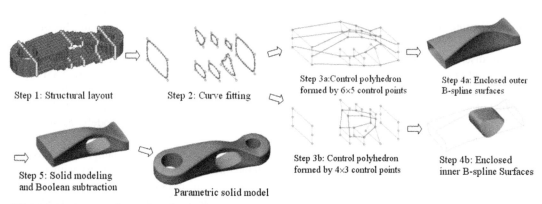

Step 1: Structural layout

Step 2: Curve fitting

Step 3a:Control polyhedron formed by 6×5 control points

Step 4a: Enclosed outer B-spline surfaces

Step 3b: Control polyhedron formed by 4×3 control points

Step 4b: Enclosed inner B-spline Surfaces

Step 5: Solid modeling and Boolean subtraction

Parametric solid model

FIGURE 2.35 Construction of B-spline Surfaces for Structural Design.

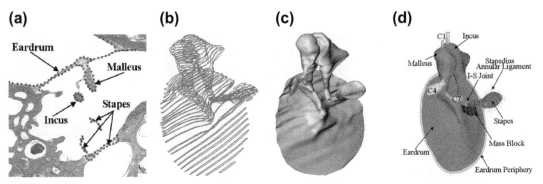

FIGURE 2.36 Human Middle Ear Surface and Finite Element Analysis Models. (a) Section Image. (b) Section Contours. (c) Surface Model. (d) Finite Element Model.

The second example is modeling a human middle ear (Sun et al., 2002). The modeling steps start with the histological section preparation of human temporal bone. Through tracing outlines of the middle ear components on the sections (Figure 2.36a), a set of discrete points is obtained and employed to construct B-spline curves that represent the exterior contours of the components using the curve fitting technique (Figure 2.36b). The surface skinning technique is then employed to quilt the B-spline curves for smooth boundary surfaces of the middle ear components using B-spline surfaces (Figure 2.36c). The solid models of the middle ear components are constructed using these surfaces and then assembled to create a complete middle ear in CAD. The geometric model constructed using the proposed method is smooth and can be used to create finite element models for mechanics study (Figure 2.36d).

The same modeling technique is applied to a human maxillary second molar, which is the third example to be presented. The main purpose of constructing a geometric model for the human tooth is to capture accurately the geometry of the critical dentino-enamel junction (DEJ), which is important for investigating stress distribution inside the tooth. The geometric modeling started with a histological section preparation of a human tooth (Figure 2.37a). Through tracing outlines of the tooth on the sections, discrete points are obtained and are employed to construct B-spline curves that represent the exterior contours and DEJ of the tooth using a least square curve fitting technique (Figure 2.37b). The surface skinning technique is then employed to quilt the B-spline curves to create a smooth boundary and DEJ of the tooth using B-spline surfaces (Figure 2.37c). These surfaces are respectively imported into SolidWorks via its Application Protocol Interface (API) to create solid models, as shown in Figure 2.37d (Chang et al., 2003).

The last example is for support of reverse engineering. An airplane tubing sample part was first scanned using an industrial CT scanner, capturing both the interior and exterior geometry with 486,107 uniformly spaced data points (Figure 2.38). A B-spline curve fitting and surface skinning approach was employed to convert the data points into B-spline surfaces (Chang et al., 2006). A physical model was produced using a stereolithography apparatus and mounted to the production fixtures to verify the accuracy of the surface model, as shown in Figure 2.38.

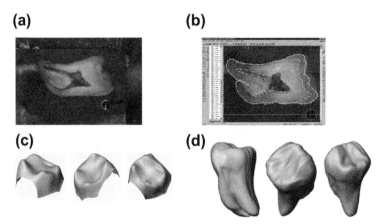

FIGURE 2.37 Geometric Model Construction for a Human Tooth. (a) A Sample Section Image. (b) Section Sketch Digitization with References. (c) Surface Model of the DEJ. (d) The Solid Model in Various Views.

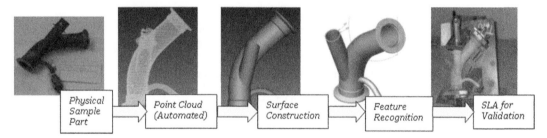

FIGURE 2.38 Reverse Engineering of Airplane Engine Tubing.

2.7 Summary

In this chapter, we discussed basic and essential topics in geometric modeling, including parametric representations for curves and surfaces. We discussed popular curve and surface formats, including the most versatile and general NURB curves and surfaces, which are widely employed for geometric modeling. We hope the discussion became directly relevant when we introduced the surfaces generated by CAD. We also include the topic of geometric transformation in this chapter, which is essential to understand how the geometric entities are transformed to support numerous needs in modeling. Detailed derivations were provided in this chapter because geometric modeling serves as the foundation for solid modeling in CAD, which is at the center of the e-Design paradigm. In addition to the mathematical forms of the curves and surfaces, we include as a case study of the curve fitting and surface skinning techniques, which are powerful for many engineering applications. We hope by now you have a fine understanding in the basics of geometric modeling, as we move to the next chapter to discuss solid modeling and CAD theory. With a good understanding of solid modeling, we will then discuss CAD assembly in Chapter 4 and then move into the heart of this book—design parameterization for part and assembly in Chapter 5.

Appendix A Basis functions of B-spline curves and surfaces

In this appendix, we provide detailed derivations that lead to the six basis functions $N_{i,3}(u)$, $i = 0, 5$, of a quadratic B-spline curve, as stated in Eqns 2.48a–f.

Recall that the basis functions $N_{i,k}(u)$ are defined recursively as

$$N_{i,k}(u) = \frac{(u - t_i)N_{i,k-1}(u)}{t_{i+k-1} - t_i} + \frac{(t_{i+k} - u)N_{i+1,k-1}(u)}{t_{i+k} - t_{i+1}} \tag{2.44}$$

and

$$N_{i,1}(u) = \begin{cases} 1, t_i \leq u \leq t_{i+1} \\ 0, \quad \text{elsewhere.} \end{cases} \tag{2.45}$$

Note that t is called knots in Eqns (2.44) and (2.45), defined as

$$t_i = \begin{cases} 0, & i < k \\ i - k + 1, & k \leq i \leq n. \\ n - k + 2, & i > n \end{cases} \tag{2.46}$$

There are $n + k + 1 = 5 + 2 + 1 = 9$ knots. Also, the knots of the curve are

$$\begin{aligned} t_{0,1,2} &= 0 \\ t_3 &= 1 \\ t_4 &= 2 \\ t_5 &= 3 \\ t_{6,7,8} &= 4. \end{aligned} \tag{2.47}$$

From Eqn (2.45), we have

$$\begin{aligned} N_{0,1}(u) &= N_{1,1}(u) = 0 \\ N_{2,1}(u) &= 1, \quad \text{for } 0 \leq u \leq 1 \\ N_{3,1}(u) &= 1, \quad \text{for } 1 \leq u \leq 2 \\ N_{4,1}(u) &= 1, \quad \text{for } 2 \leq u \leq 3 \\ N_{5,1}(u) &= 1, \quad \text{for } 3 \leq u \leq 4 \\ N_{6,1}(u) &= N_{7,1}(u) = 0 \end{aligned} \tag{A.1}$$

which are step functions, also called switch functions.

Then, for $k = 2$, from Eqn (2.44) we have

$$N_{i,2}(u) = \frac{(u - t_i)N_{i,1}(u)}{t_{i+1} - t_i} + \frac{(t_{i+2} - u)N_{i+1,1}(u)}{t_{i+2} - t_{i+1}} \tag{A.2}$$

and

$$N_{0,2}(u) = \frac{(u - t_0)N_{0,1}(u)}{t_1 - t_0} + \frac{(t_2 - u)N_{1,1}(u)}{t_2 - t_1} = 0 + 0 = 0$$

$$N_{1,2}(u) = \frac{(u - t_1)N_{1,1}(u)}{t_2 - t_1} + \frac{(t_3 - u)N_{2,1}(u)}{t_3 - t_2} = 0 + (t_3 - u)N_{2,1}(u) = (1 - u)N_{2,1}(u)$$

$$N_{2,2}(u) = \frac{(u - t_2)N_{2,1}(u)}{t_3 - t_2} + \frac{(t_4 - u)N_{3,1}(u)}{t_4 - t_3} = uN_{2,1}(u) + (2 - u)N_{3,1}(u)$$

$$N_{3,2}(u) = \frac{(u - t_3)N_{3,1}(u)}{t_4 - t_3} + \frac{(t_5 - u)N_{4,1}(u)}{t_5 - t_4} = (u - 1)N_{3,1}(u) + (3 - u)N_{4,1}(u) \tag{A.3}$$

$$N_{4,2}(u) = \frac{(u - t_4)N_{4,1}(u)}{t_5 - t_4} + \frac{(t_6 - u)N_{5,1}(u)}{t_6 - t_5} = (u - 2)N_{4,1}(u) + (4 - u)N_{5,1}(u)$$

$$N_{5,2}(u) = \frac{(u - t_5)N_{5,1}(u)}{t_6 - t_5} + \frac{(t_7 - u)N_{6,1}(u)}{t_7 - t_6} = (u - 3)N_{5,1}(u) + 0 = (u - 3)N_{5,1}(u)$$

$$N_{6,2}(u) = 0.$$

These are piecewise linear functions, as shown in Figure A.1.

Now, for k = 3, from Eqn (2.48), we have

$$N_{i,3}(u) = \frac{(u - t_i)N_{i,2}(u)}{t_{i+2} - t_i} + \frac{(t_{i+3} - u)N_{i+1,2}(u)}{t_{i+3} - t_{i+1}} \tag{A.4}$$

and

$$N_{i,3}(u) = \frac{(u - t_i)N_{i,2}(u)}{t_{i+2} - t_i} + \frac{(t_{i+3} - u)N_{i+1,2}(u)}{t_{i+3} - t_{i+1}}$$

$$N_{0,3}(u) = \frac{(u - t_0)N_{0,2}(u)}{t_2 - t_0} + \frac{(t_3 - u)N_{1,2}(u)}{t_3 - t_1} = (1 - u)^2 N_{2,1}(u)$$

$$N_{1,3}(u) = \frac{(u - t_1)N_{1,2}(u)}{t_3 - t_1} + \frac{(t_4 - u)N_{2,2}(u)}{t_4 - t_2} = \frac{1}{2}u(4 - 3u)N_{2,1}(u) + \frac{1}{2}(2 - u)^2 N_{3,1}(u)$$

$$N_{3,3}(u) = \frac{(u - t_3)N_{3,2}(u)}{t_5 - t_3} + \frac{(t_6 - u)N_{4,2}(u)}{t_6 - t_4}$$

$$= \frac{1}{2}(u - 1)^2 N_{3,1}(u) + \frac{1}{2}(-2u^2 + 10u - 11)N_{4,1}(u) + \frac{1}{2}(4 - u)^2 N_{5,1}(u) \tag{A.5}$$

$$N_{4,3}(u) = \frac{(u - t_4)N_{4,2}(u)}{t_6 - t_4} + \frac{(t_7 - u)N_{5,2}(u)}{t_7 - t_5}$$

$$= \frac{1}{2}(u - 2)^2 N_{4,1}(u) + \frac{1}{2}(-3u^2 + 20u - 32)N_{5,1}(u)$$

$$N_{5,3}(u) = \frac{(u - t_5)N_{5,2}(u)}{t_7 - t_5} + \frac{(t_8 - u)N_{6,2}(u)}{t_8 - t_6} = \frac{1}{2}(u - 3)^2 N_{5,1}(u).$$

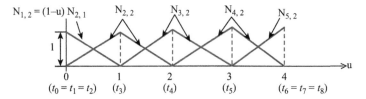

FIGURE A.1 Basis Functions $N_{i,2}(u)$.

These are the quadratic functions shown in Figure 2.12.

Appendix B Representing conics with quadratic NURB curves

In Example 2.10, we showed that a quadratic NURB curve with three control points represents a 90° circular arc analytically, in which we set the weights $h_0 = h_2 = 1$, and $h_1 = \frac{1}{\sqrt{2}}$. Now this appendix, we explain why such weights turns a NURB curve into a circular arc. We provide the explanation in a broader sense, extending the topic to include the entire conics family. We hope that by doing so we offer a more comprehensive explanation on this important topic.

First, in analytic geometry, a conic may be defined as a planar algebraic curve of degree 2, which is written as an implicit equation of degree 2 as follows:

$$f(x, y) = Ax^2 + 2Bxy + Cy^2 + 2Dx + 2Ey + 1 = 0 \tag{B.1}$$

Geometrically, a conic is the locus of a point moving on the x–y plane so that its distance from a fixed point (called the focus, point F, in Figure B.1) is proportional to its distance to a fixed line (called the directrix, usually the y-axis). As shown in Figure B.1, the focus F is located at (k,0); any point on the directrix, such as point D, can be represented as (0,y). The locus of the conics must satisfy the proportionality e, called eccentricity, defined as

$$e = \frac{FP}{PD}. \tag{B.2}$$

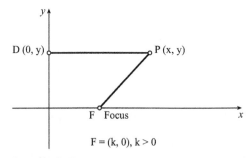

FIGURE B.1 Basics of Constructing a Conic Curve.

From Figure B.1, the eccentricity e can be written as follows:

$$e = \frac{\sqrt{(x-k)^2 + y^2}}{|x|} \tag{B.3}$$

Square both sides and arrange terms, which yields

$$(1 - e^2)x^2 - 2kx + y^2 + k^2 = 0. \tag{B.4}$$

When e = 1, Eqn (B.4) becomes

$$-2kx + y^2 + k^2 = 0 \tag{B.5}$$

which is a parabola (see example in Figure B.2a). When e < 1, the coefficient of the x^2 term in Eqn (B.4) is positive, and the equation becomes

$$(1 - e^2)x^2 - 2kx + y^2 + k^2 = 0 \tag{B.6}$$

which can be converted into a form

$$\frac{x^2}{s^2} + \frac{y^2}{t^2} = 1 \tag{B.7}$$

which represents an ellipse (see example in Figure B.2b). Equation (B.7) represents a circle when s = t. When e > 1, the coefficient of the x^2 term in Eqn (B.4) is negative, and the equation becomes

$$-(e^2 - 1)x^2 - 2kx + y^2 + k^2 = 0 \tag{B.8}$$

which can be converted into a form

$$\frac{x^2}{s^2} - \frac{y^2}{t^2} = 1 \tag{B.9}$$

which represents a hyperbola (see example in Figure B.2c).

With a basic understanding of conics, we proceed with representing a conic curve with a NURB curve.

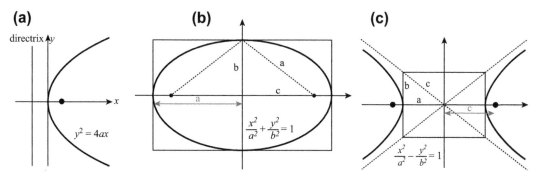

FIGURE B.2 Conic Curves. (a) A Parabola. (b) An Ellipse. (c) A Hyperbola.

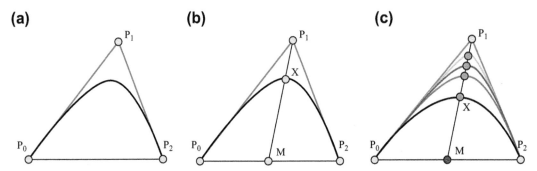

FIGURE B.3 Representing Conic Curves Using NURB. (a) A Parabola. (b) Point X on Line Segment P_1M. (c) Different Types of Conic Curves Determined by the Position of Point X.

First, a NURB of degree 2 defined by three noncollinear control points P_0, P_1, and P_2, can be a segment of a parabola, as depicted in Figure B.3a. We wish to extend this concept to define ellipse and hyperbola segments. It is well known that a conic curve that passes through P_0 and P_2 and is tangent to P_0P_1 and P_1P_2 at P_0 and P_2, respectively (see Figure B.3b), can be represented by an implicit equation of degree 2 as shown in Eqn (B.1). Note that there are five unknown in Eqn (B.1): A, B, C, D, and E. These five unknowns must be determined by five linearly independent equations. Four of these equations can be found by plugging P_0 and P_2 into Eqn (B.1) and then taking the derivative of Eqn (B.1) and applying the condition of curve tangency to line segments P_0P_1 and P_1P_2.

Each of these four equations is linear in the unknowns A, B, C, D, and E. If we could find one more condition to generate one more needed linear equation, we will have five linear equations with five unknowns. Solving this system of linear equations yields all five coefficients and the conic curve is uniquely determined.

A very natural addition would be one more point. Plugging the coordinates of this point into Eqn (B.1) will give us an equation that is similar to those for control points P_0 and P_2. This point should be inside of the triangle of the three control points so that the convex hull property can be maintained. The position of this point should also be easily changed to produce a different conic curve. One way to do this is by allowing this point to be on the line segment joining P_1 and the midpoint of P_0P_2 (point M shown in Figure B.3b). In this way, moving the point X on this line segment generates different conic curves, as shown in Figure B.3c.

Recall that the equation of a quadratic NURB curve is

$$P(u) = \frac{\sum_{i=0}^{2} h_i P_i N_{i,3}(u)}{\sum_{i=0}^{2} h_i N_{i,3}(u)} = \frac{h_0 P_0 N_{0,3}(u) + h_1 P_1 N_{1,3}(u) + h_2 P_2 N_{2,3}(u)}{h_0 N_{0,3}(u) + h_1 N_{1,3}(u) + h_2 N_{2,3}(u)}. \tag{B.10}$$

Note that moving point X has the same effect as changing the weight h_1 associated with the control point P_1 of the NURB curve. Also, we assumed $h_0 = h_2 = 1$ to ensure the curve tangency at P_0 and P_2, respectively.

If we put P_0 and P_2 on the opposite sides of the x-axis, with the midpoint of P_0P_2 being the coordinate origin (by a simple translation followed by a rotation), we have $P_0 = -P_2$. Let the NURB curve meet the line segment MP_1 at X as shown in Figure B.3b. A simple calculation using the quadratic NURB curve equation shown in Eqn (B.10) yields the following:

$$P(0.5) = \frac{h_1}{1 + h_1} P_1 \quad \text{or} \quad \frac{P(0.5)}{P_1} = \frac{h_1}{1 + h_1} \tag{B.11}$$

In other words, from Eqn (B.3b), we have the following important relationship:

$$\frac{MX}{MP_1} = \frac{h_1}{1 + h_1} \tag{B.12}$$

Now, for the quarter circle shown in Figure 2.16b, or any circular arc like that of Figure B4, we have

$$MX = OX - OM = r - r \sin a = r(1 - \sin a)$$

and

$$MP_1 = OP_1 - OM = \frac{r}{\sin a} - r \sin a = \frac{r(1 - \sin^2 a)}{\sin a}.$$

From the above two equations, we have

$$\frac{h_1}{1 + h_1} = \frac{MX}{MP_1} = \frac{\sin a}{1 + \sin a}$$

which implies $h_1 = \sin a$ as desired.

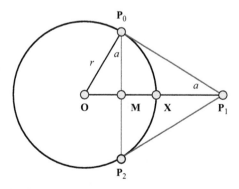

FIGURE B.4 A Circular Arc Being Defined by a Quadratic NURB with Three Control Points.

Questions and exercises

1. Verify that the functions in Eqn (2.3) are indeed representing a circle. Plot points $(x(u), y(u))$ using a program, such as Matlab, or write a computer program to do so.

2. Given two points, $\mathbf{P}_0$, $\mathbf{P}_1$, and a tangent vector at the start point $\mathbf{P}_{0,u}$, derive equations for a parametric quadratic curve that passes through these two points at $u = 0$ and 1, respectively, with tangent vector $\mathbf{P}_{0,u}$ at $u = 0$.

   Graph the basis functions in Matlab, make observations on the characteristics of the function, and comment on their influence on the curve geometry.

   Graph the curve for

   $\mathbf{P}_0 = [0,1]$, $\mathbf{P}_1 = [3,2]$, $\mathbf{P}_{0,u} = [2,-7]$.

   In addition to the derivations, submit screen captures of Matlab graphs to show the curve and basis functions.

3. Given three points, $\mathbf{P}_0$, $\mathbf{P}_1$, $\mathbf{P}_2$, and the tangent vector at the end point $\mathbf{P}_{2,u}$, derive equations for a parametric cubic curve that passes through these three points at $u = 0$, ½, and 1, respectively, with tangent vector $\mathbf{P}_{2,u}$ at $u = 1$.

   Graph the basis functions and comment on their influence on the curve geometry.

   Graph the curve for:

   $\mathbf{P}_0 = [0,1]$, $\mathbf{P}_1 = [2,0]$, $\mathbf{P}_2 = [3,2]$, $\mathbf{P}_{2,u} = [2,-7]$.

   Submit screen captures of the Matlab graph to show the curve and basis functions.

4. Continue from Problem 3. Calculate the tangent vectors of the curve at both start and end points, following curve format conversion. Calculate the positions of the curve points at $u = 1/3$ and 2/3, following curve format conversion. Calculate the position of the interior control points $\mathbf{P}_1$ and $\mathbf{P}_2$ of the equivalent Bézier curve using curve format conversion.

5. Four control points on the x–y plane are given as follows:

   $\mathbf{P}_0 = [0,0]$, $\mathbf{P}_1 = [1,4]$, $\mathbf{P}_2 = [2,-5]$, $\mathbf{P}_3 = [3,8]$.

   a. Construct a Bézier curve enclosed by the control polygon formed by the four given points;

   b. Graph the curve in Matlab and submit screen captures to show the curve and basis functions.

6. Show that the 4×4 $\mathbf{M}^4$ matrix of a cubic uniform B-spline curve defined as

$$\mathbf{P}^i(u) = \begin{bmatrix} u^3 & u^2 & u & 1 \end{bmatrix}_{1\times4} \mathbf{M}_{4\times4}^4 \mathbf{M}_4 \begin{bmatrix} \mathbf{P}_{i-1} \\ \mathbf{P}_i \\ \mathbf{P}_{i+1} \\ \mathbf{P}_{i+2} \end{bmatrix} = \mathbf{U}_{1\times4} \mathbf{M}_{4\times4}^4 \begin{bmatrix} \mathbf{P}_{i-1} \\ \mathbf{P}_i \\ \mathbf{P}_{i+1} \\ \mathbf{P}_{i+2} \end{bmatrix}, \quad u \in [0, 1], i \in [1, n-2]$$

is

$$\mathbf{M}_{4\times4}^4 = \frac{1}{6} \begin{bmatrix} -1 & 3 & -3 & 1 \\ 3 & -6 & 3 & 0 \\ -3 & 0 & 3 & 0 \\ 1 & 4 & 1 & 0 \end{bmatrix}.$$

7. Show that the closed uniform B-spline curve of cubic basis functions is C^2-continuous. Calculate the Cartesian coordinates of the start and end points of all curve segments of Example 2.9, both quadratic and curve B-spline curves.

8. Derive a parametric equation for the surface of the quarter cone, using the following:

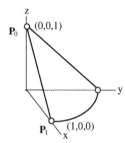

(i) Surface of revolution. Plot the surface using Matlab, and
(ii) Sweep surface. Note that
    $P_0 = [0,0,1]$, $P_1 = [1,0,0]$.
    Submit the following:
(i) Detailed equations that describe the surface of revolution and sweep surface;
(ii) Matlab scripts and screen captures of the surface plotted in Matlab.

9. Derive a parametric equation for the surface of the quarter cone shown below, using the surface of revolution. Plot the surface using Matlab. Note that

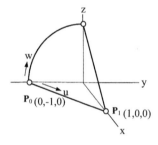

    $P_0 = [0,-1,0]$, $P_1 = [1,0,0]$.
    Submit the following:
(i) Detailed parametric equations that describe the surface;
(ii) Matlab script and screen capture of the surface plotted in Matlab.

10. Derive a parametric equation for a 1/8 sphere of radius 1 shown below formed by revolving a quadratic NURB curve $P(u)$ on the x–z plane along the z-axis.

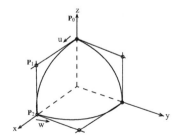

11. Derive a parametric equation for a blend surface formed by four curves. These four curves are:

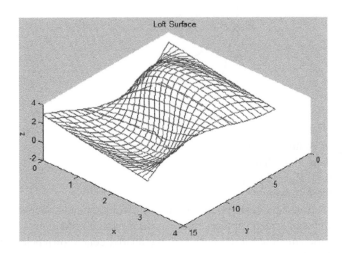

(a) Straight with end points:
$G_0 = [0,0,3]$, $G_1 = [3,0,1]$;
(b) Spline curve with three points:
$P_0 = [0,5,0]$, $P_1 = [1,5,3]$, $P_2 = [3,5,2]$;
(c) Bézier curve with four control points:
$Q_0 = [0,10,1]$, $Q_1 = [2,10,2]$, $Q_2 = [2.5,10,0.5]$, $Q_3 = [3,10,3]$.
(d) Straight line with end points:
$R_0 = [0,15,3]$, $R_1 = [3,15,1]$.
Also, create a solid (or surface) feature using these four curves in Pro/ENGINEER or SolidWorks. Submit the following:
  (i) Screen capture of the Pro/ENGINEER or SolidWorks model and sketch view with all four sections;
  (ii) Detailed equations that describe the surface;
  (iii) Matlab script and screen capture of the surface plotted in Matlab.

12. Derive a parametric equation for a sweep surface formed by sweeping a cubic Bézier curve $P(u)$ on the x–z plane along a trajectory of the same curve $Q(u)$ on the x–y plane. The control points of these two curves are, respectively:
$P_0 = [0,0,0]$, $P_1 = [-1,0,3]$, $P_2 = [-2,0,0.5]$, $P_3 = [-3,0,2]$;
$Q_0 = [0,0,0]$, $Q_1 = [1,3,0]$, $Q_2 = [2,0.5,0]$, $Q_3 = [3,2,0]$.
Also, create a sweep solid feature using these two curves in Pro/ENGINEER or SolidWorks. Submit the following:
  (i) Screen capture of the Pro/ENGINEER or SolidWorks solid model and a sketch view of both curves;
  (ii) Detailed equations that describe the surface;
  (iii) Matlab script and screen capture of the surface plotted in Matlab.

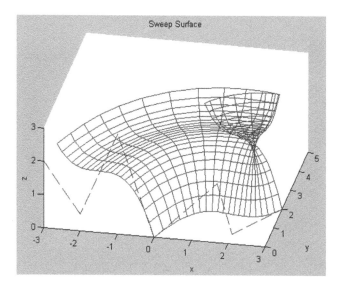

References

Bendsoe, M.P., Sigmund, O., 2003. Topology Optimization, Theory, Methods, and Applications. Springer-Verlag, Berlin, Heidelberg, New York, ISBN 3-540-42992-1.

Chang, K.H., Siddique, Z., Edke, M., Chen, Z., 2006. An integrated testbed for reverse engineering of aging systems and components. Computer-aided Design and Applications 3 (1–4), 21–30.

Chang, K.H., Magdum, S., Khera, S., Goel, V.K., May 2003. An advanced computer modeling and prototyping method for human tooth mechanics study. Annals of Biomedical Engineering 31 (5), 621–631.

Chang, K.H., Choi, K.K., 1992. A geometry-based parameterization method for shape design of elastic solids. Mechanics of Structures and Machines 20 (2), 215–252.

Mortenson, M.E., 2006. Geometric Modeling, third ed. Industrial Press.

Piegl, L., Tiller, W., 1987. Curve and surface construction using rational B-splines. Computer-aided Design 19 (9), 485–498.

Sun, Q., Chang, K.H., Dormer, K., Dyer, R., Gan, R.Z., November 2002. An advanced computer-Aided geometric modeling and fabrication method for human middle ear. Medical Engineering and Physics 24 (9), 595–606.

Tang, P.-S., Chang, K.H., 2001. Integration of topology and shape optimizations for design of structural components. Journal of Structural Optimization 22 (1), 65–82.

Solid Modeling

3

CHAPTER OUTLINE

With the basic understanding of geometric modeling discussed in Chapter 2, we are moving closer to the core of product design modeling—that is, solid modeling, especially feature-based parametric solid modeling, which is the key topic to be discussed in this chapter. In recent decades, the term *solid modeling* has been associated with the technology of using computer-aided design (CAD) systems to create the shape and form of part geometry and associated physical properties with a computer for the purpose of engineering designs. Today, CAD models with built-in essential product design information play a central role in e-Design.

Product Design Modeling using CAD/CAE.

Solid modeling in CAD applications has evolved through a series of phases in order to improve the geometric representation of physical artifacts or design concepts that are being developed in the engineering design process. It started in early 1960s when the first wireframe computer graphic was invented at the Massachusetts Institute of Technology (MIT) Lincoln Laboratory. In the mean time, design automated by computer (DAC-1), the first production interactive graphics manufacturing system, was developed by General Motors. Since then, with further development, surface modeling became a reality in the 1960s and solid modeling began in the 1970s, followed by feature modeling in the mid-1980s and parametric modeling by Parametric Technology in the late 1980s. With the development of feature modeling and parametric modeling in the 1980s, feature-based parametric solid modeling has since become the mainstream CAD theory in support of engineering design. It is well recognized that the most significant development appeared in the mid-1990s, in which major CAD tools were made available in personal computers (PCs) that allowed end users in mid- and small-size companies to be able to bring designs from the drawing board into digital form. More recently, direct modeling technology brought CAD one step further in support of engineering design by allowing designers to directly manipulate solid models by pulling or squeezing solid features on the computer screen using a mouse.

Today, major CAD systems employ feature-based parametric modeling techniques to support engineering designs through respective interactive user interfaces. Because solid modeling is the heart and soul of CAD, we devote this chapter to introducing basic knowledge in solid modeling methods and theory. This is the knowledge that readers must have in order to proceed with the study of the e-Design paradigm and gain practical skills in practicing e-Design, in which product geometry is represented in CAD solid models throughout the product development process.

This chapter is organized with the assumption that the reader has used CAD software (e.g., SolidWorks, Pro/ENGINEER) for creating solid models but has no or little background in solid modeling theory. If you are not familiar with CAD software, you are strongly encouraged to review excellent references for tutorial lessons, such as Toogood and Zecher (2012) for Pro/ENGINEER or Planchard and Planchard (2013) for SolidWorks. With the assumption that you are familiar with CAD software, we offer discussion on numerous topics involved in solid modeling, with examples extracted mostly from SolidWorks and Pro/ENGINEER.

Overall, the objectives of this chapter are (1) to provide an introduction to the basic solid modeling theories that help readers understand how the product design is realized in CAD, and (2) to help readers become familiar with the behind-the-scenes operations in CAD modeling so as to effectively use these tools for design. We also provide a short discussion on commercial CAD software tools, with the hope of offering readers guidance on selecting proper tools that are suitable for their specific needs.

3.1 Introduction

In the 1970s, nearly every engineering drawing produced in the world was done with pencil or ink on paper. A drafter leaned on the drawing board and used a T-square ruler, protractors, a compass, and templates to carefully sketch the lines, arcs, letters, and symbols that constitute an engineering drawing. Any changes or mistakes required erasing and redrawing, whereas major changes often necessitated recreation of the drawing from scratch. In manually created drawings, one of the most challenging tasks is that a drafter must envision the intersection of solid entities, unwrap the intersecting curves, and sketch the curves accurately on the drawing paper. Engineering drawing has been the backbone of product design and development for many years (Bozdoc, 2003).

Although engineering drawing still plays an important role in product design and manufacturing in many industrial sectors around the world, manual sketching for creating drawings has been gradually replaced by CAD (computer-aided design) software using computers. Beginning in the 1980s, CAD software reduced the need for draftsmen significantly, especially in small to mid-sized companies. The software's affordability and ability to run on personal computers in the mid-1990s allowed engineers to do their own drafting and analytic work to some extent.

In fact, instead of just creating drawings, CAD has fundamentally changed the way design is done. As in the manual drafting of technical and engineering drawings, the output of CAD conveys information, such as materials, processes, dimensions, and tolerances, according to application-specific conventions in solid models. Instead of drafting in digital form, designers use CAD to create product models in solid model forms with adequate product data, then they create drafting if necessary. CAD solid models offer flexibility and efficiency when making design changes; provide geometric and physical data that support product performance evaluations using computer-aided engineering (CAE); support virtual manufacturing, prototyping, manufacturing process planning, and product cost estimating; and offer product life cycle and product knowledge repository for archiving. Most important, product model in CAD serves as the centerpiece for e-Design.

The backbone of CAD is solid modeling. It is indispensable for designers to acquire adequate knowledge in CAD and solid modeling in order to effectively practice e-Design in support of engineering design. We introduce numerous theories and schemes that support product (or more specifically, parts) representation in solid models, with a focus on feature-based parametric solid modeling, which is the mainstream solid modeling method offered in major CAD systems. The main theme of the chapter is understanding the behind-the-scenes operations while you are using CAD for creating solid models. It is also important for readers to understand how CAD rebuilds solid models when a design change is made by changing dimension values associated with solid features.

We start in Section 3.2 by introducing the basic theories of solid modeling, including constructive solid geometry (CSG) and boundary representation (B-rep), which are the two most widely used schemes for solid modeling. With a basic understanding of solid modeling, we discuss the main topic of the chapter in Section 3.3—that is, the feature-based parametric solid modeling method. In Section 3.4, we offer the practical aspects of creating solid models by discussing model construction plans. We then provide a short overview of commercial CAD software in Section 3.5.

3.2 Basics of solid modeling

Before getting into the main topic of this chapter—feature-based parametric solid modeling—we discuss a few important basic topics in solid modeling in this section. We start by discussing three basic methods for representing solid models: wireframe, surface, and solid forms. We include the advantages and disadvantages of each form, as well as the use of the models represented in the form for design and manufacturing applications. We will then narrow our focus to solid modeling, for which we introduce two major modeling methods: CSG and B-rep.

3.2.1 Wireframe models

Wireframe is the simplest and the earliest form of representing physical objects; it was first introduced in 1963 at MIT's Lincoln Laboratory. The wireframe form represents a shape by its characteristic

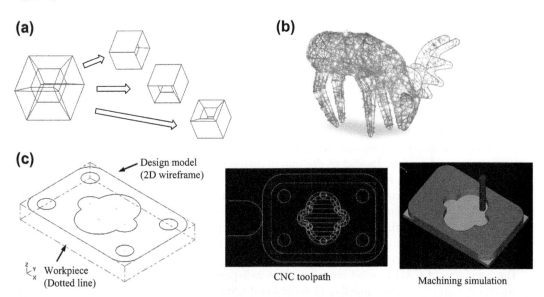

(a)

(b)

(c)

Design model
(2D wireframe)

Workpiece
(Dotted line)

CNC toolpath

Machining simulation

FIGURE 3.1 Wireframe Models. (a) Ambiguity in Representing a Solid Object. (b) Outdoor Christmas Decoration of a Rattan Reindeer (courtesy of http://www.brookstone.com/pre-lit-outdoor-christmas-decorations-rattan-reindeer). (c) Supporting Toolpath Generation in MasterCAM.

curves (lines, arcs, splines, and so on) and points, as illustrated in Figure 3.1a. A reindeer frame decoration, as shown in Figure 3.1b, which is displayed in residential front yards during Christmas, is a good example of wireframe model in real-world applications. The major advantages of this method are that it requires simple input from users and the modeling software is relatively easy to implement. One of the examples in its applications in design and manufacturing is the two-dimensional (2D) wireframe models in MasterCAM that supports numerical control (NC) toolpath generation for machining simple prismatic features, such as pocket milling or profile milling (Figure 3.1c), in which all contours exist in flat planes and only planar geometric information is required.

Although there are some applications that a wireframe model is able to support, there are major issues involved in representing solid models in wireframe. First, it is ambiguous for representing a solid object in a wireframe, as illustrated in Figure 3.1a, due to its inability to determine the inside or outside of a solid object. Second, a wireframe is not able to represent objects with nonpolygonal boundaries due to a lack of curvature information on surfaces. In addition, it is impossible to calculate mass properties of a solid object represented in a wireframe form. A wireframe model is not capable of supporting a finite element mesh for structural analysis of a physical object other than beam or truss structures. Generating a toolpath on a nonpolygon surface of a solid model represented in a wireframe is impossible due to its lack of surface geometric information.

Because of these reasons, no CAD tool uses wireframe alone to represent part geometry. Wireframes are only used in CAD as one of the options for visualizing solid objects—that is, the wireframe mode—due to its quick response in displaying objects on the computer screen without rendering.

3.2.2 **Surface models**

A surface can be thought of as an infinitely thin shell stretched over a wireframe. In addition to lines and points, surface models represent a shape by its surface geometry, as illustrated in Figure 3.2. A surface model includes information about the faces and edges of a part. Modeling methods for a surface offered in CAD include protrusions, such as extrude, sweep, loft, and revolve; interpolating points; and knitting and trim, as shown in Figure 3.2. To represent a solid object, the surfaces must form an airtight cavity that replicates the geometry of the object without any gap between them or any dangling surface or line. A hot air balloon shown in Figure 3.2f, is a good example of a surface model in real-world applications. Surfaces are commonly used to model complex, freeform (or organic) shapes that are commonly found in applications in the automotive, aircraft, mold, and consumer goods industries.

A surface model is good for visualizing complex surfaces and supports NC toolpath generation. As illustrated in Figure 3.3a, a toolpath was generated on a Coons patch in MasterCAM. The surface model is also widely used in finite element analysis (FEA) for thin-shell structures, as shown in Figure 3.3b, in which an FEA was carried out for a surface model that represents the mid-plane of a thin-shell solid object created in CAD. In addition, a stereolithography (STL) model, which is a surface model consisting of triangular facets that form an airtight cavity representing a solid object, is the de-facto standard for three-dimensional (3D) printing (also called rapid prototyping or solid freeform fabrication). An example of the STL model is given in Figure 3.3c.

As for modeling solid objects, the surface model generally works well. However, the mass or volume information of a solid object represented in a surface model is hard to determine, partly

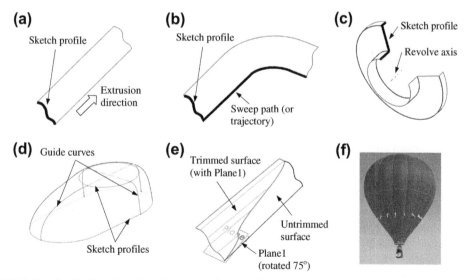

FIGURE 3.2 Creating Surface Models in Computer-aided Design. (a) Extrusion. (b) Sweep. (c) Revolve. (d) Loft (e) Trimmed. (f) A Hot Air Balloon.

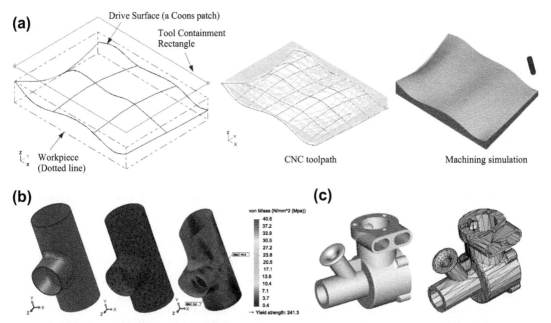

FIGURE 3.3 Surface Models for Support of Design and Manufacturing. (a) Toolpath and Machining Simulation in MasterCAM. (b) FEA of a Thin-shell Structure as a Solid Model in CAD (Left), Surface Model of Mid-plane (Middle) with Finite Element Mesh, and FEA Stress Fringe Plot (Right). (c) Engine Block as a Solid Model (Left) and Stereolithography (STL) Model (Right).

because a surface model lacks the mathematic representation of the solid object. Additional information must be added to a surface model in order to specify in/out and top/bottom of the physical object that the surface model represents.

3.2.3 Solid models

Solid models contain information about the edges, faces, and the interior of the part. The mathematical description contains information that determines whether any location is inside, outside, or on the boundary surface. Modeling a solid object in the solid model form generally includes primitive creation and Boolean operations, surface operations, protrusion operations, pick-and-place, feature-based modeling, and parametric modeling. It is important to note that individual CAD systems only use some of these methods for the modeling capabilities they respectively offer.

Primitives are basic solid objects with simple mathematical surfaces, as depicted in Figure 3.4a. These primitives can be controlled by a small number of parameters and positioned using a transformation matrix (as discussed in Chapter 2). Boolean operations, such as union, intersection, and difference, are used to make more complicated objects by combining the basic objects. This method is often referred to as CSG. More about this approach will be discussed in Section 3.2.4.

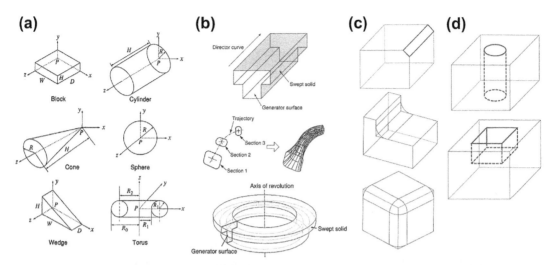

FIGURE 3.4 Solid Model Construction Methods. (a) Primitives and Boolean Operations. (b) Protrusion Operations. (c) Pick-and-place. (d) Feature Operations.

Surface operations trim and knit surfaces to form an airtight cavity that represents a solid object. This method is called boundary representation or B-rep. More details of this method are discussed in Section 3.2.4.

Protrusion operations use 2D sketch profiles to generate a 3D solid by extruding, revolving, sweeping, and loft. Examples are shown in Figure 3.4b.

Pick-and-place operates directly on the solid model surfaces, edges, and vertices to create a desired modification. Some examples include chamfering, rounding/filleting, drafting, and shelling, as illustrated in Figure 3.4c.

Feature modeling mainly supports manufacturing operations. Manufacturing features are shapes having engineering manufacturing significance. They usually are the geometric embodiment of machining operations, such as hole, pocket, slot, and boss, as illustrated in Figure 3.4d. Note that *feature* (instead of *manufacturing feature*) is a generic term used by CAD users and developers to refer to almost all kinds of geometric entities in solid modeling, sometimes including nongeometric entities, such as datum features (including planes, axis, points, coordinate systems).

Parametric modeling manipulates parameters to control the geometric shape of a solid object. Parameters come from dimensions in 2D profiles in sketch, dimensions on 3D solid features, and variables in user-defined equations. If defined properly, the entire part geometry can be controlled by a small number of key parameters. Design intents can therefore be captured through the change of the small set of parameters. Parametric modeling supports design parameterization; therefore, it becomes an indispensable part of the product design modeling in the context of e-Design. More about feature-based parametric modeling is discussed in Section 3.3 as a key section of the chapter. Design parameterization, which is an important topic of this book, is discussed in Chapter 5.

A solid model is the ultimate way to represent general objects, which are physically solid objects. Solid models support NC toolpath generation of complex surfaces and meshing with solid elements for finite element analysis. In addition, solid models are adequate for the calculation of mass properties in support of motion simulations. Solid models also support collision and interference checking, which are critical in assembly and kinematic analyses.

3.2.4 **Major modeling schemes**

As mentioned, there are several methods for constructing solid models. From a CAD user's perspective, protrusion and pick-and-place methods are most often employed. In terms of mathematically representing a solid object, two major modeling methods, CSG and B-rep, are widely employed by geometric modeling kernels, which are the core of CAD systems. More about kernels is discussed in Section 3.2.5. In this subsection, we discuss these two modeling methods in greater detail to offer readers a more in-depth understanding of CAD theory and behind-the-scenes mathematic operations.

3.2.4.1 Constructive solid geometry

CSG is a modeling method that supports the construction of solid objects through operations on solid primitives. CSG records both the information of operations and information of the primitives. The major components of the CSG method are primitives and instances, as well as the Boolean set operations and CSG tree.

3.2.4.1.1 Primitives and instances

A typical CSG system uses primitives, such as cylinders, boxes, cones, spheres, as shown in Figure 3.4a, as well as their instances, as the building blocks for constructing solid models. The idea is to use the primitives that can be manipulated easily in the computer system. Each individual primitive is stored as a geometric family together with a set of parameters. An instance of the primitive family is a scaled and/or transformed replica of its original. The scaling may be uniform or differential. It is a method of keeping primitives in their basic minimum condition, such as unit cube and unit sphere. For example, a cube belongs to the family of box with all three parameters—length, width, and height—having the same value.

3.2.4.1.2 Boolean set operations and CSG tree

To construct a complex object, the CSG approach decomposes it as a compound object composed of a number of primitives. CSG typically uses Boolean set operations, including union, difference, and intersection, to construct objects. These are mathematical operations taken from set theory. The primitives involved in an operation are referred to as operands.

A union of primitives gives a volume occupied by each operand minus the volume shared (or overlapped) by them, as illustrated in Figure 3.5b, in which union operation is carried out for two primitives P and Q (Figure 3.5a). In CAD, such a union operation is realized in many ways. For example, in Figure 3.5b, you may sketch a profile and extrude it for Q, and extrude (both sides) a square block from Q for P. Implicitly, union operation is employed in CAD.

Difference operations require two operands playing different roles. The base operand defines the source volume (Q in Figure 3.5c) and the second operand defines the volume to be removed (P in Figure 3.5c). The resulting object contains the volume of the base operand but not the second operand. For example, the object shown in Figure 3.5c represents difference operation Q − P. In CAD, you may

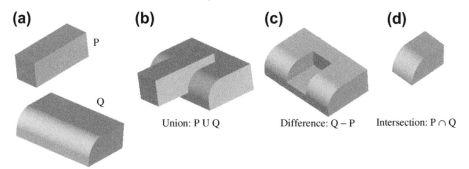

(a) P Q   **(b)** Union: P U Q   **(c)** Difference: Q – P   **(d)** Intersection: P ∩ Q

FIGURE 3.5 Boolean Operations. (a) Solid Primitives P and Q. (b) Union of P and Q. (c) Difference of P and Q. (d) Intersection of P and Q.

create a sketch profile and extrude it for Q, and extrude a square blind cut P to Q to yield the object shown in Figure 3.5c.

An intersection of primitives gives a volume commons to all operands, for example, the object shown in Figure 3.5d representing an intersection P ∩ Q. Note that there is no direct intersection operation in major CAD systems.

In fact, as a user, we do not see any direct Boolean operation capabilities offered by CAD systems. However, there are capabilities for users to cut and union features as illustrated in Figure 3.5. In general, location and orientation of primitives involved in the Boolean operations determine largely the resulting object. A primitive or its instance is usually scaled, translated, and rotated to a prescribed location and orientation before carrying out a Boolean operation.

As seen clearly, Boolean operations are binary, involving two (or two sets of) primitives. The primitives involved in each operation and the sequence of operations create a so-called CSG tree. A CSG tree, as shown in Figure 3.6 schematically, is a binary tree with leaf nodes as the primitives and interior nodes (or branch) represent Boolean set operations. The root node represents the final part.

The CSG tree creates a procedural model that specifies how the solid features are combined to form the final solid model. In general, the solid model must be "evaluated" by computing intersecting curves

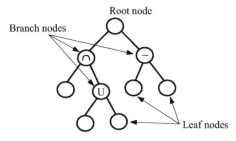

Root node

Branch nodes

Leaf nodes

FIGURE 3.6 A Schematic of a Constructive Solid Geometry Tree.

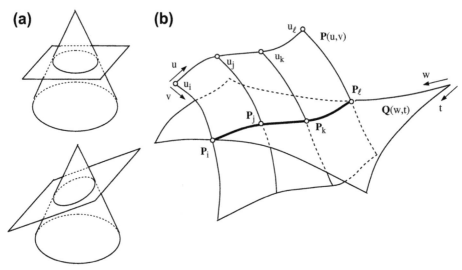

FIGURE 3.7 Surface to Surface Intersection. (a) Intersecting Two Standard Primitives. (b) Intersecting Two Parametric Surfaces.

from the parametric surface equations of the geometric features, based on the position and orientation of the primitives. Quantitative information, such as intersecting curves, must be generated and stored to define and to display the solid model.

The curve intersecting two surfaces can be analytical only for basic primitives, such as a circular cone intersecting with a plane. Conic curves, such as a circle or ellipse, are generated as a result of the intersection, as illustrated in Figure 3.7a, which can be represented analytically. However, analytical representations of intersecting curves are generally not available, so approximation methods must be used. We discuss one approximation method that calculates intersecting curves of two parametric surfaces as an example.

Consider two parametric surfaces $P(u,v) = [P_x(u,v), P_y(u,v), P_z(u,v)]$ and $Q(w,t) = [Q_x(w,t), Q_y(w,t), Q_z(w,t)]$, where u and v, and w and t are parametric coordinates of the two surfaces, respectively (see Figure 3.7b). The intersecting curve is constructed by generating parametric curve that passes through a number of intersection points.

The intersecting curve must satisfy the following equation:

$$P(u, v) - Q(w, t) = 0. \tag{3.1}$$

There are four unknowns u, v, w, and t, but only three equations (parametric equations of Eqn (3.1) in the x-, y-, and z-directions, respectively). Basically, the intersecting curve cannot be obtained by solving Eqn (3.1).

One approach to construct the intersecting curve:

1. Fix a value of, for example, the u parameter of the surface $P(u,v)$, to generate a curve on the surface (i.e., $P(u_j,v)$), as shown in Figure 3.7b.

2. Compute the intersection points of this curve with the surface $\mathbf{Q}(w,t)$ by solving numerically.

$$\mathbf{P}\left(u_j, v\right) - \mathbf{Q}(w, t) = 0 \tag{3.2}$$

We have now three equations (parametric equations of Eqn (3.2) in the x-, y-, and z-directions, respectively), and three unknowns (v, w, and t).

3. Repeat Steps 1 and 2 for as many u_j as needed. For example, if a cubic curve is desired, then four intersecting points at $u = 0$, 1/3, 2/3, and 1 can be calculated for a cubic spline curve to approximate the intersecting curve, as illustrated in Figure 3.7b.

As can be seen from this discussion, CSG is intuitive because the concept of solid model construction is in some sense parallel to manufacturing operations. The solid model constructed is always valid (except if one unions two cones only at their vertices, for example). Another advantage of the CSG is that the solid model requires a small data set (i.e., the primitives involved and CSG tree). In general, the CSG model is an unevaluated model, which must be "evaluated" for numerous purposes (e.g., display on computer screen) and calculating engineering data (e.g., mass properties). As discussed, evaluating a CSG model can be inefficient because the process involves computations, such as calculations of intersecting curves.

3.2.4.2 Boundary representation

B-rep is an important method of 3D modeling for solid objects. A B-rep model represents a solid object by assembling (or gluing) surfaces to form an "airtight" boundary that encloses the 3D space occupied by the object, as illustrated in Figure 3.8.

In B-rep, a solid model is bounded by faces, a face is bounded by edges, and an edge is bounded by vertices. Essentially, there is a hierarchy of four levels of geometric entities: volume, face, edge, and vertex. Face, edge, and vertex are topology entities that specify connectivity information. In addition to topological entities, the corresponding geometric entities are surface, curve, and point that define the shape, location, and orientation of the entities. Both geometric and topology data must be defined to construct the solid model. Both must be stored in a database.

For example, as shown in Figure 3.9, a block is defined in a volume with its six boundary faces. Each rectangular face is defined by its own edges to form an enclosed surface. All the surfaces are

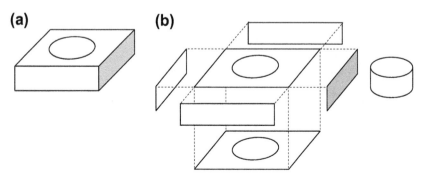

FIGURE 3.8 A Boundary Representation Model of a Solid Object. (a) Solid Model. (b) Boundary Surfaces.

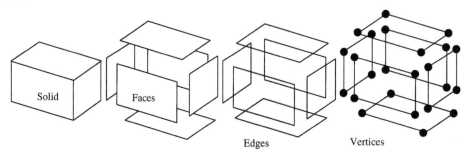

FIGURE 3.9 The Boundary Representation Model of a Rectangular Block.

joined along the common edges of their respective neighboring faces. Every edge—a straight line in this case—is bounded by its end points.

One advantage of the B-rep model is that the model is fully evaluated; that is, all geometric entities are explicitly defined and are ready for display. However, a complex B-rep model requires a relatively large database. It is less intuitive to create solid models using the B-rep method as compared with CSG because users must deal with points, curves, and surfaces instead of primitives that are much more relevant to physical objects. More importantly, a B-rep model constructed by a designer may be invalid, and a B-rep model must be verified for its topology before putting it to use.

How can we (or the computer, in this case) tell if a B-rep model is topologically valid? In a topologically valid model, all faces are properly "glued" to wrap the solid object airtight, all edges are properly "joined" that fence the face, and all edges are properly bounded by end vertices. In addition, there must be no dangling faces or edges, and no split solid object.

Which objects in Figure 3.10 are topologically valid? Apparently, objects in Figure 3.10a and b are valid, but not Figure 3.10c. It is apparent that normal objects found in nature have the property that, at every point on the boundary, a small enough sphere around the point is divided into two pieces: one inside and one outside the object. This property can be easily verified for the objects in Figure 3.10a and b. The so-called nonmanifold models break this rule. For example, in the object shown in Figure 3.10c, a small sphere around any point on the four edges of the rectangle at the bottom face of the cube is divided into four pieces: two inside and two outside. Essentially, the object in Figure 3.10c is two solids "welded" along the edges of the rectangle on the bottom face. The welded edges are

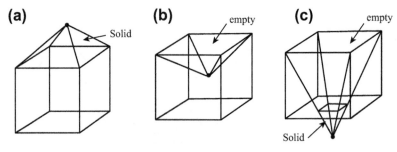

FIGURE 3.10 Examples of Manifold (a, b) and Nonmanifold (c) Objects.

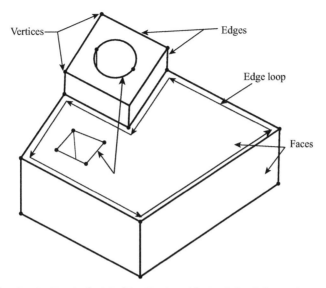

FIGURE 3.11 Illustration for the Topological Entities Employed in the Euler-Poincare Law.

infinitely thin without a cross-sectional area, which is physically impossible. A nonmanifold object, such as the one on Figure 3.10c, is considered to be topologically invalid.

To ensure the topological validity of a B-rep model, the number of its topological entities must satisfy the Euler-Poincare law, which is stated as follows:

$$v - e + f = 2(s - h) + r \qquad (3.3)$$

where v, e, f, s, h, and r are the numbers of vertices, edges, faces, solids, through holes, and rings, respectively. Note that r can be a ring or an inner loop of edges that are completely within a face. Figure 3.11 offers an illustration for the topological entities mentioned in Eqn (3.3).

The following simple examples illustrate the law and verify the topological validity of the respective physical objects.

EXAMPLE 3.1

Use the Euler-Poincare law to verify if the following objects are topologically valid: (a) a rectangular block with a rectangular through hole, (b) a circular cylinder, and (c) a rectangular block with a circular through hole.

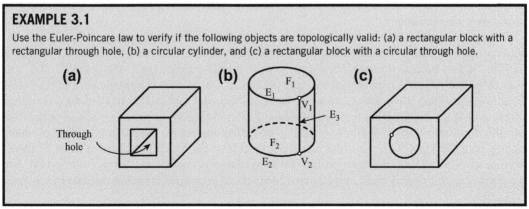

Continued

EXAMPLE 3.1—CONT'D

Solutions

For part (a), we have f = 10 (four exterior, four interior, and two ends), v = 16, e = 24, s = 1, h = 1, and r = 2. From Eqn (3.3), we have

$$v - e + f = 16 - 24 + 10 = 2$$

and

$$2(s - h) + r = 2(1 - 1) + 2 = 2.$$

Therefore, part (a) is topologically valid. Physically, the block is a valid object.

For part (b), we have f = 3, v = 2, e = 3, and s = 1. Note that a silhouette edge must be added (and the associated vertices) to a cylindrical surface when counting the number of topological entities. From Eqn (3.3), we have

$$v - e + f = 2 - 3 + 3 = 2$$

and

$$2(s - h) + r = 2(1 - 0) + 0 = 2.$$

Therefore, part (b) is topologically valid.

We are applying the same principle for part (c)—that is, adding a silhouette edge to the circular cylindrical surface inside the rectangular block.

Hence, for part (c), we have f = 6 + 1 = 7, v = 8 + 2 = 10, e = 12 + 3 = 15, s = 1, h = 1, and r = 2. From Eqn (3.3), we have

$$v - e + f = 10 - 15 + 7 = 2$$

and

$$2(s - h) + r = 2(1 - 1) + 2 = 2.$$

Therefore, part (c) is topologically valid.

Boundary representation is essentially a local representation connecting faces, edges, and vertices. An extension of this is to group the primitive geometric entities of the shape into logical units called geometric features. Features are the basis of many other developments, allowing high-level "geometric reasoning" about shape for comparison, process planning, manufacturing, etc. Feature-based modeling is discussed next in Section 3.3.

Compared to the CSG representation, which uses only primitive objects and Boolean operations to combine them, boundary representation is more flexible and has a much richer operation set. This makes boundary representation a more appropriate choice for CAD systems. CSG was used initially by several commercial systems because it was easier to implement. The advent of reliable commercial B-rep kernel systems, such as Parasolid and ACIS, has led to widespread adoption of B-rep for CAD. B-rep kernels systems offer CAD-like operations, such as protrusion, chamfer, blending, drafting, shelling, tweaking, and other operations. Note that most CAD systems employ both CSG and B-rep for solid modeling, or at least the major principles of these methods. In general, CSG keeps the relationship between features, and B-rep stores topological and geometric data for display and computations. More about geometric modeling kernels can be seen in Section 3.3.7.

3.3 Feature-based parametric solid modeling

Most modern CAD software employs a methodology called feature-based parametric solid modeling as the major interface for users to interactively create solid models. Feature-based modeling approach is more desirable in constructing solid models, in which designers use features that correspond to physical entities to construct solid models, instead of dealing with primitive geometric entities, such as points, curves, and solid primitives. The features available in CAD are usually designed to relate to how engineers think in their design and manufacturing work. The parametric modeling method allows designers to create solid models in such a way that by varying a few parameters (e.g., geometric dimensions), the solid models rebuild automatically as intended (i.e., capturing design intent). For example, a hole in the block shown in Figure 3.12a is intended to stay at the middle of the block when the width of the block changes. To capture this design intent, first the sketch profile of the base block must be fully defined (Figure 3.12b), with dimension d2 as a design variable that is to vary. The hole must be placed to the profile of the base block with its width dimension (d1 in Figure 3.12c) related to that of block width d2, as $d1 = 0.5d2$. This one-way parameter assignment is essentially parametric modeling.

In most CAD, the base block is a protrusion feature generated by extruding the sketch profile shown in Figure 3.12b along an extrusion direction that is normal to the sketch profile. The hole is an extrude cut feature that does not require users to sketch its profile, which is called a pick-and-place feature.

The question is how CAD software captures our design intent. How does CAD handle the sketch profile of the block when you make a change? How does CAD rebuild the part and what are the potential pitfalls you, as a designer, should avoid?

In this section, we intend to answer these questions. We start by revisiting "features" to have a better understanding of the terminology so that the "feature-based modeling" becomes more vivid. We then discuss the variational modeling method that CAD employs to support the needed calculation for determining a sketch profile. We discuss parent–child relationships, which are generated

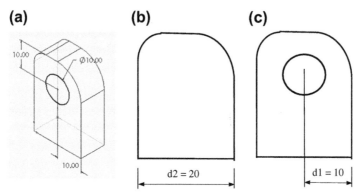

(a)   **(b)**   **(c)**

FIGURE 3.12 (a) The Block Example for the Illustration of Design Intent Capturing Using Feature-based Parametric Solid Modeling Method. (b) Sketch Profile of the Base Block with Width Dimension d2 Shown. (c) Position of the Hole by Dimension d1, Which can be Parametrically Related to Dimension d2.

when features are added to the solid model. Finally, we discuss the parametric modeling method and the solid modeling procedure we often exercise in CAD, in which we show how CAD rebuilds a part by walking through the steps using a simple example. We also discuss a newly developed modeling method called direct modeling. This section is wrapped up by a short introduction to geometric modeling kernels.

3.3.1 Geometric features

The term "feature" implies different meanings in different engineering disciplines. This has resulted in many ambiguous definitions for feature. A feature in computer-aided design (CAD) usually refers to a region of a part with certain geometric or topological properties (Pratt and Wilson, 1985). These are more precisely called geometric (or form) features. Geometric features contain both shape information and parametric information of a region of interest. They are now ubiquitous in most current CAD software, where they are used as the primary means of creating 3D solid models.

Another frequently used feature is the manufacturing feature, which can be defined simply as a geometric shape and its manufacturing information to create the shape. Manufacturing features support the generation of process plans in a feature-based process planning system. Machining features are an important subset of manufacturing features. A machining feature can be regarded as the volume swept by a "cutting" tool, which is always a negative (subtracted) volume. Some CAM software, such as CAMWorks, offers an automatic (manufacturing) feature recognition capability that recognizes manufacturing features embedded in CAD solid models and generates a toolpath accordingly. There are also tolerance features that specify deviations from the nominal form (or shape), size, or locations, such as surface flatness, circularity, and concentricity. Finally, there is the concept of assembly feature, which encodes the assembly method between connected components.

The features mentioned previously are highly related to part geometry. There are also non-geometric features, such as material features that specify material composition and heat treatment for a part.

In this subsection, we discuss geometric features from a design perspective. Instead of providing a generic and philosophical discussion, we narrow the focus to CAD solid modeling. Nowadays, almost everything that contributes to the construction of a solid model in CAD is called a feature. Basically, geometric features involved in creating a solid object can be categorized into the five groups: construction (or datum) features, shape (or protrusion) features, pick-and-place (or hard-coded) features, mirror and pattern features, and thickened features, as shown in Figure 3.13. Therefore, features can be thought of as the individual shapes that, when combined, make up the part.

Construction or datum features—such as coordinate systems, planes, axes, or points—are auxiliary entities that aid solid model creation. Default construction features, including a coordinate system and

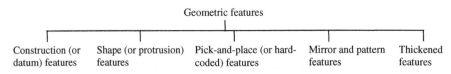

FIGURE 3.13 Classification of Geometric Features.

three perpendicular planes (front, top, and right), are provided in CAD as starting point for part solid modeling (and assembly).

Protrusion features (some also call them "sweep" features) are the most important set of features that support solid modeling. Such features include extrude, sweep, loft (or blend), and revolve, in which sketch profiles are required, as illustrated in Figure 3.14. In addition to protrusion that adds volume, protrusion can also be used to create cut features that remove volume from existing objects. Attributes, such as protrusion direction (one or both sides), are options offered to designers to complete a protrusion feature conveniently.

Pick-and-place features are hard-coded features, including chamfer, fillets, rounds, draft, and holes, which are placed on a face or an edge of existing objects without sketching a profile. Such features are often added in the final stage of the solid modeling process.

Mirror and pattern features are created from existing features, as illustrated in Figure 3.15. Mirror copies the selected features or all features, mirroring them about the selected plane or face (Figure 3.15c). Pattern, as shown in Figures 3.15a and b, repeats the selected features in an array based on a seed feature. The array can be linear (a linear pattern), a circular (a circular pattern), or following a curve. Some CAD, such as SolidWorks, offers feature copy and paste capabilities, in which designers can pick an existing feature (e.g., a through hole), then copy and paste it on a different face of the solid object in a different orientation.

The thickened feature creates a solid feature by thickening one or more adjacent surfaces. For example, the tracked vehicle roadarm surface model discussed in Chapter 2 (Section 2.7.2) was created in B-spline surfaces, imported into SolidWorks, and then thickened for a solid model in support of structural analysis using FEA. Note that except for the construction features, geometric features are solid features.

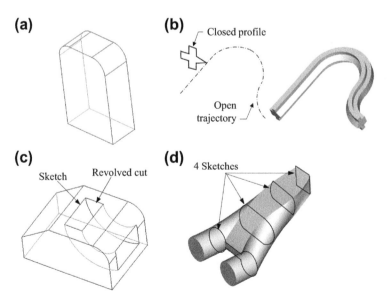

FIGURE 3.14 Protrusion Features. (a) Extrusion Feature. (b) Sweep Feature. (c) Revolved Cut Feature. (d) Blend Feature.

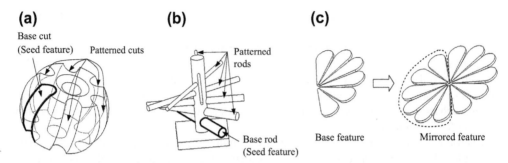

FIGURE 3.15 Copy, Pattern, and Mirror Features. (a) Patterned Cut Features. (b) Patterned Extrusion Features. (c) Mirrored Feature.

Among these features, protrusions are probably the most important ones in supporting designers to create most features in solid modeling. Protrusions features (including cuts) require designers to sketch a profile, in which a section view of the feature is defined. How does CAD support designers to create sketch profiles interactively? How does CAD receive inputs from the designer, then formulate and solve equations to define the sketch profile mathematically? We discuss sketch profiles in the next subsection.

3.3.2 Sketch profiles

When we start a new part, we pick a sketch plane and create a profile. The profile is the basis for a 3D model. We usually create a profile on one of the default construction planes (front, top, and right), or a created plane. In sketching a profile, as a CAD user, we create an open or closed profile with lines, arcs, and so on. A CAD system, such as SolidWorks or Pro/ENGINEER, automatically adds sketch relations (also called sketch constraints) to relate or constrain entities; for example, a straight line connects to a circular arc with a tangent relation at the junction point. After completing the profile, we add dimensions and enter proper values to adjust the profile that meets our design requirements. After entering or modifying a dimension value, CAD is able to adjust the profile as a logical consequence of the change. How does the CAD system do that? CAD employs the so-called variational modeling theory in sketch mode.

3.3.2.1 Sketch relations

In this section, we use examples to illustrate the sketch relations and variational modeling technique. We assume SolidWorks sketch mode to be more specific. Other CAD systems follow a similar approach.

We often start the sketch profile at the origin; that is, we use the origin as the anchor for the profile. SolidWorks (and other modern CAD systems) creates relations for the geometric entities in the profile, based on how these entities are created. Designers add dimensions (or more relations) to make the profile fully defined (or fully constrained).

In SolidWorks, before creating any dimensions, the geometric entities, including lines and vertices, are either in black or (mostly) in blue color. As illustrated in Figure 3.16 with a simple example, black

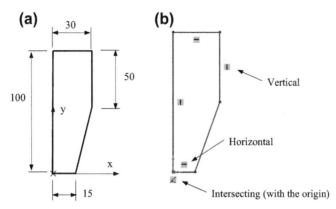

FIGURE 3.16 Sketch Profile as Designed (a) and with Underconstrained Relations (b).

indicates that the entity is fully defined. Blue indicates that the entity is not fully defined and is free to change in certain way. You may drag a vertex or line in blue color to see how it can be changed in SolidWorks.

When you add dimensions or relations, affected entities will change from blue to black color, indicating they become fully defined as a result. When all the entities of the sketch are in black, the entire sketch is fully defined. In the model tree (called Browser in SolidWorks), the (−) sign in front of the sketch is removed. Sometimes, the sketch is overconstrained when a conflict occurs or more dimensions (or relations) are created than required.

Note that some CAD systems, such as Pro/ENGINEER, offer "smart" sketching tools. When turned on, design intent is inferred, and sketch relations and dimensions are added automatically to make the sketch profile fully defined. In some cases, line and curve entities are slightly adjusted with imposed relations. For examples, two straight lines that are nearly perpendicular may "snap" perpendicular with a perpendicularity relation (Figure 3.17a), and two fillets with about the same radii may be adjusted to have equal radii with an equal radii relation added (Figure 3.17b). For more details on the sketch relations in SolidWorks and Pro/ENGINEER, please refer to Appendix A.

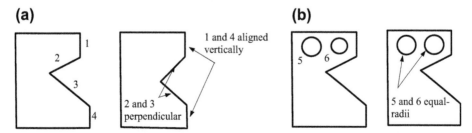

FIGURE 3.17 Effects of Imposed Sketch Relations to the Profile. (a) Perpendicular and Alignment Relations were Added. (b) Equal-radii Relation was Added.

The question is how SolidWorks knows what dimensions and relations are just right for the sketch (i.e., to make it fully defined). How does SolidWorks figure out if your sketch is overdefined or underdefined?

3.3.2.2 Variational modeling

CAD software employs the variational modeling theory in sketch mode. The first step involved in variational modeling is identifying a set of characteristic points (or vertices) on the sketch profile. For example, the sketch shown in Figure 3.16 has five characteristics points that are at the corners of the polygon. Then, system of equations are derived that incorporate the relations and dimensions defined on the sketch to relate the x- and y-locations of these characteristic points. The system of equations are solved to determine the locations of the characteristic points.

In formulating the system of equations, the number of equations must be identical to the number of unknowns and equations must be linearly independent for a unique solution. This is when a sketch is called fully defined. When the dimension values are changed, the same system of equations is solved again for the locations of the characteristic points.

When the number of equations is greater than the number of unknowns, we have an overdefined sketch. When the number of equations is less than the number of unknowns, we have an underdefined case, where a ($-$) sign will stay in front of the sketch in the Browser.

EXAMPLE 3.2

Determine if the following sketch profile with the relations and dimensions is fully defined. In this sketch, **P1** is fixed to the origin, and there are four relations (two horizontal and two vertical) and four dimensions. If we change the dimension d3 from 50 to 100 and d4 from 15 to 30, what will happen to the profile? Would CAD accept such changes and be able to regenerate the profile? If we add a dimension, such as the length of the line segment **P4P5**, would the sketch profile still be fully defined?

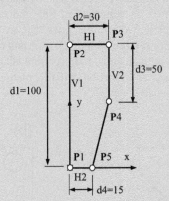

Solutions
There are five characteristic points in the sketch profile, **P1**, **P2**, **P3**, **P4**, and **P5**; therefore, we have $5 \times 2 = 10$ unknowns and we need to have 10 linearly independent equations to solve for the unknowns. The following equations are derived from the relations and dimensions defined on the sketch.

EXAMPLE 3.2—CONT'D

Because point **P**1 coincides with the origin, we have the following two equations:

$$\text{Coincident:} \quad P1x = 0 \tag{1}$$

and

$$\text{Coincident:} \quad P1y = 0 \tag{2}$$

The remaining equations are:

$$V1: \quad P2x - P1x = 0 \tag{3}$$

$$d1: \quad P2y - P1y = d1 \tag{4}$$

$$H1: \quad P3y - P2y = 0 \tag{5}$$

$$d2: \quad P3x - P2x = d2 \tag{6}$$

$$V2: \quad P4x - P3x = 0 \tag{7}$$

$$d3: \quad P3y - P4y = d3 \tag{8}$$

$$H2: \quad P5y - P1y = 0 \tag{9}$$

$$d4: \quad P5x - P1x = d4 \tag{10}$$

These ten equations are linearly independent. How can you tell? You may arrange these ten equations into a matrix form and check the rank of the matrix.

If we change the dimension d3 from 50 to 100 and d4 from 15 to 30, points **P**4 and **P**5 coincide, making the length of the line segment **P**4**P**5 zero. Most CAD software, such as SolidWorks, will not accept any line or curve entity with zero length.

Because we have already ten linearly independent equations that solve uniquely for the ten unknowns, adding more dimensions, such as the length dimension for the line segment **P**4**P**5, causes the profile to become overdefined.

Now, let us take a look at a bit more complex problem, in which the profile consists of circular arcs.

EXAMPLE 3.3

Add adequate dimensions to the following sketch profile with the given relations to make the profile fully defined. In this sketch, **P**1 is fixed to the origin, and there are another eight relations, as shown below. Formulate system of equations and solve them for the locations of the characteristics points that determine the shape of the profile.

Continued

EXAMPLE 3.3—CONT'D

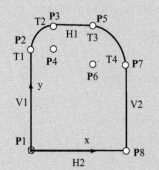

Tangent (T1)
Tangent (T2)
Tangent (T3)
Tangent (T4)
Vertical (V1)
Horizontal (H1)
Vertical (V2)
Horizontal (H2)

Solutions

There are eight characteristic points (including the arc centers); therefore, there are $8 \times 2 = 16$ unknowns, and we need to have 16 linearly independent equations to solve for the unknowns. The following equations are derived from the relations given to the profile shown above.

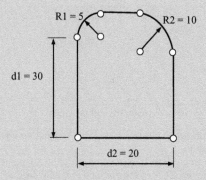

and

$$\text{Fix:} \quad P1x = 0, \tag{1}$$

$$\text{Fix:} \quad P1y = 0 \tag{2}$$

$$\text{V1:} \quad P2x - P1x = 0 \tag{3}$$

$$\text{T1 (and V1):} \quad P4y - P2y = 0 \tag{4}$$

$$\text{T2:} \quad P3x - P4x = 0 \tag{5}$$

$$\text{H1:} \quad P5y - P3y = 0 \tag{6}$$

$$\text{T3 (and H1):} \quad P5x - P6x = 0 \tag{7}$$

EXAMPLE 3.3—CONT'D

$$T4: \quad P7y - P6y = 0 \tag{8}$$

$$V2: \quad P7x - P8x = 0 \tag{9}$$

$$H2: \quad P1y - P8y = 0 \tag{10}$$

We need six more equations. These six additional equations come from the dimensions we are about to add to the profile.

If you add the four dimensions d1, d2, R1, and R2, as shown on previous page, the profile becomes fully defined. Why? Let us take a look at the equations that the dimensions will provide.

$$d1: \quad P2y - P1y = d1 \tag{11}$$

$$R1(T1 \text{ and } T2): \quad P4x - P2x = R1 \tag{12}$$

$$P3y - P4y = R1 \tag{13}$$

$$R2(T1 \text{ and } T2): \quad P5y - P6y = R2 \tag{14}$$

$$P7x - P6x = R2 \tag{15}$$

$$d2: \quad P8x - P1x = d2 \tag{16}$$

Now we have all 16 equations identified. Among them, Eqns 1–3, and 10 are trivial; therefore, they are removed together with the four unknowns (i.e., P1x = P1y = P2x = P8y = 0).

We assemble the remaining 12 equations for the 12 unknowns in a matrix form as follows.

													unknown	= mid	= right
(11)	1	0	0	0	0	0	0	0	0	0	0	0	P2y	d1	30
(5)	0	1	0	-1	0	0	0	0	0	0	0	0	P3x	0	0
(13)	0	0	1	0	-1	0	0	0	0	0	0	0	P3y	R1	5
(12)	0	0	0	1	0	0	0	0	0	0	0	0	P4x	R1	5
(4)	-1	0	0	0	1	0	0	0	0	0	0	0	P4y	0	0
(7)	0	0	0	0	0	1	0	-1	0	0	0	0	P5x	0	0
(6)	0	0	-1	0	0	0	1	0	0	0	0	0	P5y	0	0
(15)	0	0	0	0	0	0	0	1	0	-1	0	0	P6x	-R2	-10
(14)	0	0	0	0	0	0	-1	0	1	0	0	0	P6y	-R2	-10
(9)	0	0	0	0	0	0	0	0	0	1	0	-1	P7x	0	0
(8)	0	0	0	0	0	0	0	0	-1	0	1	0	P7y	0	0
(16)	0	0	0	0	0	0	0	0	0	0	0	1	P8x	d2	20

The system of equation can be solved using, for example, Matlab. We are solving five cases: Case 1: d1 = 30, R1 = 5, R2 = 10, d2 = 20 (base case); Case 2: d1 = 60, R1 = 5, R2 = 10, d2 = 20 (taller profile); Case 3: d1 = 30, R1 = 5, R2 = 5, d2 = 20 (equal fillet radii); Case 4: d1 = 30, R1 = 5, R2 = 15, d2 = 20 (zero length profile); and Case 5: d1 = 30, R1 = 5, R2 = 25, d2 = 20 (penetrating profile).

The Matlab script for solving the equations is shown (next page) is, followed by the resulting profiles.

Continued

EXAMPLE 3.3—CONT'D

$$EDU \gg a = \begin{bmatrix} 1 & 0 & 0 & 0 & 0 & 0 & 0 & 0 & 0 & 0 & 0 & 0 \\ 0 & 1 & 0 & -1 & 0 & 0 & 0 & 0 & 0 & 0 & 0 & 0 \\ 0 & 0 & 1 & 0 & -1 & 0 & 0 & 0 & 0 & 0 & 0 & 0 \\ 0 & 0 & 0 & 1 & 0 & 0 & 0 & 0 & 0 & 0 & 0 & 0 \\ -1 & 0 & 0 & 0 & 1 & 0 & 0 & 0 & 0 & 0 & 0 & 0 \\ 0 & 0 & 0 & 0 & 0 & 1 & 0 & -1 & 0 & 0 & 0 & 0 \\ 0 & 0 & -1 & 0 & 0 & 0 & 1 & 0 & 0 & 0 & 0 & 0 \\ 0 & 0 & 0 & 0 & 0 & 0 & 0 & 1 & 0 & -1 & 0 & 0 \\ 0 & 0 & 0 & 0 & 0 & 0 & -1 & 0 & 1 & 0 & 0 & 0 \\ 0 & 0 & 0 & 0 & 0 & 0 & 0 & 0 & 0 & 1 & 0 & -1 \\ 0 & 0 & 0 & 0 & 0 & 0 & 0 & 0 & -1 & 0 & 1 & 0 \\ 0 & 0 & 0 & 0 & 0 & 0 & 0 & 0 & 0 & 0 & 0 & 1 \end{bmatrix}$$

Case 1		Case 2	Case 3	Case 4	Case 5
EDU» c=[30	d1	EDU» c=[60	EDU» c=[30	EDU» c=[30	EDU» c=[30
0	0	0	0	0	0
5	R1	5	5	5	5
5	R1	5	5	5	5
0	0	0	0	0	0
0	0	0	0	0	0
0	0	0	0	0	0
-10	-R2	-10	-5	-15	-25
-10	-R2	-10	-5	-15	-25
0	0	0	0	0	0
0	0	0	0	0	0
20]	d2	20]	20]	20]	20]
EDU» b=inv(a)*c		EDU» b=inv(a)*c	EDU» b=inv(a)*c	EDU» b=inv(a)*c	EDU» b=inv(a)*c
b =	b =	b =	b =	b =	b =
30	p2y	60	30	30	30
5	p3x	5	5	5	5
35	p3y	65	35	35	35
5	p4x	5	5	5	5
30	p4y	60	30	30	30
10	p5x	10	15	5	-5
35	p5y	65	35	35	35
10	p6x	10	15	5	-5
25	p6y	55	30	20	10
20	p7x	20	20	20	20
25	p7y	55	30	20	10
20	p8x	20	20	20	20

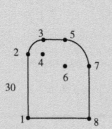

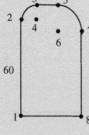

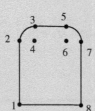

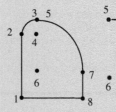

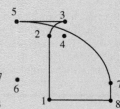

EXAMPLE 3.3—CONT'D

A number of important points are observed from Example 3.3:

1. Solutions exist for all five cases, which is because the number of equations and number of unknowns are identical, and the equations are linearly independent. As a result, the matrix equation is nonsingular and can be solved for a unique solution.
2. It is apparent that if we add a dimension d3 shown below, the sketch becomes overconstrained. Now, if we add d3 and remove d2, would the sketch be fully defined? The answer is no. Adding d3 creates a reddendum dimension because d3 is determined by $d1 + R1 - R2$. Removing the equations contributed by d2 and adding those contributed by d3 will make the system of equations linearly dependent; therefore, the resulting matrix equation becomes singular.

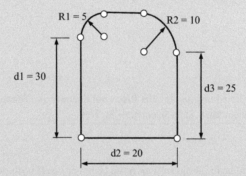

3. Cases 1–3 are regular profiles that CAD generates. For Case 4, because there is a zero-length line segment **P3P5**, some CAD programs, such as SolidWorks, prompt an error message and will not generate the profile. For Case 5, although the entities penetrate to each other, indicating a physically infeasible profile, some CAD systems, such as Pro/ENGINEER, check the validity of the profile and prompt with a warning message; however, others, such as SolidWorks, do not catch the problem and generate an invalid sketch profile anyway.
4. In both examples, we have all linear equations. In some cases, for instance, if angle dimensions are present, nonlinear equations are required. In these cases, an iterative numerical method, such as Newton's method, may be employed for solving the nonlinear equations.

3.3.3 Parent–child relationships

Once a sketch profile is completed, a solid feature can be created by using, for example, one of the protrusion capabilities. The first solid feature serves as the first building block, called the base feature, in the model construction process. Follow-on features are added to the base feature or existing solid features.

Depending on the sequence of feature construction, there is a parent–child relationship created between solid features. For example, the part shown in Figure 3.18a consists of five solid features. The first solid feature is the base block, which was created by extruding a sketch profile like that of Figure 3.12b along the extrusion direction perpendicular to the sketch plane of the profile. A center through hole was then added as a cut extrusion feature (or a pick-and-place feature), with a sketch

(a) **(b)**

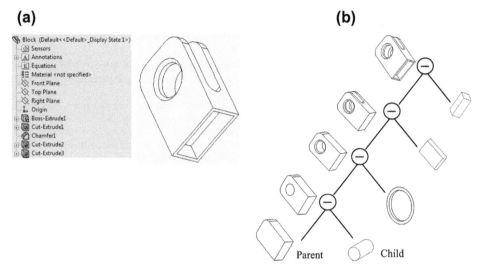

FIGURE 3.18 The Parent–child Relationships. (a) The Block Solid Model and Feature Tree in Browser. (b) Feature Construction Sequence in a Constructive Solid Geometry-like Tree.

placed on the front face of the base block, in which the hole is placed with position dimensions referred to the right and bottom edges of the base block, respectively. As a result, hole became a child feature of the base block, as illustrated in Figure 3.18b. The third feature is a chamfer (a pick-and-place feature) placed on the outside circle of the hole on the front face of the block. As a result, the chamfer is a child feature of the hole. The fourth and fifth features are the side cut and bottom slot, respectively, as shown in Figure 3.18b; both are child features of the base block.

In addition to parent–child relationships between solid features, a solid feature may be a child feature of a construction feature, and vice versa. For example, the base block of Figure 3.18b is a child feature of a default construction plane because its sketch profile was created on the plane.

Parent–child relationships are critical in the construction and rebuild of the solid model. Changes made to the parent feature will propagate to all child features during the rebuild. On some occasions, the part may not be rebuilt successfully due to numerous reasons, mainly because the part is not properly parameterized. Also, operations such as deleting, suppressing, or hiding a parent feature will affect all its child features. Therefore, it is extremely critical that designers arrange the sequence of feature construction and the way the child feature is related to its parent feature. A desirable solid model should have less coupled parent–child relationships between features. Solid model construction sequence or history is critical in the feature-based modeling approach. More about the design parameterization is discussed in Chapter 5.

3.3.4 Parametric modeling

Unlike the variational modeling technique that formulates and solves a system of equations, parametric modeling adopts a one-way assignment approach. For example, two dimensions d0 and d1 can be parametrically related as d0 = d1 × 2, in which d0 is a dependent parameter and d1 is an independent parameter that is free to change.

Such "assignment"-type equations are explicit and they are solved sequentially, in which each assigned value is computed as a function of previously assigned or computed values. Therefore, parametric modeling is also called "unidirectional" modeling or "procedural" modeling. In the above example, d1 must be defined first, and then d0 can be evaluated. The solid model must be rebuilt (regenerated) by propagating the changed parameter (dimension) through all equations that involve the parameter.

Computation in solving the explicit equation sequentially is efficient and straightforward. However, this approach lacks flexibility in relating parameters. For example, in the variational modeling method, d0 = d1 × 2 can be written as d0 − d1 × 2 = 0, in which either d0 or d1 can be independent.

3.3.5 Solid modeling procedure in CAD

After reviewing the solid modeling methods discussed so far, we revisit the general process of creating solid models in CAD. In the meantime, we walk through how CAD rebuilds the part when we make a design change, using a simple example. By going through this exercise, we hope you gain a better understanding of the behind-the-scenes operations while using CAD for solid modeling.

In general, as illustrated in Figure 3.19, when we start a new solid model, we are given datum features, such as datum planes and datum coordinate systems. It is in general a good idea to develop a modeling plan before beginning the actual modeling work. More about modeling plans is discussed in Section 3.4. At the beginning of creating a new part, we usually pick a datum plane and create a sketch profile for the first (or base) solid feature using one of the protrusion capabilities. As discussed before, a variational modeling technique is employed in CAD to determine the locations of characteristic points of the sketch profile. After the base solid feature is created, we either add more solid features by repeating the same process, sketch and make a cut feature, or place a pick-and-place feature on the existing features. We repeat some or all steps to create more features. In the meantime, CAD records the feature creation sequence in the model tree and parent–child relationships between features. Once a solid model is completely created, we often make a few adjustments to make sure the solid model accurately represents the design of the part. CAD rebuilds the part based on the changes we made by updating features (both datum and solid features) following the feature creation sequence, one feature at a time. For each feature, the CAD system does the following:

1. Takes the new dimension values, from user input and computation through parameter relations, to update the sketch profile first. In the sketch profile, a variational modeling technique is exercised. The system of equations that govern the profile of the section are solved again for the new parameter values.
2. Rebuilds the geometric features using the new sketch profile by
   a. New parameter values in the protrusion direction, including extrude, revolve, sweep, or loft; and
   b. Feature attributes, such as one side or both sides.
3. If the feature cannot be rebuilt (e.g., when an invalid geometric feature is encountered), an error message will appear. For example, in SolidWorks, a Rebuild Errors window appears, as shown in Figure 3.20.

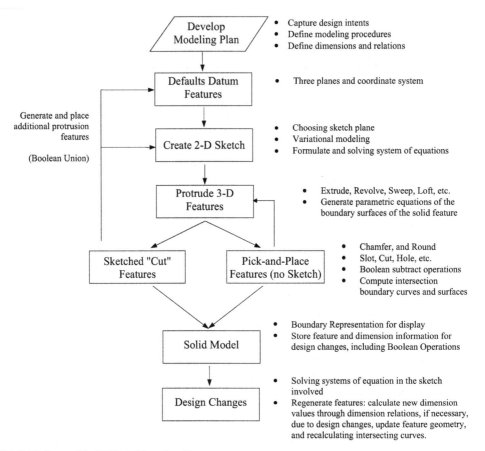

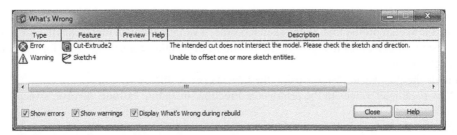

FIGURE 3.19 General Solid Model Creation Process.

FIGURE 3.20 The Rebuild Error Message Window in SolidWorks.

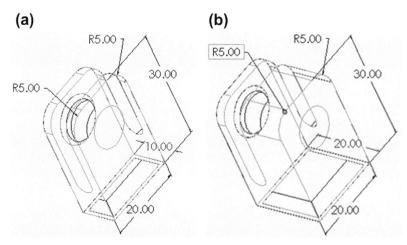

FIGURE 3.21 Change of the Depth Dimension of the Base Block from 10 to 20. (a) Solid Model before Design Change. (b) Solid Model after Design Change.

4. When this error message appears, we must read the messages carefully and identify which feature is problematic. Once the problematic feature is identified, it is a good practice to try to figure out what the problem was, and then undo the change to restore the dimension values, fix the problem, and try a similar change again. In SolidWorks or Pro/ENGINEER, you may choose *Edit > Undo Change Dim* (or CTL + Z) to undo the change.
5. Find child features through the parent-child relations. Note that while rebuilding child features, intersecting curves of feature boundary surfaces may need to be recomputed.
6. Repeat steps 1–5 until all the features are rebuilt.

In the following, we walk through the part rebuild process using the block example shown in Figure 3.18. We first change the depth dimension of the base block (d5) from 10 to 20, as shown in Figure 3.21. The regeneration follows the feature creation sequence, as shown in Figure 3.18b, which is described below.

1. First feature: base block
   The sketch profile is unchanged. There is no need to resolve the system of equations. Therefore, the only action for CAD is to update the width of the base block on one side along the extrude direction, which is the attribute of the base block.
2. Second feature: big hole
   Both placement data (placement plane, placement references, and dimensions) and hole dimension are unchanged. The hole is rebuilt following the feature attribute (i.e., through all and one side). The intersecting curve (in this case, the circle in the back face) is computed.
3. Third feature: side cut
   Check sketch profile. Is the sketch profile changed? Why and why not? The cut feature is regenerated following the feature attribute (i.e., through all and cut directions), as

illustrated in Figure 3.22a. The intersecting curves of the base block and the cut features are computed.

4. Fourth feature: chamfer

This feature is not affected because the placement edges (the circle on the front face) and the size of the chamfer are unchanged.

5. Fifth feature: cut on the bottom face

Check the sketch profile. Is the sketch profile changed? Yes, because the rectangular profile is defined with an offset from the exterior rectangle, as illustrated in Figure 3.22b. The system of equations will be resolved for the four characteristic points (corner points of the rectangle) due to the change of d5. The cut feature is regenerated following the feature attribute—that is, blind with depth 5 and one side with the same cut direction. The intersecting curves are computed.

If we change the width of the base block (d5) from 10 to 0.3, which feature(s) will fail to rebuild? Both the side cut and bottom cut features will not be generated because there is not enough room for the sketch profiles to be generated. However, CAD usually stops at the first unsuccessful feature; therefore, in this case, the rebuild error on the side cut will be reported.

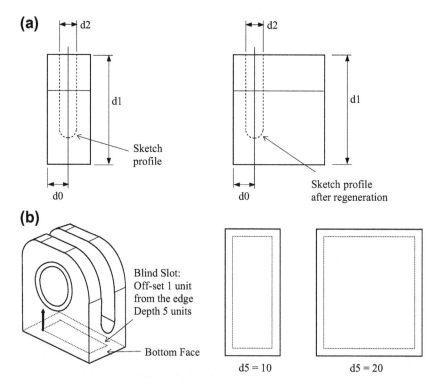

FIGURE 3.22 Feature Rebuilds. (a) Side Cut. (b) Bottom Cut.

3.3.6 **Direct modeling**

As can be seen from the discussion above, the parametric modeling approach requires the designer to anticipate design changes and accordingly define features, add relations to sketch entities, and add parameter relations between features. As a result, the solid model is created in such a way that a design modification (e.g., change in a dimension value) triggers rebuild in solid features in a prescribed manner. Feature-based parametric modeling is a structured modeling process, in which feature creation sequence or history tree masters the model rebuild process and design intent is captured implicitly through sketch relations, parent–child relationships, and parametric relations between dimensions.

Although the feature-based parametric modeling is indispensable in support of product design in the e-Design paradigm, capturing design intents in complex solid models is not always straightforward, to say the least. It requires the designer's effort, considerable planning, and careful implementation in achieving such parametric solid models. In general, parametric CAD tools lack ease of use, speed, and modeling flexibility. It requires a relatively steep learning curve and modeling effort upfront for the designer, and the solid models created suffer from model interoperability issues; that is, a CAD model created in software A cannot be understood or imported to software B with features and dimensions due to the nature of "history-based" model.

The newly developed direct modeling approach provides a geometric-based modeling strategy that gives designers the power to quickly define and edit geometry by simply clicking on the model geometry and moving it. Designers can focus on creating geometry rather than building features, adding constraints and design intent into their models and therefore speeding up design, saving time and development costs, and increasing productivity. The direct modeling paradigm is especially suited to the needs of designers working with legacy and heterogeneous CAD data. The direct modeling eliminates the need to access feature-level information to implement design changes. Designers can easily edit, modify, and repurpose solid models from any CAD sources.

Both Pro/ENGINEER (Creo™ 2.0 and higher) and SolidWorks (2012 and newer) are equipped with direct modeling (also called direct model editing) capabilities, which is built on top of existing feature-based parametric modeling technique. With the added direct modeling capability, designers are able to copy, move, split, replace, offset, push, and drag geometry to create the result as desired, instead of clicking on a dimension, entering a different value, and asking for model rebuild. In addition, with direct modeling capability, CAD automatically imports nonnative, imported model geometry without a model tree. The imported geometric model can be modified through direct geometry manipulation.

In general, parametric modeling is a history-based modeling method that enables design automation and creates product platforms for a product family, which are suitable for product design strategy that is aimed to be family-based or platform-driven. On the other hand, direct modeling is a geometry-centered and history-free approach that supports quick and easy 3D solid model construction, allows design change through direct manipulation of geometric models, and supports direct geometry-editing from any CAD source. There are pros and cons of these two methods. They are not exclusive but in general complement each other.

3.3.7 **Geometric modeling kernels**

No matter what kind of modeling method is offered by a CAD system, the core of any CAD software is its geometric modeling kernel. The kernel is key to support underline computing and modeling capabilities of solid objects, as well as output or export solid models, including 2D drawings, from 3D geometry. All commercially available solid modeling systems today are built on top of a geometric modeling kernel (also referred to as a modeling engine or geometry library). This is the library of core mathematical functions that defines and stores 3D solid objects in response to users' commands. The kernel library processes commands input through the application's user interface, stores the results, and submits the output to the graphics package for display, as illustrated in Figure 3.23.

Following Figure 3.23, it is commonly understood that there are two layers of information created when designers work with CAD. The top layer records user interaction with the CAD through either feature-based or direct modeling capabilities in the form of geometric features, including sketches, attributes, parameters, and equations, and feature construction sequence or history tree. On the bottom layer is the resulting geometric entities or objects. A history-based CAD system is basically recording every function it sends to the kernel into the history tree. For example, a sketch with an extrusion distance creates an extruded feature. Constraints control the size and position of the new feature. Then, a Boolean function is added to specify whether the feature is added or removed from the parent geometry. The Boolean function with the feature and related parameters are passed to the kernel and the resulting geometry is processed and revealed. This kernel function is processed every time this feature is rebuilt. The kernel function along with its required parameters is very specific to the kernel as discussed above. It is highly unlikely that another kernel will understand this very specific function, and even if it did the geometrical results could be very different. This is one of the major issues to address in solid model interoperability among CAD systems.

Some geometric modeling kernels such as ACIS (Spatial Inc.), Parasolid (Unigraphics Solutions, Inc.), and SMLib (Solid Modeling Solutions) are licensed by their respective developers for use in many different CAD systems. Others, such as thinkernel (think3), Granite One (Pro/ENGINEER), and UPG2 (Varimetrix Corporation), are proprietary kernels developed exclusively for a specific CAD system.

Apart from the underlying functionality supported, both licensed and proprietary kernels offer distinct advantages. CAD systems that license the same kernel can directly exchange model files that the kernel generates. For example, you can load SAT (ACIS) files directly into a CAD system that uses the ACIS kernel. On the other hand, developers who use different kernels in their CAD systems must write specific translators to read and write model files for import and export. Exporting and importing parametric solid models is not straightforward. More about CAD model translation is discussed in Chapter 6.

The geometric modeling kernels adopted by major commercial CAD systems are summarized in Table 3.1. In the following, we briefly introduce the two kernels that are widely employed in CAD: ACIS and Parasolid.

ACIS is an object-oriented C++ geometry library that integrates wireframe, surface, and solid modeling with both manifold and nonmanifold topology. It gives application developers a rich set of geometric operations for constructing and manipulating complex models. These include extruding,

FIGURE 3.23 Relationship of a CAD Interface to Geometric Modeling Kernels.

Table 3.1 Geometric Modeling Kernels and CAD Systems

CAD Systems	Software Developer	Geometric Modeling Kernel
AutoCAD 2000	AutoDesk	ACIS
Pro/ENGINEER	PTC	Granite one
I-DEAS	SDRC	Geomod
Unigraphics	EDS	Parasolid
SolidWorks	Dassault systems	ACIS
CATIA	Dassault systems	CGM (Convergence geometric modeler)
NX	Siemens	Parasolid

sweeping, lofting, skinning, offsetting, slicing, stitching, sectioning, fitting, and interpolating surfaces. ACIS also offers a complete set of Boolean operations, and length, area, and mass property inquiry functions. The ACIS kernel outputs a SAT file format that any ACIS-enabled application can read directly.

Parasolid from Unigraphics Solutions is an exact B-rep modeler that supports solid modeling and integrated free-form surface and sheet modeling. Parasolid's Extreme Modeling is a set of tightly integrated, proprietary technologies that enable modeling of complex geometry. Parasolid comprises more than 600 object-oriented functions for applications running on Windows, UNIX, and LINUX.

3.4 **Solid model build plan**

One of the objectives in creating solid models for product design using the e-Design paradigm is to capture design intents so that design changes can be made by simply modifying a few dimension values and rebuilding the solid models. To achieve the objective, it is important for the designer to plan ahead and spell out detail steps in terms of the type of features, their sketch profiles, relations and dimensions, parent–child relationships, and the feature creation sequences, including the construction of datum features. This is especially true when we use a CAD system with a feature-based parametric modeling method, which is still the mainstream technology implemented in major CAD systems.

Why do we need to spend time in developing such a part build plan? First, we always want to complete our work in the minimum time with the best result. If we plan ahead and think through the best possible way to create a design in using a CAD system, we often foresee possible pitfalls and are able to take precautions, which save us time in the end by not throwing out the problematic and incomplete models and starting over. So, please think before you do it! Planning ahead saves you time and makes you a better CAD user.

Let us take a look at the two examples in Figure 3.24. How should we construct the hose support and bracket shown in Figures 3.24a and b, especially when a design change is anticipated for the bracket, as illustrated in Figures 3.25a and b? After all, we are not just creating a part, we are creating a quality part that is clean, organized, and well thought of.

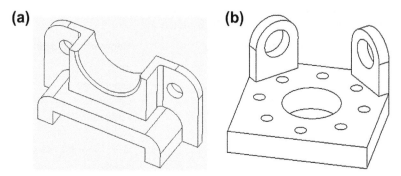

(a)   **(b)**

FIGURE 3.24 Sample Parts for Illustrating Part Build Plan: (a) Hose Support and (b) Bracket.

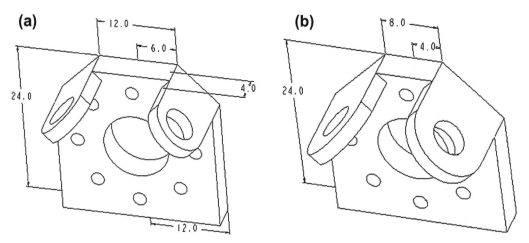

(a)   **(b)**

FIGURE 3.25 Design Intent for the bracket shown in Figure 3.24(b). (a) Dimension Design Variable. (b) Dimension Changed from 12 to 8.

What do we mean by a quality part? A quality part must be accurate in revealing geometric features in support of product design and manufacturing. Its feature construction sequence must be logical so that other team members can understand the solid model. The part should have minimum number of features. However, it does not imply that you must create complex sketches. It is a tradeoff in minimizing the number of geometric features and complexity level of individual sketch profiles. Moreover, a quality part must have minimum number of dimensions, implying as many sketch relations as possible. Most important, a quality model must be correctly parameterized and capture design intents. More about design parameterization is discussed in Chapter 5.

What should be included in the part construction plan? At least three things must be included:

1. Features and feature creation sequences (also include construction or datum features)
2. Sketch of each feature, including sketch plane, geometric entities, dimensions, and relations
3. Equations between dimensions as needed

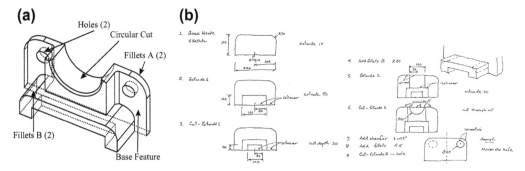

FIGURE 3.26 Build Plan for the hose support shown in Figure 3.24(a). (a) Major Features. (b) Hand-Sketched Sample Build Plan.

After a plan is jotted down on paper, it is a good idea to review it and try to optimize it before implementing it in CAD. The plan does not have to be fancy; it does have to facilitate your work. Note that a part construction plan is not unique. There is no "best plan".

Let us take a look at sample the example part: hose support shown in Figure 3.24a. Usually the first question to come to mind is how many solid features need to be created and which one the base feature (first solid feature) should be. For this part, it seems to be logical to create the back plate as the base feature (see Figure 3.26a). Is it a good idea to create the semicircle in the sketch of the base block? Is it a good idea to create fillets A in the sketch? How about the holes? Is it a good idea to take the advantage of part symmetry by focusing on only half of the part and then mirror the first half for the remaining half? Again, there is no "correct" answer to these questions. The key principle is creating a quality part with a minimum effort. A hand-sketched sample build plan for the hose support example is given in Figure 3.26b for your reference.

Now, let us discuss the bracket example. The detailed dimensions are provided in Figures 3.27a and b.

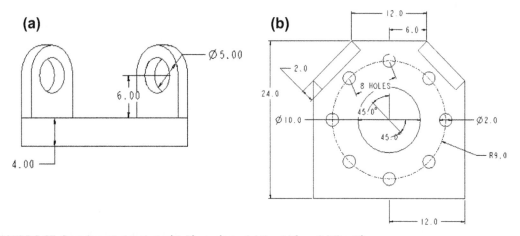

FIGURE 3.27 Sample part: bracket with Dimensions. (a) Front View. (b) Top View.

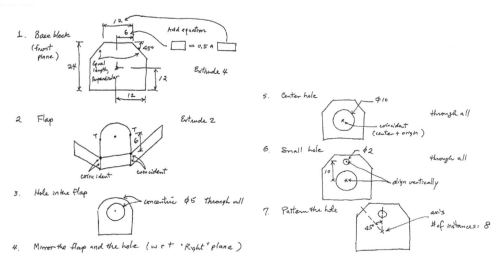

FIGURE 3.28 Hand-Sketched Build Plan for Sample Part: Bracket.

Note that the design variable is the length dimension of the top edge (current value: 12.0, shown in Figure 3.27b). When the design variable is changed, we expect to see the changes shown in Figure 3.25b; that is:

1. The 45° edges of the base block remained 45°, but with their lengths changed in accordance with new design variable value.
2. The back faces of the flaps align with the 45° edge of the base block.
3. The hole in the flaps stays in the middle with size unchanged.

How do we create such a part with the requirements described above? A sample build plan is shown in Figure 3.28 for your reference. To a less experienced CAD user, for a part like the bracket shown in Figure 3.24b, if you do not think ahead, you may end up with throwing away several "wrong" models before actually creating one that works. So please think before you actually construct a solid model in front of a computer.

3.5 Commercial CAD systems

Since the early 1960s when the first wireframe computer graphics was invented at MIT's Lincoln Laboratory, CAD has advanced significantly and has become the de facto design tool for the industry around the world. The first commercial applications of CAD were in large companies of automotive and aerospace industries, as well as in electronics. Only large corporations could afford the computers capable of performing the calculations. Notable company projects were at GM with DAC-1 (Design Augmented by Computer) in 1964 and at Renault–UNISURF 1971 car body design and tooling.

As computers became more affordable, the application areas have gradually expanded. The development of CAD software for personal desktop computers was the impetus for almost universal application in all areas of engineering. The most significant development appeared in the mid-1990s, in which major CAD tools were made available in PCs that allowed end users in mid- and small-size companies to be able to bring design from drawing board into digital form.

3.5.1 General purpose codes

Several general purpose commercial CAD systems available today were developed decades ago. These systems include the solid modeling packages Romulus (ShapeData) and Uni-Solid (Unigraphics), and the release of the surface modeler CATIA (Dassault Systemes) in 1981. Autodesk was founded in 1982, which led to the 2D system AutoCAD. Integrated Design and Engineering Analysis Software (I-DEAS) was produced by Structural Dynamics Research Corporation in 1982, and used primarily in the automotive industry, most notably by Ford Motor Company and General Motors. The next milestone was the release of Pro/ENGINEER in 1988, which heralded greater usage of feature-based modeling methods and parametric linking of the parameters of features. Also of importance to the development of CAD was the development of the B-rep solid modeling kernels Parasolid (ShapeData) and ACIS (Spatial Technology Inc.) at the end of the 1980s and beginning of the 1990s. This led to the release of mid-range packages such as SolidWorks in 1995, Solid Edge (then Intergraph) in 1996, and Autodesk Inventor in 1999. Several major mergers occurred throughout the years. SDRC was bought in 2001 by Electronic Data Systems, which had also acquired UGS Co. (maker of Unigraphics). EDS merged these two products into NX. UGS was purchased by Siemens AG in 2007 and was renamed Siemens PLM Software.

All major CAD systems offer not only solid modeling capabilities, but also CAE and CAM. Some CAD systems are equipped with in-house CAE/CAM, such as CATIA and Pro/ENGINEER. Some CAD partners with third-party software developers and fully integrates the third-party codes to the system; for example, CAMWorks integrated with SolidWorks. Although all major CAD systems offer excellent solid modeling and CAE/CAM, they serve different industrial sectors with slightly different focuses. CATIA is widely used by aerospace and automotive industry because of its superior surface modeling capabilities. AutoDesk is popular in small and mid-size companies due to its excellent capability in 2D drafting and its availability on PC in early years. Pro/ENGINEER serves heavy equipment industry, such as Caterpillar, due to its pioneer parametric modeling technology and strong CAE in the 1990s. SolidWorks became popular in almost all industrial sectors, as well as academia, because the software is intuitive and easy to use.

Several review articles on CAD software tools, such as those offered by 10 Top Ten Reviews (cad-software-review.toptenreviews.com), Cadalyst (www.cadalyst.com/listing/9/3d-modeling), and Wikipedia (http://en.wikipedia.org/wiki/List_of_computer-aided_design_editors, and http://en.wikipedia.org/wiki/Comparison_of_3D_computer_graphics_software), provide in-depth reviews and comparisons among major commercial systems. Readers are strongly encouraged to take a look at these articles for a better understanding of commercial CAD software.

3.5.2 Special codes

Besides general-purpose CAD software tools, there are at least two special codes worth mentioning. They are Rhinoceros (www.rhino3d.com) and SpaceClaim Engineer (www.spaceclaim.com/en/default.aspx).

Rhinoceros (Rhino) is a stand-alone, commercial nonuniform rational B-spline (NURB)-based 3D modeling software, commonly used for industrial design, architecture, marine design, jewelry design, automotive design, as well as the multimedia and graphic design industries. Rhino specializes in free-form NURB modeling. Rhino is gaining popularity due to its diversity, multidisciplinary functions, low learning curve, relatively low cost, and its ability to import and export many file formats, which allows Rhino to act as a "converter" tool between programs in a design workflow.

SpaceClaim Engineer is a 3D direct modeler. It enables engineers to easily create concepts and prepare 3D designs for prototyping, analysis, and manufacturing without becoming experts in traditional feature-based CAD systems. SpaceClaim helps engineers interact with CAD geometry in new ways. Without becoming a CAD expert, users can edit models, conceptualize on-the-fly, and communicate quickly and easily with prototyping and manufacturing. Direct modeling changes the way designers think about working with 3D solid models by letting them focus on what they are designing. Intuitive tools such as Pull and Move let users directly select portions of the model and move them where users want. The Combine tool slices and divides parts into pieces and lets users merge in portions from other designs. The Fill tool cleans up small features and fills holes. Together, these direct modeling tools let designers get job done without resorting to traditional CAD.

3.6 Summary

There is no doubt that CAD offers a better visualization of the design, easier creation of drawings once the model is completed, and better integration with CAE and CAM for product development. We discussed in this chapter the fundamentals in solid modeling, including CSG and B-rep, the two most commonly employed methods for underline solid modeling in CAD. We introduced the mainstream solid modeling technique—feature-based parametric solid modeling—that is employed in major CAD systems. We discussed key concept and theories, including variational and parametric modeling techniques, the parent–child relationship, and feature construction sequence or history tree. We walked through the steps of model rebuild in CAD using a simple example. We also briefly introduced geometric modeling kernels and newly developed direct modeling method. In addition, we offered a brief overview of commercial CAD systems. We hope you have gained adequate knowledge of CAD and solid modeling techniques and understand the behind-the-scenes operations that CAD carries out when you interact with it.

It is important to point out that although CAD becomes essential for product design, especially using the e-Design paradigm, it has a few issues. First, CAD can be slow for conceptual design. In the early stages, we tend to think faster than anybody could model in 3D. The direct modeling method may offer good alternatives to this issue. Also, CAD may require a lot of computing power to handle complex parts and assemblies. Display and rendering such models can be slow and model rebuild due to design changes can be too sophisticated to handle. Finally, model interchange between 3D parametric CAD systems is still an open issue. More about CAD interoperability is discussed in Chapter 6. We are now ready to move on to the next chapter to discuss the theory and methods employed by CAD for assembly. Our goal again is to understand the behind-the-scenes operations in CAD when we use it for creating assemblies.

Appendix A **Sketch relations**

Sketch relations play an important role in solid modeling. In this appendix, we offer tables that illustrate the commonly seen relations (also called sketch constraints) in SolidWorks and Pro/ENGINEER in Table A.1. Examples of such relations in SolidWorks are provided in Table A.2.

Table A.1 Commonly Seen Sketch Relations (or Constraints) in *Pro/ENGINEER* and *SolidWorks*

Pro/E	SolidWorks	Name	Entities	Descriptions
H or V	— I	Horizontal or vertical	One or more lines or two or more points	The lines become horizontal or vertical.
I —	I₁	Alignment	Two or more vertices	The items are aligned vertically or horizontally.
⊥	⊥	Perpendicular	Two lines	The two items are perpendicular to each other.
//		Parallel	Two or more lines	The items remain parallel.
T		Tangent	An arc, ellipse, or spline, and a line, or arc	The two items remain tangent.
⊕	◎	Concentric	Two or more arcs, or a point and an arc	The circles and/or arcs share the same center point.
-○-	⦵	Coincident	A point and a line, arc, or ellipse	The point lies on the line, arc, or ellipse.
L1 L1 R1 R1	=	Equal	Two or more lines, or two or more arcs	The line lengths or radii remain equal.
→ ←	N/A	Symmetric	A centerline and two points, lines, arcs, or ellipses	The items remain equidistant from the centerline, on a line perpendicular to the centerline.

Table A.2 Examples of Sketch Relations Seen in *SolidWorks*

Relations	Icons	Notes
Horizontal		Horizontal line sketched.
Perpendicular		Second line was sketched perpendicular to the first. Sketch tool is active, so midpoint sketch snap is displayed on line.
Parallel		Two lines sketched with parallel relation.
Horizontal and tangent		Tangent arc added to horizontal line.
Horizontal and coincident		Second circle. Sketch tool is active, so quadrant sketch snaps display on the second arc.
Vertical, horizontal, intersection, and tangent		Circle sketched with center inferred to sketch origin (vertical). Horizontal line intersects circle quadrant. Tangent relation added.

Continued

Table A.2 Examples of Sketch Relations Seen in *SolidWorks*—cont'd

Relations	Icons	Notes
Horizontal, vertical, and equal		Horizontal and vertical relations inferred. Equal relation added.
Concentric		Concentric relation added.
Horizontal		Horizontal relation added to spline handles.

Questions and exercises

1. Are the following sketches *fully defined*? Why or why not? Please answer the question by formulating equations similar to those discussed in Section 3.3.

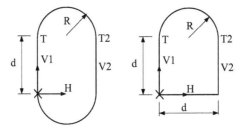

2. Create a solid model shown below. Make sure your model has identical dimensions as shown.

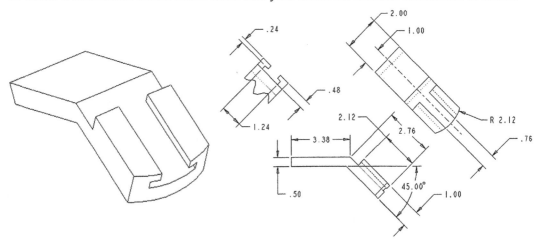

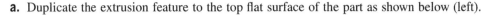

a. Duplicate the extrusion feature to the top flat surface of the part as shown below (left).

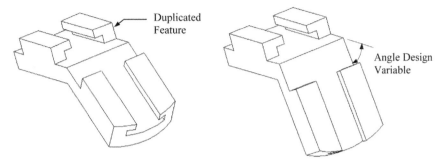

Duplicated Feature

Angle Design Variable

b. Define necessary relations so that when the angle design variable changes, the copied feature stays on the surface (above, right).

c. Submit four views, front, top, side, and isometric of the final part (including hidden lines; angle design variable $= 45°$). Submit a screen capture that shows equation employed to capture the angle design variable.

3. Create a solid model with a smooth loft feature using the exact dimensions shown in the sketch. Note that there are three sections in the loft, and the distances between them are 5 units for each pair. Note that you will have to state how guide curves if any are defined to create such a loft.

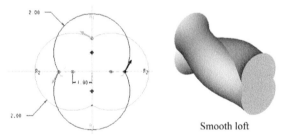

Smooth loft

a. Define relation(s) so that modifying a single dimension can change the diameter of all arcs. Note that the relation(s) must be implemented in the solid model and you must show screen captures of the solid model before and after changes, with a single dimension change.

b. What is the range of the diameter dimension that you can change that results in valid solid models?

4. Create a steering wheel solid model using the exact dimensions shown in the figures below.

Note that the width of the spoke is 0.4; the triangle on top of the hub is an equal-lateral triangle. The bottom sketch profile of the hub is a circle with diameter 2.5. The distance between the bottom edge of the triangle and the origin is 0.52, and the height of the hub is 2.25 as shown below.

a. What is the outer diameter of the wheel? _____

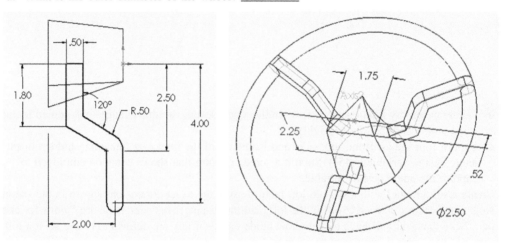

b. Define relations and equation(s) so that when the height dimension of the spoke is changed— for example, from 0.5 to 0.75—the sketch profile of the spoke can be regenerated as shown.
  Relations: Please show relations on the sketch with brief explanation.
  Equation(s): Please list all equations and point the dimensions involved in the sketch or part.
  What is the range of the outer diameter of the wheel that you can change that results in valid solid models?
    Minimum = _____, Maximum: _____

Height of the Spoke

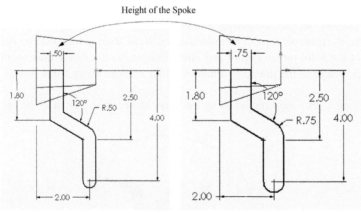

c. Please submit four views, front, top, side, and isometric of the final part (including hidden lines).

References

Bozdoc, M., 2003. The History of CAD. www.mbinfo.mbdesign.net/CAD-History.htm.

Lee, W., 1999. Principles of CAD/CAM/CAE Systems. Addison Wesley Longman, Inc.

Planchard, D.C., Planchard, M.P., 2013. Official Certified SolidWorks Professional (CSWP) Certification Guide with Video Instruction. SDC Publications.

Pratt, M.J., Wilson, P.R., 1985. Requirements for Support of Form Features in a Solid Modeling System. CAM-I, R-85-ASPP-01.

Toogood, R., Zecher, J., 2012. Creo Parametric 1.0 Tutorial and MultiMedia DVD. SDC Publications.

References

Assembly Modeling

CHAPTER OUTLINE

Assembly could mean very different things to different engineers. Mechanical engineers often consider mechanical assembly at the shop floor or assembly line, for which topics relevant to the physical assembly of a product—such as manual assembly vs automatic assembly, force and mass of parts, tools and equipment involved in assembly, tolerance analysis, and interference checking—are often the emphasis. Assembly process planning and assembly/disassembly are popular considerations for industrial engineers, who are often in charge of designing and running a product assembly line. Overall, it is essential for design engineers to acquire knowledge in these areas so that the practical aspects of product assembly can be incorporated in product design.

These topics, although important, will not be the focus of the chapter. For those who are interested in learning more about mechanical assembly or assembly process planning to enter this area for thesis work, there are excellent references that provide in-depth reviews and discussions on various topics related to mechanical assembly, such as Dawari and Sen (2007) and Whitney (2004).

In this chapter, we focus on assembly modeling that addresses methods employed in computer-aided design (CAD) to represent assembly. Topics such as mating constraints, degrees of freedom (DOFs), and fully constrained vs underconstrained assemblies are included. In addition, we discuss methods that support design changes and kinematic analysis in CAD assembly, which are the two most common activities encountered in assembly modeling using CAD. We discuss both open-loop and closed-loop systems. Note that the methods discussed in this chapter are mainstream methods adopted in the CAD community; they do not necessarily represent a specific CAD system.

In addition to theoretical discussion, we include virtual reality (particularly the applications that support product design) as a case study to illustrate and demonstrate the application of CAD assembly for practical engineering designs. In addition, a single-piston engine assembly is employed as a tutorial example to illustrate the detailed steps in creating the assembly using both Pro/ENGINEER and SolidWorks. Detailed instructions for bringing up these models and steps for carrying out the assembly discussed in this chapter can be found in Projects P1 and S1 for Pro/ENGINEER and SolidWorks, respectively. Example models are available for download at the book's companion website http://booksite.elsevier.com/9780123985132.

This chapter was written with the assumption that readers are familiar with basic CAD operations in part modeling, especially using Pro/ENGINEER or SolidWorks. If this is not the case, we encourage you to go over examples presented in other books (e.g., Toogood and Zecher, 2012; Shih, 2013; Lombard 2013, Reyes, 2013) before going over this chapter.

The overall objectives of this chapter are to (1) provide you with a general understanding of the methods that support assembly modeling in CAD, (2) familiarize you with the behind-the-scenes operations of CAD when a change is made or a part is dragged in an assembly, and (3) help you use Pro/ENGINEER or SolidWorks for creating basic assembly models (after going through the tutorial lessons).

4.1 Introduction

In the physical assembly of rigid parts, they are positioned (including location and orientation) relative to one another. The positioning of parts causes some of the low-level geometric entities, such as faces, edges, and vertices of the parts, to be in contact. The entities in contact between parts constrain the relative motion between them because a rigid part cannot deform or penetrate through other parts in the assembly.

The position of a part in space is uniquely defined by specifying its location and orientation with respect to some reference system. Three parameters are required to specify the location and another three parameters are required to specify the orientation. A rigid body in space has six degrees of freedom (DOFs) representing the allowable motions of the part. Assembly models are created by fixing the location and orientation of individual parts relative to one another through mating constraints, whereas kinematic models are created by specifying the allowed motions between the parts by defining kinematic joints.

CAD assembly has been commonly employed for product design. It is well known that the assembly design has a significant impact on many downstream activities, such as production process planning and control, tolerance analysis, and packaging. Assembly design involves the creation of assembly models that specify the relative location and orientation of components. In the design activity, component geometry is assembled together to create an assembly model. Mating constraints

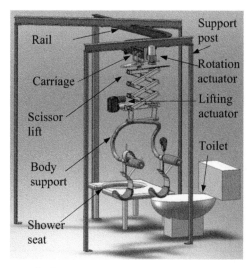

FIGURE 4.1 The Bathroom Transport Device.

also called assembly mates (or placement constraints) are used to locate and orient components with respect to one another. With a CAD assembly, basic yet essential questions in design can be readily answered by the product assembled in CAD. For example, will the parts fit into the designated space as an assembly? Will the components of the assembly collide or interfere in operation? Will the assembly operate as intended?

A bathroom transport device shown in Figure 4.1, which was designed and manufactured by a team of undergraduate students as their capstone project, is used as an example to illustrate some of the points mentioned above. This device was created for the purpose of transporting a disabled woman from her wheelchair to the toilet and shower seat without human assistance. It also transports the person from the toilet or shower seat back to the wheelchair. The device is compact (to fit into a very small bathroom), durable, and tailored to help a person to overcome a physical disability. The design features a three-button remote control that will move the person to the toilet, shower, and back to the wheelchair; a scissor lift with a linear actuator that provides lift; a carriage on a rail system that carries the person to designated locations; and a body support that safely holds the person while the system transports her to designated locations. A second actuator mounted on top of the scissor lift provides a 90-degree rotation to the body support when the carriage is moved to the toilet so that the user will be properly oriented on top of the toilet. A motor and a cable system are employed to pull the carriage.

The design of the device was extremely challenging because it was made to accommodate a severely disabled person who can only use her right hand to operate the device. The person would pull her wheelchair to the entrance of the bathroom, right in front of the device, as shown in Figure 4.2. She would use her right hand to move the two leg supports under her thigh, place the two arm supports under her arms, and press a button on the remote control mounted on top of the right arm support. The button pressing triggers the actuator of the scissor lift to contract, creating a lift to move her out of the wheelchair. Then, a motor is activated to pull a cable that draws the carriage along the curve rail and

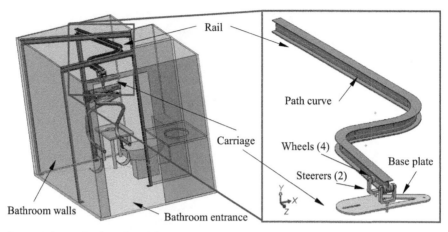

FIGURE 4.2 The Rail and Carriage Subsystems.

transport her to the toilet or shower seat. Position sensors are mounted on top of the rail to detect the location of the carriage and activate motor or actuators for the desired motions.

While designing the device, path mates were employed to assemble the carriage to the rail, allowing the carriage to move along the rail. The rail and carriage are important features of the device. The rail is a curve I-beam, created by sweeping an I-cross section sketch along an open loop curve composed of three straight lines and two circular arcs, as shown in Figure 4.2. The carriage consists of a base plate, two steerers, and four wheels, as shown in Figure 4.2. The wheels are sitting on the top faces of the bottom flange of the rail. A cable connecting to a motor is pulling the steerers to move the carriage along the rail. A universal joint under the base plate connects the body support. This assembly model with motion animation helps verification of the design concept, facilitates communications within the design team, and supports demonstration of the device design to the sponsors and user.

In this chapter, we start with a short and brief introduction in Section 4.2 on the mating constraints and kinematic joints commonly offered by CAD systems. After becoming familiar with the constraints and joints, Section 4.3 discusses a method that supports the calculation of a transformation matrix that positions a mating part to the base part in space. This illustrates how CAD supports part assembly as designers bring individual components into an assembly and define mating constraints. In Section 4.4, we discuss a kinematic modeling technique, in which we introduce the conversion of a CAD assembly to a kinematic model, the mapping of mating constraints to kinematic joints, and the mathematical representation of a kinematic assembly in CAD. We include both open-loop and closed-loop systems. The chapter wraps up by introducing a case study that involves applications of virtual reality technology for product design. In addition, a tutorial example of a single-piston engine is provided.

4.2 Assembly modeling in CAD

In CAD, an assembly model is created by specifying the relative location and orientation of parts. In general, an assembly model is static, in which all parts are completely constrained (also called fully constrained). On many occasions, the desired relative motion is required in product design to meet certain design requirements or verify functionalities. In these cases, an assembly is underconstrained, in which parts are allowed to move with respect to one another in order for the designer to explore or verify the kinematic characteristics of the assembly design.

A task common to both assembly modeling and kinematic analysis is the determination of part location and orientation satisfying certain constraints between these parts. There are two categories of geometric assembly relationships: geometry mating and joint mating. The former is usually static, whereas the latter allows relative motion and holds despite changes in the components' dimensions.

In general, geometry mating constrains geometric entities between mating parts. There are usually multiple pairs of entities constrained between the mating parts. On the other hand, joint mating constrains the relative motion between mating parts, instead of between geometric entities. As a result, there is one single joint between the two mating parts.

Some CAD systems, such as SolidWorks, support designers in creating an assembly model that is underconstrained so that the kinematic characteristics of the assembly can be explored by dragging individual parts. In other CAD systems, designers are required to complete the product assembly using mating constraints, and then convert the assembly model to a motion model by defining kinematic joints on top of the assembly in order to verify the kinematic characteristics of the assembly. This was the case, for example, in SolidWorks versions before 2008. In some CAD systems, such as Pro/EN-GINEER, designers are given choices in either selecting mating constraints or kinematic joints or a mixed set to create the assembly model. If the assembly is intentionally created with an underconstrained status, components can be dragged and moved.

In Section 4.2.1, we introduce commonly employed mating constraints in CAD, especially SolidWorks and Pro/ENGINEER. Then, in Section 4.2.2, we provide readers with a list of standard and advanced mating constraints offered by SolidWorks for a more complete picture in terms of the kind of mating constraints you may expect to use. We use a slider-crank mechanism as an example to go over the assembly in both SolidWorks and Pro/ENGINEER. We also introduce kinematic joints and the associated DOFs they constrain. This section serves as a prelude to the theoretical discussion on the subject in Sections 4.3 and 4.4.

4.2.1 Mating constraints

There are six DOFs for each component in space: three translations and three rotations. In the geometry mating approach, users specify the relative positions of parts by interactively defining spatial relationships between the geometric elements of mating parts. The geometric elements used in geometry mating include points, planar faces, surfaces, and axes of cylinders and holes. Commonly employed mating constraints (or placement constraints in Pro/ENGINEER and assembly mates in SolidWorks) include coincident-mate, mate offset, coincident-aligned, concentric (or fit), angle, parallel, and align. These mating constraints are usually applied to the same type of geometric entities, such as a pair of planar faces for a coincident-mate, or different entities, such as a point on a curve for a path mate.

In CAD, the first part brought into the assembly is fixed to the default datum features with all six DOFs constrained. In Pro/ENGINEER, the first part can be assembled to the assembly datum features, such as datum planes or the datum coordinate system, using placement constraints (e.g., by aligning their respective coordinates). In SolidWorks, the first component is fixed by aligning the component coordinate system with the default coordinate system provided in the assembly (also called the world

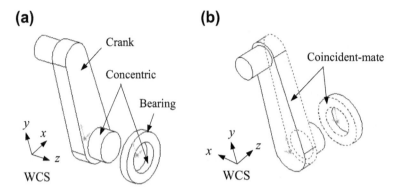

FIGURE 4.3 Mating Constraints for the Bearing and Crank Assembly. (a) Concentric, and (b) Coincident-mate.

coordinate system or WCS). The first part serves as the base part for assembling other parts. When an existing part is brought into the assembly, there are an additional six DOFs associated with it for the designer to work with.

Most mating constraints restrict part motion between regular surfaces, such as flat surfaces and cylindrical surfaces. As a result, a mating part is allowed to translate or rotate along a fixed direction if it is underconstrained. For example, the lower shaft of the crank is to be inserted into the hole of the bearing, as shown in Figure 4.3. The bearing is fixed. The crank is assembled to the bearing using two mating constraints, concentric (called Mate: Concentric in SolidWorks and Insert in Pro/ENGINEER) and coincident-mate (called Mate: Coincident in SolidWorks and Mate or Align Surfaces in Pro/ENGINEER). The concentric mating constraint eliminates two translational DOFs and two rotational DOFs. The coincident-mate mating constraint eliminates one translational DOF and two rotational DOFs. As a result, only one DOF, R_z, remains, as summarized in Table 4.1. SolidWorks allows designers to move (rotate) the crank by simply dragging the part, according to the free DOF. The designer is able to check the kinematics of the product in the assembly mode. In Pro/ENGINEER, such a rotational DOF is allowed to be undefined; similarly in SolidWorks, components can be dragged to check the kinematic behavior of the assembly.

Note that in Figure 4.3b, the coincident-mate mating constraint is more precisely called coincident with antialigned condition. In SolidWorks, you can set the alignment condition. The alignment conditions for a coincident mating constraint are either aligned, in which vectors normal to the selected faces point in the same direction; or antialigned, in which vectors normal to the selected faces point in

Table 4.1 Degrees of Freedom Eliminated by the Two Mating Constraints in Figure 4.3

	Tx	Ty	Tz	Rx	Ry	Rz
Mate: Concentric	×	×		×	×	
Mate: Coincident			×	×	×	

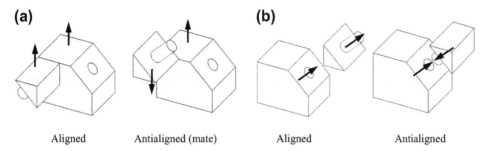

(a)   Aligned   Antialigned (mate)   **(b)**   Aligned   Antialigned

FIGURE 4.4 Aligned and Antialigned Conditions. (a) Between Two Flat Faces, and (b) Between Two Cylindrical Surfaces.

opposite directions, as illustrated in Figure 4.4a. For cylindrical surfaces, the axis vector is aligned or antialigned, as illustrated in Figure 4.4b.

The most commonly used mating constraints in Pro/ENGINEER and SolidWorks are listed in Table 4.2. In addition, a complete list of standard mating constraints in SolidWorks with mate symbols is provided in Table 4.3. You may expect to use these mating constraints to create an assembly in SolidWorks and most modern CAD systems. In SolidWorks, mating constraints (standard) are imposed to surfaces, which are physically intuitive. In Pro/ENGINEER, in addition to surfaces, some mating constraints are applied to abstract geometric entities, such as point-on-surface and edge-on-surface constraints. In some cases, Pro/ENGINEER and SolidWorks will not accept the mate constraints as defined if they conflict with existing ones.

Table 4.2 Mating Constraints in Pro/ENGINEER and SolidWorks

Pro/ENGINEER	SolidWorks	Descriptions
Mate surfaces Align surfaces	Mate: Coincident, antialigned Mate: Coincident, aligned	Positions selected faces or planes so they coincide. Antialigned implies that the two faces or planes mate and the normal vectors of the two faces or planes point in the opposite directions, and aligned implies that the normal vectors of the two faces or planes point in the same directions.
Align axes or insert surfaces	Mate: Concentric	Places the selected cylindrical surfaces so that they share the common axis.
Orient	Mate: Parallel	Places the selected items so they lie in the same direction and remain a constant distance apart from each other.
Coordinate system	Default	Place the first part to the default coordinate system in assembly.
Tangent	Mate: Tangent	Places the selected items in a tangent mate (at least one item must be a cylindrical surface)

Table 4.3 Standard Mates in SolidWorks

Standard Mates	Descriptions from SolidWorks Help
Coincident ⟨	Positions selected faces, edges, and planes (in combination with each other or combined with a single vertex) so they share the same infinite plane. Positions two vertices so they touch.
Parallel ⟍	Places the selected items so they remain a constant distance apart from each other.
Perpendicular ⊥	Places the selected items at a 90° angle to each other.
Tangent ⟨	Places the selected items tangent to each other (at least one selection must be a cylindrical, conical, or spherical face).
Concentric ◎	Places the selections so that they share the same center line.
Lock 🔒	Maintains the location and orientation between two components.
Distance ↗	Places the selected items with the specified distance between them.
Angle ⌂	Places the selected items at a specified angle to each other.
Default	Places the first part to the default coordinate system in assembly.

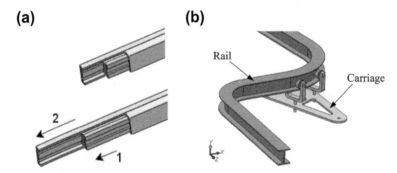

(a) **(b)**

FIGURE 4.5 Examples of Advanced Mates in SolidWorks. (a) Linear Coupler, and (b) Path Mate.

In this chapter, we adopt SolidWorks terminologies for mating constraints, except that we use *coincident-mate* instead of *coincident antialigned*.

In addition to standard mates, such as concentric and coincident, some CAD systems, such as SolidWorks, offer advanced mates, as listed in (Table 4.4). Advanced mates provide additional ways to constrain or couple movements between parts. A coupler removes one additional degree of freedom from the kinematic model. For example, a linear coupler shown in Figure 4.5a removes one translational DOF by coupling the respective translational DOF between components 1 and 2. Also, path mate (one of the advanced mates in SolidWorks) allows a part to move along a curve slot, a groove, or fluting, varying its moving direction specified by the path curve. For example, in the rail

Table 4.4 Advanced Mates in SolidWorks

Advanced Mates	Descriptions
Symmetric	Forces two similar entities to be symmetric about a plane or planar face.
Width	Centers a tab within the width of a groove.
Path	Constrains a selected point on a component to a path.
Linear/Linear coupler	Establishes a relationship between the translation of one component and the translation of another component.
Limit	Allows components to move within a range of values for distance and angle mates.

and carriage assembly of the transport device shown in Figures 4.1 and 4.2, a vertex in the carriage is moving along the sweep curve (which can be either open- or closed-loop, composed of several curves) of the rail, as shown in Figure 4.5b. As a result, path mate allows the carriage to move along the curve groove of the rail, varying its moving direction specified by the path curve. In addition, the pitch, yaw, and roll of the moving part can be defined to resemble the physical conditions. Such a capability supports animation and kinematic analysis for a whole new set of applications that involves curvilinear motion.

Some CAD systems, such as SolidWorks and Pro/ENGINEER, also offer mechanical mates, such as cam follower, gear, hinge, rack and pinion, screw, and universal joint. These are essential for kinematic analysis of the product design. More about kinematic and dynamic analysis can be found in Chapter 3 of *Product Performance Evaluation using CAD/CAE* in this book series. Tutorial lessons can be found in Projects P2 and S2 for Pro/ENGINEER and SolidWorks, respectively. More tutorial lessons can also be found in Chang (2010).

Next, we use a slider-crank example shown in Figure 4.6 to illustrate the mating constraints employed for the assembly in SolidWorks. We will use the same example in Section 4.2.2 to illustrate

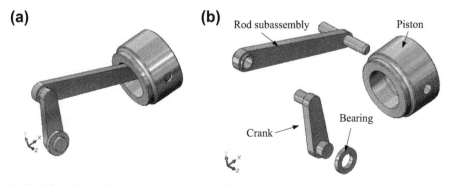

(a) **(b)** Rod subassembly Piston

Crank Bearing

FIGURE 4.6 The Slider-Crank Example. (a) Unexploded View, and (b) Exploded View.

the joint constraint approach for assembly as in, for example, Pro/ENGINEER. Note that model files of both examples are available for download at the book's companion website http://booksite.elsevier.com/9780123985132.

The slider-crank mechanism consists of five parts and one subassembly. They are bearing, crank, rod, pin, piston, and rod subassembly (consisting of rod and pin rigidly connected). An exploded view of the mechanism is shown in Figure 4.6b. There are eight assembly mates, including five coincident and three concentric, defined in the assembly.

The first three mates (*Concentric1*, *Coincident1*, and *Coincident2*) assemble the crank to the fixed bearing, as shown in Figure 4.7a. As a result, the crank is completely fixed. Note that the mate *Coincident2* orients the crank to the upright position, defining the configuration of the mechanism. Suppressing this mate will allow the crank to rotate with respect the bearing.

The next two mates (*Concentric2* and *Coincident3*) assemble the rod to the crank, as shown in Figure 4.7b. Unlike the crank, the rod is allowed to rotate with respect to the crank. The next two mates (*Concentric3* and *Coincident4*) assemble the piston to the pin, allowing the piston to rotate about the pin. The final mate (*Coincident5*) eliminates the rotation by mating two planes, *Plane3* of the piston and the *Plane2* of the bearing, as shown in Figure 4.7c.

At this point, the entire assembly is fully constrained. No relative motion between any components is allowed. If we suppress *Coincident2* defined between the right plane of crank and right plane of the bearing, the crank is allowed to rotate along the z-direction of the WCS. If you drag the crank (or any component), the entire assembly is moving, as illustrated in Figure 4.8.

4.2.2 Kinematic joints

In some CAD systems, such as Pro/ENGINEER, designers are given an option in choosing either mating constraints between geometric entities (like those of Tables 4.2 and 4.3) or defining kinematic joints between components.

A kinematic joint is a connection between two components that imposes constraints on their relative movement. There are in general two kinds of joints—a lower pair and higher pair. Physically, a lower pair joint is used to describe the connection between a pair of rigid components when the relative motion is characterized by two common surfaces sliding over one another. Commonly employed lower pair joints include revolute (also called hinge or pin), prismatic (also called slider or translation), cylindrical, planar, spherical, and screw, as shown in Figure 4.9. On the other hand, higher pair joints describe joints with points or lines, such as a cam-follower joint.

A prismatic, slider, or translational joint (Figure 4.9a) requires that a line in the moving component (or mating part) remains colinear with a line in the fixed component (or base part), and a plane parallel to this line in the moving component maintains contact with a similar parallel plane in the fixed component. This restricts five DOFs on the relative movement of the links—two translational and three rotational—which therefore has one translational degree of freedom.

A revolute, hinge, or pin joint (Figure 4.9b) requires a line in the moving component to remain colinear with a line in the fixed component, and a plane perpendicular to this line in the moving component maintain contact with a similar perpendicular plane in the fixed component. This restricts five DOFs on the relative movement of the parts—three translational and two rotational—which therefore allows only one rotational degree of freedom.

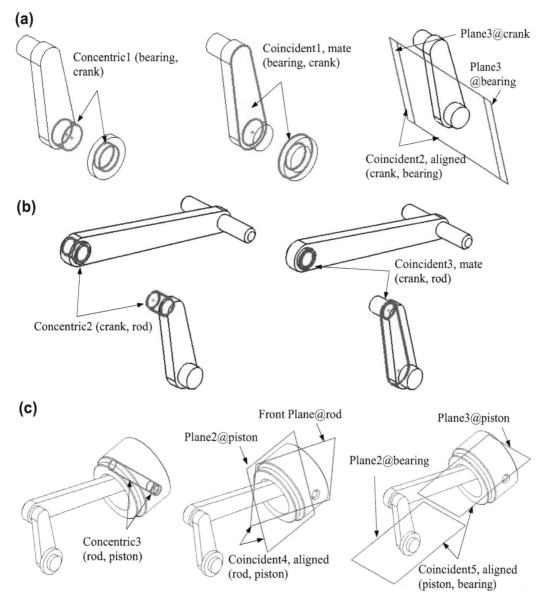

FIGURE 4.7 Assembly Mating Constraints Defined for the Slider-Crank Mechanism. (a) Mating Constraints for Crank (exploded view), (b) Mating Constraints for Rod (exploded view), and (c) Mating Constraints for Piston (unexploded view).

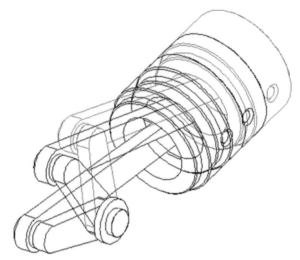

FIGURE 4.8 Drag the Crank to Explore the Kinematic Characteristics of the Slider-Crank Mechanism in SolidWorks.

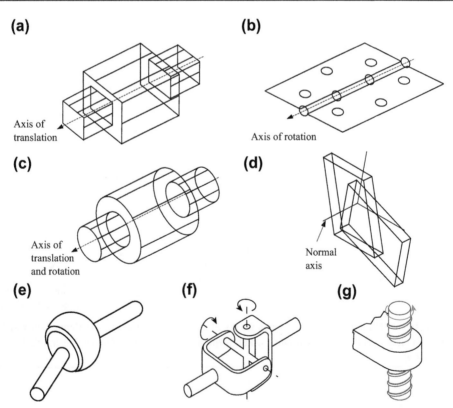

FIGURE 4.9 Lower Pair Kinematic Joints. (a) Revolute, Hinge, or Pin, (b) Prismatic, Slider, or Translational, (c) Cylindrical, (d) Planar, (e) Spherical or Ball, (f) Universal, and (g) Screw.

A cylindrical joint (Figure 4.9c) requires that a line in the moving component remain co-linear with a line in the fixed component. It is a combination of a revolute joint and a prismatic joint. This joint has two DOFs—one translational and one rotational.

A planar joint (Figure 4.9d) requires that a plane in the moving component maintain contact with a plane in the fixed component. This joint has three DOFs—two translational and one rotational.

A spherical joint, or ball joint (Figure 4.9c), requires that a point in the moving component maintain contact with a point in the fixed component. This joint has three DOFs—all rotational.

A universal joint (Figure 4.9f) allows the rotation of one component to be transferred to the rotation of another component. This joint is particularly useful to transfer rotational motion around corners or to transfer rotational motion between two connected shafts that are permitted to bend at the connecting point (such as the drive shaft in an automobile transmission system).

A screw joint (Figure 4.9g) requires cut threads in two components, so that there is a turning as well as sliding motion between them. This joint has one degree of freedom—coupled rotational and translational.

The DOFs that the lower joints constrained are summarized in Table 4.5.

Next, we use the same slider-crank example discussed in Section 4.2.1 to illustrate the kinematic joints employed for assembly in Pro/ENGINEER. Note that when using kinematic joints for assembly in Pro/ENGINEER, designers must use less physically intuitive entities, such as axis and points, to define joints.

Kinematically, the slider-crank example shown in Figure 4.6a is a four-bar linkage, as illustrated in Figure 4.10 schematically. They are commonly found in mechanical systems, such as internal combustion engines and oil-well drilling equipment. For the internal combustion engine, the mechanism is driven by a firing load that pushes the piston (slider), converting the reciprocal motion into rotational motion at the crank.

In the oil-well drilling equipment, a torque is applied at the crank. The rotational motion is converted to a reciprocal motion at the slider or piston that digs into the ground. Note that in any case the length of the crank must be smaller than that of the rod in order to allow the mechanism to operate. This is called Grashof's law (Erdman et al., 2001).

Table 4.5 Lower Pair Joints and the DOF Constrained

Joint Type	DOF Constrained			Remarks
	Translation	**Rotation**	**Total**	
Revolute	3	2	5	Rotates about an axis
Translational	2	3	5	Translates along an axis
Cylindrical	2	2	4	Translates along and rotates about an axis
Planar	2	1	3	Components connected by a planar joint move in a plane with respect to each other. Rotation is about an axis perpendicular to the plane.
Spherical	3	0	3	Rotates in any direction
Universal	3	1	4	Rotates about two axes
Screw	0.5	0.5	1	Coupled rotation and translation along one axis

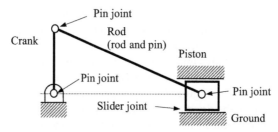

FIGURE 4.10 The Schematic View of the Kinematic Model of the Slider-Crank Mechanism.

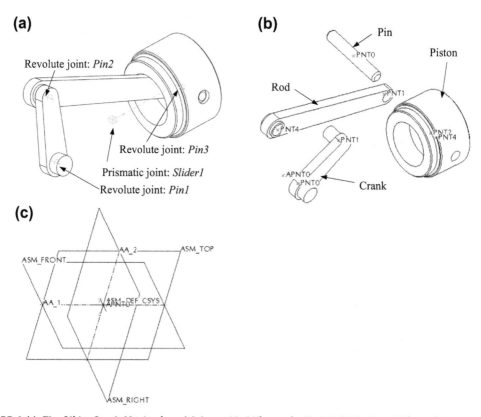

FIGURE 4.11 The Slider-Crank Mechanism. (a) Assembled Kinematic Model, (b) Exploded View with Datum Points for Defining Joint Locations, and (c) Assembly Datum Features Serving as the Ground Part.

The slider-crank assembly shown in Figure 4.11a consists of four parts: crank (*crank.prt*), rod (*rod.prt*), pin (*pin.prt*), and piston (*piston.prt*), as shown in the exploded view in Figure 4.11b. Instead of using a bearing part as the ground, we use the assembly datum features shown in Figure 4.11c as the ground. Datum points (such as APNT0 in assembly and PNT0 of crank shown in Figure 4.11b and datum axes (such as AA_1 of assembly shown in Figure 4.11c) created in parts and assembly will be used to define joints between parts.

The assembly datum features shown in Figure 4.11c include datum planes, datum axes, and datum points. Note that the datum axes AA_1 and AA_2 and datum point APNT0 will be used for creating joints—specifically, the pin joint between the ground and the crank, as well as the slider joint between the ground and the piston.

We define a pin joint (Pin1) that allows one rotational motion between the crank and the ground. The second pin joint (Pin2) is created to allow rotation motion between the crank and the rod. After assembling the crank and the rod, the system should have two DOFs, allowing the crank and rod to rotate along their respective pin joints independently.

Next, the pin is assembled to the rod rigidly using placement constraints, still maintaining two DOFs. Then, the piston is assembled to the pin by defining a third pin joint. Therefore, the piston will be free to rotate along the common axes A_1 (pin) and A_5 (piston). The total number of DOFs now increases to three.

Finally, the piston is assembled to the ground by defining a prismatic joint. The prismatic joint is created by aligning two parallel axes (A_6 in piston and AA_1 in the assembly) and two datum planes (DTM3 in piston and ASM_TOP in the assembly). The prismatic joint allows only one translational movement between piston and ground—that is, along the common axes without rotation. The slider-crank mechanism is now restricted to planar motion, with three rotations (Pin1, Pin2, and Pin3) and one translation (Slider1) motion. However, all three rotations and the translational motion are coupled to form a closed-loop mechanism, leaving only one free DOF, which can be any one of the three rotations or the translational motion. Note the joint symbols of Pro/ENGINEER shown in Figure 4.11a.

The total number of DOFs of the slider-crank mechanism can also be calculated as follows by using Gruebler's count:

3 (Bodies) × 6 (DOFs/body) − 3 (revolute joints) × 5 (DOFs/revolute) − 1 (prismatic joint) × 5 (DOFs/prismatic) = 18 − 20 = −2.

We know that for this slider-crank mechanism there is only one DOF. However, the count yields −2. This is because there are three redundant DOF created in the system. This is fine because the CAD system, such as Pro/ENGINEER, filters out the redundant DOF for kinematic analysis. Joints defined in this simulation model are summarized in Table 4.6. The pairs of datum points and datum axes

Table 4.6 Joints Defined in the Simulation Model

	Ground Body	**Crank**	**Rod/Pin**	**Piston**
Crank	*Pin1 A_1* (crank)/ AA_2 and PNT0/APNT0		*Pin2 A_2* (crank)/ A_1 (rod) and *PNT1* (crank)/*PNT4* (rod)	
Rod/pin		*Pin2 A_2* (crank)/ A_1 (rod) and *PNT1* (crank)/*PNT4* (rod)		*Pin3 A_5* (piston)/ A_1 (pin) and *PNT2* (piston)/*PNT0* (pin)
Piston	*Slider1 A_6* (piston)/ AA_1 and DTM3 (piston)/*ASM_TOP*		*Pin3 A_5* (piston)/ A_1 (pin) and *PNT2* (piston)/*PNT0* (pin)	

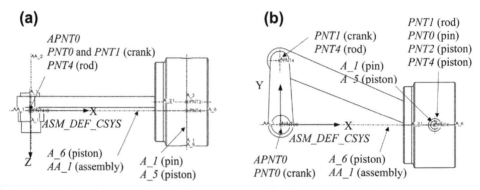

FIGURE 4.12 Locations of Datum Points and Datum Axes. (a) Top View, and (b) Front View.

created in the parts and the assembly for defining these four joints can be seen in the top and front views of the mechanism, as shown in Figure 4.12.

Note that the way the joints are defined is not unique. One of the pin joints may be replaced with a bearing joint, which describes an identical slider-crank mechanism kinematically, in which the total DOF becomes 1.

After completing the assembly using kinematic joints, you may click the Drag Components button at the top of the graphics window in Pro/ENGINEER, and click and drag a component to see how parts move. You may also bring the assembly into Mechanism Design by choosing from the pull-down menu: Applications > Mechanism, in which you may create a driver (e.g., a rotary motor) to drive the mechanism or define a force that pushes the piston to conduct a dynamic simulation.

4.3 Assembly modeling technique

An assembly model in CAD can be created by specifying assembly constraints between parts. As discussed in Section 4.2, there are mating constraints and joint constraints. In this section, we discuss the technique that determines the location and orientation of a mating part in an assembly with respect to the base part by defining mating constraints. Joint constraints will be discussed in Section 4.4.

In most mechanical assemblies, part positioning is carried out sequentially, with only two parts (or subassemblies) positioned at a time. Using this strategy, a smaller number of relations, and hence constraints, must be satisfied at each stage, even for a large assembly. This can offer significant computational advantages in comparison with a simultaneous strategy.

Part positioning in assembly involves specifying part location and orientation. It can be expressed relative to some global reference or with respect to other parts. In either case, part location and orientation are specified by a 4 × 4 homogeneous transformation matrix. In this section, we first discuss the transformation method and solution scheme proposed by Kim et al. (2000) in Section 4.3.1. Then, we introduce a technique for degree of freedom analysis based on the mating constraints in Section 4.3.2. Again, we adopt the terminologies of mating constraints defined in SolidWorks.

4.3.1 **Transformation matrix**

The method we discuss in this subsection takes well-constrained mating conditions between a base and a mating part and directly transforms them into a 4×4 matrix that determines the relative location and orientation of the mating part with respect to the base part. Well-constrained mating conditions imply that mating constraints are not in conflict in positioning the mating part to the base part.

In the example shown in Figure 4.13, a mating part (Figure 4.13b) is assembled first to the base part (Figure 4.13a), with a concentric mate applied to the inner surface of the hole in the base part and the cylindrical surface of the mating part (Figure 4.13c), in which axes of the hole and cylinder align and the mating part is free to rotate and translate along the common axis. Then, a coincident-aligned is applied to the top face of the base part and bottom face of the mating part, resulting in a fully constrained assembly (Figure 4.13d). To assemble the mating part to the base part, a 4×4 matrix (similar to that in Chapter 2), which is determined by directly computing a rotation matrix $\mathbf{T_R}$ and a translation matrix $\mathbf{T_L}$ that define the relative orientation and location of the mating part, respectively, must be calculated.

In determining the transformation matrix, we compute the rotation matrix $\mathbf{T_R}$ first by solving a set of linear constraint equations associated with the orientation of two mating parts. After orienting the mating part by applying the rotation matrix $\mathbf{T_R}$, the translation matrix $\mathbf{T_L}$ is calculated by solving a set of linear constraint equations associated with location. This method is computationally very effective because the transformation matrix for relative location and orientation of the mating part is algebraically derived directly from the linear equations associated with the mating conditions. We assume that the mating part is fully constrained.

The mating conditions considered in this subsection are concentric and coincident. We adopt the conventions in (Kim et al. 2005), in which the superscripts b, m, mr, and ma in the following equations indicate the base part, the mating part, the mating part after rotation, and the mating part after assembly, respectively.

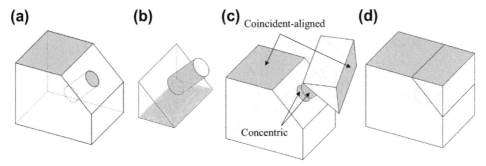

(a)   **(b)**   **(c)** Coincident-aligned   **(d)**

Concentric

FIGURE 4.13 An Example of a Two-Part Assembly. (a) Base Part, (b) Mating Part, (c) Concentric Mating Constraint Applied to the Hole in the Base Part and the Cylindrical Surface in the Mating Part, and (d) Coincident-aligned Applied to the Top Face of the Base Part and Bottom Face of the Mating Part, Resulting in a Fully Constrained Assembly.

4.3.1.1 Coincident

The coincident-mate holds between two planar faces and requires the two faces to touch each other (Figure 4.14a). The designated faces, shaded in Figure 4.14a, are the faces to be mated. Each face is specified by its unit normal vector $\mathbf{n}$ and one point $\mathbf{P}$ on the face in terms of its local coordinate system. This condition is accomplished by constraining the two normal vectors to be opposite to each other, and the two points that are noncoincident to lie on the same plane at which the two faces mate. Thus, equations of the coincident-mate constraint can be expressed as follows:

$$\mathbf{n}^b = -\mathbf{n}^{ma} \tag{4.1a}$$

where $\mathbf{n}^b = [n_x^b, n_y^b, n_z^b]^T$ and $\mathbf{n}^{ma} = [n_x^{ma}, n_y^{ma}, n_z^{ma}]^T$ are also called direction vectors; and

$$\mathbf{n}^{b^T} \cdot \left(\mathbf{P}^b - \mathbf{P}^{ma}\right) = 0 \tag{4.1b}$$

where $\mathbf{P}^b = [P_x^b, P_y^b, P_z^b]^T$, and $\mathbf{P}^{ma} = [P_x^{ma}, P_y^{ma}, P_z^{ma}]^T$. Note that $\mathbf{P}^b$ and $\mathbf{P}^{ma}$ must not coincide.

The coincident-aligned condition is assigned between two planar faces when they lie in the same plane, as shown in Figure 4.14b. Equations of the coincident-aligned constraint are similar those of the coincident-mate constraint, except that the two normal vectors $\mathbf{n}^b$ and $\mathbf{n}^m$ are required to be in the same direction. Thus, a coincident-aligned constraint can be expressed mathematically by

$$\mathbf{n}^b = \mathbf{n}^{ma} \tag{4.2a}$$

and

$$\mathbf{n}^{b^T} \cdot \left(\mathbf{P}^b - \mathbf{P}^{ma}\right) = 0. \tag{4.2b}$$

4.3.1.2 Concentric

The concentric condition holds between two cylindrical faces: a shaft face, and a hole face, as shown in Figures 4.14c and d. The concentric condition is accomplished by requiring the center axes of shaft and hole components to be parallel and a point $\mathbf{P}^m$ on the axis of the mating part lies on the axis of the base part. An axis is defined by a unit direction vector and a point on it. The hole axis is specified by a point $\mathbf{P}^b$ and a unit direction vector $\mathbf{n}^b$ defined in terms of its local coordinate system. Similarly, the shaft axis is specified by a point $\mathbf{P}^m$ and a unit direction vector $\mathbf{n}^m$ in terms of its local coordinate system. Thus, the constraint equations for concentric conditions can be written as

$$\mathbf{n}^b = \mathbf{n}^{ma} \tag{4.3a}$$

for aligned (see Figure 4.14c), $\mathbf{n}^b = -\mathbf{n}^{ma}$ for antialigned (see Figure 4.14d), and

$$\frac{P_x^{ma} - P_x^b}{n_x^b} = \frac{P_y^{ma} - P_y^b}{n_y^b} = \frac{P_z^{ma} - P_z^b}{n_z^b} = C \neq 0 \tag{4.3b}$$

because vectors $\mathbf{n}^b$ and $\mathbf{P}^{ma} - \mathbf{P}^b$ are collinear.

Equations (4.1)–(4.3) specify partially the relative rotation and translation of the mating part with respect to the base part, associated with the respective mating constraints. We assume that the origins of the coordinate systems of the mating and base parts coincide (not necessarily aligned), although they are sketched separately in Figure 4.14 for clarity. This point is illustrated in Example 4.1.

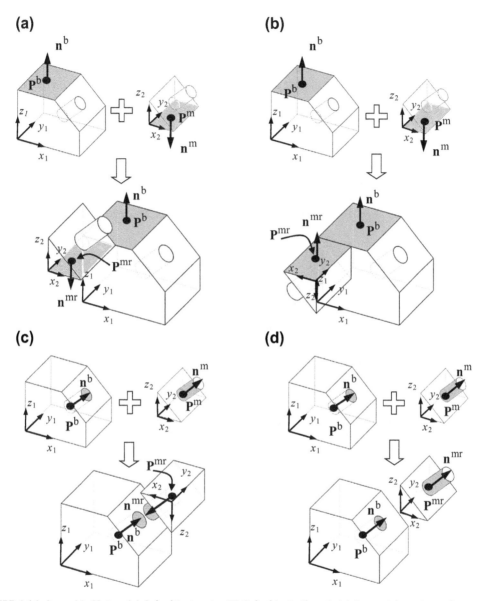

FIGURE 4.14 Assembly Mates. (a) Coincident-mate, (b) Coincident-aligned, (c) Concentric-mate, and (d) Concentric-aligned.

Although we only present equations for concentric and coincident mating constraints, equations of remaining mating constraints can be derived following the same ideas presented. For examples, Eqn (4.1a) or Eqn (4.2a) (for mate or align, respectively) is sufficient to support a parallel mating constraint. In addition, a coincident offset constraint can be represented by using the same equation as either Eqn (4.1a) or Eqn (4.2a) (for mate or align, respectively) for part orientation, and the following equation for location:

$$\mathbf{n}^{b^T} \cdot \left[\mathbf{P}^b - (\mathbf{P}^{ma} - \mathbf{O}) \right] = 0 \tag{4.3c}$$

where $\mathbf{O}$ is the vector of the offset specified by the designer.

4.3.1.3 Computation of the transformation matrix

The relative orientation and location of the mating part with respect to the base part is represented by a 4×4 transformation matrix. The transformation matrix can be written in homogeneous coordinates, defined as

$$\mathbf{T} = \begin{bmatrix} R_{1x} & R_{1y} & R_{1z} & L_x \\ R_{2x} & R_{2y} & R_{2z} & L_y \\ R_{3x} & R_{3y} & R_{3z} & L_z \\ 0 & 0 & 0 & 1 \end{bmatrix} = \begin{bmatrix} \mathbf{R} & \mathbf{L} \\ 0 & 1 \end{bmatrix} \tag{4.4a}$$

in which the 3×3 matrix $\mathbf{R}$ defines the rotation transformation, and the 3×1 column vector $\mathbf{L} = [L_x \ L_y \ L_z]^T$ determines the translation. Physically, this transformation can be viewed as a representation of a coordinate system in a fixed reference coordinate system. Each unit vector of the coordinate frame $\mathbf{n}_1$, $\mathbf{n}_2$, and $\mathbf{n}_3$ is mutually perpendicular, as illustrated in Figure 4.15. With this, the transformation matrix can be rewritten as

$$\mathbf{T} = \begin{bmatrix} \mathbf{n}_{1_x} & \mathbf{n}_{2_x} & \mathbf{n}_{3_x} & L_x \\ \mathbf{n}_{1_y} & \mathbf{n}_{2_y} & \mathbf{n}_{3_y} & L_y \\ \mathbf{n}_{1_z} & \mathbf{n}_{2_z} & \mathbf{n}_{3_z} & L_z \\ 0 & 0 & 0 & 1 \end{bmatrix} = \begin{bmatrix} \mathbf{R} & \mathbf{L} \\ 0 & 1 \end{bmatrix}. \tag{4.4b}$$

With such a transformation matrix, any given vector $\mathbf{v} = [v_x \ v_y \ v_z]^T$ defined in the mating part with respect to its local coordinate system can be transformed to $\mathbf{V} = [V_x \ V_y \ V_z]^T$ with respect to the coordinate system of the base part by the following,

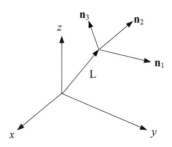

FIGURE 4.15 Representation of a Frame in a Frame.

$$\begin{bmatrix} \mathbf{V} \\ 1 \end{bmatrix} = \begin{bmatrix} \mathbf{R} & \mathbf{L} \\ 0 & 1 \end{bmatrix} \begin{bmatrix} \mathbf{v} \\ 1 \end{bmatrix} = \begin{bmatrix} \mathbf{R}v + \mathbf{L} \\ 1 \end{bmatrix} \tag{4.5}$$

which clearly shows the rotational and translational portions of the transformation. Therefore, the transformation matrix $\mathbf{T}$ in Eqn (4.4) can be represented by the product of a translation matrix $\mathbf{T_L}$ and a rotation matrix $\mathbf{T_R}$ as

$$\mathbf{T} = \begin{bmatrix} \mathbf{I} & \mathbf{L} \\ 0 & 1 \end{bmatrix} \begin{bmatrix} \mathbf{R} & 0 \\ 0 & 1 \end{bmatrix} = \mathbf{T_L T_R} \tag{4.6a}$$

where

$$\mathbf{T_R} = \begin{bmatrix} \mathbf{R} & 0 \\ 0 & 1 \end{bmatrix} = \begin{bmatrix} R_{1x} & R_{1y} & R_{1z} & 0 \\ R_{2x} & R_{2y} & R_{2z} & 0 \\ R_{3x} & R_{3y} & R_{3z} & 0 \\ 0 & 0 & 0 & 1 \end{bmatrix} \tag{4.6b}$$

and

$$\mathbf{T_L} = \begin{bmatrix} \mathbf{I} & \mathbf{L} \\ 0 & 1 \end{bmatrix} = \begin{bmatrix} 1 & 0 & 0 & L_x \\ 0 & 1 & 0 & L_y \\ 0 & 0 & 1 & L_z \\ 0 & 0 & 0 & 1 \end{bmatrix}. \tag{4.6c}$$

These matrices are determined sequentially. We first derive $\mathbf{T_R}$ from the rotational relationships between the mating parts, and then derive $\mathbf{T_L}$ from the translational relationships between the base part and the mating part after it is oriented by applying the rotation matrix $\mathbf{T_R}$.

When we are given two independent pairs of direction vectors, $(\mathbf{n}_i^b, \mathbf{n}_i^m)$, $i = 1, 2$, from the well-constrained mating conditions between a base part and a mating part, as shown in Figure 4.16, the equation associated with rotation of components are expressed as

$$\mathbf{n}_i^{mr} = \mathbf{R} \, \mathbf{n}_i^m, \quad i = 1, 2 \tag{4.7}$$

where $\mathbf{n}_i^{mr} = \mathbf{n}_i^b$, $i = 1, 2$ is for aligned and $\mathbf{n}_i^{mr} = -\mathbf{n}_i^b$, $i = 1, 2$ is for mate.

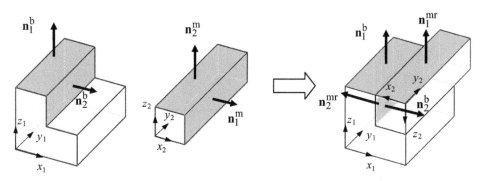

FIGURE 4.16 Mating Conditions for Assembly Modeling.

Here, $\mathbf{n}_i^{mr}$ is a mating direction vector after the mating part is reoriented by applying the rotation matrix $\mathbf{T}_R$. If $\mathbf{n}_3^m$ and $\mathbf{n}_3^{mr}$ are defined as $\mathbf{n}_3^m = \mathbf{n}_1^m \times \mathbf{n}_2^m$ and $\mathbf{n}_3^{mr} = \mathbf{n}_1^{mr} \times \mathbf{n}_2^{mr}$, respectively, then, the relation between $\mathbf{n}_3^m$ and $\mathbf{n}_3^{mr}$ is derived as $\mathbf{n}_3^{mr} = \mathbf{R}\,\mathbf{n}_3^m$. These equations are rewritten as the matrix product

$$\begin{bmatrix} \mathbf{n}_1^{mr} & \mathbf{n}_2^{mr} & \mathbf{n}_3^{mr} \end{bmatrix} = \mathbf{R} \begin{bmatrix} \mathbf{n}_1^m & \mathbf{n}_2^m & \mathbf{n}_3^m \end{bmatrix}. \tag{4.8a}$$

Hence, the rotational submatrix $\mathbf{R}$ is obtained by

$$\mathbf{R} = \begin{bmatrix} \mathbf{n}_1^{mr} & \mathbf{n}_2^{mr} & \mathbf{n}_3^{mr} \end{bmatrix} \begin{bmatrix} \mathbf{n}_1^m & \mathbf{n}_2^m & \mathbf{n}_3^m \end{bmatrix}^{-1}. \tag{4.8b}$$

The translation submatrix $\mathbf{L}$ is computed algebraically by solving the equations associated with translation after reorienting the mating part by applying the rotation matrix $\mathbf{T}_R$. After reorienting, the mating direction vectors are parallel and a point on the mating part after assembly $\mathbf{P}^{ma}$, is expressed as

$$\mathbf{P}^{ma} = \mathbf{P}^{mr} + \mathbf{L} \tag{4.9}$$

where $\mathbf{P}^{mr}$ is a point on the mating part after reorienting and is obtained by $\mathbf{P}^{mr} = \mathbf{R} \cdot \mathbf{P}^m$. Thus, the constraint equations associated with the translation of mating parts are expressed next.

First, we consider coincident-mate, in which the direction vectors of the mating parts are parallel after reorientation. The condition requires that one point on the mating face of the mating part lies on the mating face of the base part and they are not coincident. Thus, the translational constraint equation for the coincident-mate is expressed, following Eqn (4.3b), as:

$$\mathbf{n}^{b^T} \cdot \left(\mathbf{P}^b - \mathbf{P}^{ma} \right) = \mathbf{n}^{b^T} \cdot \left[\mathbf{P}^b - \left(\mathbf{P}^{mr} + \mathbf{L} \right) \right] = 0. \tag{4.10}$$

For concentric mate, we know that the center axes of the mating parts are parallel after repositioning; the constraint requires that one point $\mathbf{P}^m$ on the axis of the mating part lies on the axis of the base part. Thus, the translational constraint equations for the concentrate mate are expressed, following Eqn (4.2b), as:

$$\frac{\left(P_{1x}^{mr} + L_x \right) - P_{1x}^b}{n_{1x}^b} = \frac{\left(P_{1y}^{mr} + L_y \right) - P_{1y}^b}{n_{1y}^b} = \frac{\left(P_{1z}^{mr} + L_z \right) - P_{1z}^b}{n_{1z}^b} = C \neq 0 \tag{4.11a}$$

or

$$\begin{aligned}
\left(P_{1x}^{mr} + L_x \right) - P_{1x}^b &= C n_{1x}^b \\
\left(P_{1y}^{mr} + L_y \right) - P_{1y}^b &= C n_{1y}^b \\
\left(P_{1z}^{mr} + L_z \right) - P_{1z}^b &= C n_{1z}^b.
\end{aligned} \tag{4.11b}$$

EXAMPLE 4.1

Consider the two assembly components shown in Figure 4.13. Each component is defined in its local coordinate system, with dimensions shown on the next page. The two components are assembled by applying a concentric mate to the inner surface of the hole in the base part and the outer surface of the cylinder in the mating part, and imposing a coincident-aligned on the top face of the base part and the bottom face of the mating part, as discussed before. Therefore, the vectors $\mathbf{n}_1^b$ and $\mathbf{n}_1^m$ align with the axes of the hole and cylinder, respectively, and the vectors $\mathbf{n}_2^b$ and $\mathbf{n}_2^m$

EXAMPLE 4.1—CONT'D

are normal to the faces on the respective components, as shown below. The reference point $\mathbf{P}_1^b$ is at the center of the bottom face of the hole, and point $\mathbf{P}_1^m$ is located at the center of the bottom face of the cylinder on the mating part. Also, reference points $\mathbf{P}_2^b$ and $\mathbf{P}_2^m$ are located at the center of their respective mate planes, as shown in the figures below. These points must be chosen such that they do not coincide after assembly. The table below lists the geometric data of the vectors and reference points.

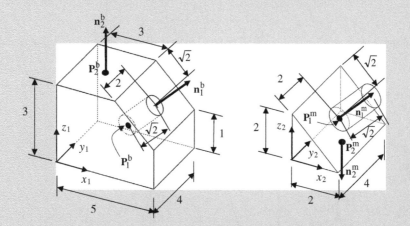

	Base Part	Mating Part	Mating Part After Assembly
Concentric	$\mathbf{n}_1^b = \begin{bmatrix} \frac{1}{\sqrt{2}} \\ 0 \\ \frac{1}{\sqrt{2}} \end{bmatrix}, \mathbf{P}_1^b = \begin{bmatrix} 3 \\ 2 \\ 1 \end{bmatrix}$	$\mathbf{n}_1^m = \begin{bmatrix} \frac{1}{\sqrt{2}} \\ 0 \\ \frac{1}{\sqrt{2}} \end{bmatrix}, \mathbf{P}_1^m = \begin{bmatrix} 1 \\ 2 \\ 1 \end{bmatrix}$	$\mathbf{n}_1^{mr} = -\mathbf{n}_1^b = \begin{bmatrix} -\frac{1}{\sqrt{2}} \\ 0 \\ -\frac{1}{\sqrt{2}} \end{bmatrix}, \mathbf{P}_1^{ma} = \begin{bmatrix} 4 \\ 2 \\ 2 \end{bmatrix}$
Coincident	$\mathbf{n}_2^b = \begin{bmatrix} 0 \\ 0 \\ 1 \end{bmatrix}, \mathbf{P}_2^b = \begin{bmatrix} 1.5 \\ 2 \\ 3 \end{bmatrix}$	$\mathbf{n}_2^m = \begin{bmatrix} 0 \\ 0 \\ -1 \end{bmatrix}, \mathbf{P}_2^m = \begin{bmatrix} 1 \\ 2 \\ 0 \end{bmatrix}$	$\mathbf{n}_2^{mr} = \mathbf{n}_2^b = \begin{bmatrix} 0 \\ 0 \\ 1 \end{bmatrix}, \mathbf{P}_2^{ma} = \begin{bmatrix} 4 \\ 2 \\ 3 \end{bmatrix}$

When the two direction vectors $\mathbf{n}_1^m$ and $\mathbf{n}_2^m$ are given, the third direction vector can be computed as

$$\mathbf{n}_3^m = \mathbf{n}_1^m \times \mathbf{n}_2^m = \begin{bmatrix} \frac{1}{\sqrt{2}} \\ 0 \\ \frac{1}{\sqrt{2}} \end{bmatrix} \times \begin{bmatrix} 0 \\ 0 \\ -1 \end{bmatrix} = \begin{bmatrix} 0 \\ \frac{1}{\sqrt{2}} \\ 0 \end{bmatrix}.$$

After normalizing it, we have $\mathbf{n}_3^m = \begin{bmatrix} 0 & 1 & 0 \end{bmatrix}^T$. After applying these two mating constraints, the mating part is assembled to the base part shown on the next page (left). The vectors and reference points after assembly are listed in the table above.

Continued

EXAMPLE 4.1—CONT'D

Therefore, we have

$$
\mathbf{n}_3^{mr} = \mathbf{n}_1^{mr} \times \mathbf{n}_2^{mr} = \begin{bmatrix} -\dfrac{1}{\sqrt{2}} \\ 0 \\ -\dfrac{1}{\sqrt{2}} \end{bmatrix} \times \begin{bmatrix} 0 \\ 0 \\ 1 \end{bmatrix} = \begin{bmatrix} 0 \\ \dfrac{1}{\sqrt{2}} \\ 0 \end{bmatrix}.
$$

After normalizing it, we have $\mathbf{n}_3^m = \begin{bmatrix} 0 & 1 & 0 \end{bmatrix}^T$. Then the rotation matrix $\mathbf{R}$ can be calculated using Eqn (4.8b) as

$$
\mathbf{R} = \begin{bmatrix} \mathbf{n}_1^{mr} & \mathbf{n}_2^{mr} & \mathbf{n}_3^{mr} \end{bmatrix} \begin{bmatrix} \mathbf{n}_1^{m} & \mathbf{n}_2^{m} & \mathbf{n}_3^{m} \end{bmatrix}^{-1} = \begin{bmatrix} -\dfrac{1}{\sqrt{2}} & 0 & 0 \\ 0 & 0 & 1 \\ -\dfrac{1}{\sqrt{2}} & 1 & 0 \end{bmatrix} \begin{bmatrix} \dfrac{1}{\sqrt{2}} & 0 & 0 \\ 0 & 0 & 1 \\ \dfrac{1}{\sqrt{2}} & -1 & 0 \end{bmatrix}^{-1} = \begin{bmatrix} -1 & 0 & 0 \\ 0 & 1 & 0 \\ 0 & 0 & -1 \end{bmatrix}.
$$

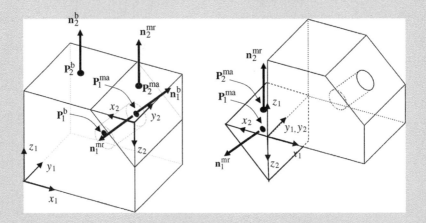

Note that the rotation matrix $\mathbf{R}$ rotates the mating part to an orientation shown in the figure above (right). The columns of the submatrix $\mathbf{R}$ represent respectively the three coordinate axes of the coordinate system in the mating part with respect to the coordinate system of the base part. For example, the first column of the submatrix $\mathbf{R}$ shows that the axis x_2 is now aligned with x_1 but in the opposite direction.

Now we find the translation matrix $\mathbf{L}$. Using the rotation submatrix $\mathbf{R}$ obtained in the previous step, two points $\mathbf{P}_1^{mr}$ and $\mathbf{P}_2^{mr}$ on the reoriented mating part are computed first by

$$
\mathbf{P}_1^{mr} = \mathbf{R}\mathbf{P}_1^m = \begin{bmatrix} -1 & 0 & 0 \\ 0 & 1 & 0 \\ 0 & 0 & -1 \end{bmatrix} \begin{bmatrix} 1 \\ 2 \\ 1 \end{bmatrix} = \begin{bmatrix} -1 \\ 2 \\ -1 \end{bmatrix}
$$

and

$$
\mathbf{P}_2^{mr} = \mathbf{R}\mathbf{P}_2^m = \begin{bmatrix} -1 & 0 & 0 \\ 0 & 1 & 0 \\ 0 & 0 & -1 \end{bmatrix} \begin{bmatrix} 1 \\ 2 \\ 0 \end{bmatrix} = \begin{bmatrix} -1 \\ 2 \\ 0 \end{bmatrix}.
$$

EXAMPLE 4.1—CONT'D

Then, the translational constraint equations are derived from the mating conditions as follows. From the coincident condition in Eqn (4.10), we have

$$\mathbf{n}_2^{b^T} \cdot \left[\mathbf{P}_2^b - (\mathbf{P}_2^{mr} + \mathbf{L}) \right] = 0, \quad \text{or} \quad \begin{bmatrix} 0 & 0 & 1 \end{bmatrix} \left\{ \begin{bmatrix} 1.5 \\ 2 \\ 3 \end{bmatrix} - \left(\begin{bmatrix} -1 \\ 2 \\ 0 \end{bmatrix} + \begin{bmatrix} L_x \\ L_y \\ L_z \end{bmatrix} \right) \right\} = 0$$

which gives $L_z = 3$. From the concentric condition in Eqn (4.11a), we have

$$\frac{(P_{1x}^{mr} + L_x) - P_{1x}^b}{n_{1x}^b} = \frac{\left(P_{1y}^{mr} + L_y\right) - P_{1y}^b}{n_{1y}^b} = \frac{(P_{1z}^{mr} + L_z) - P_{1z}^b}{n_{1z}^b} = C \neq 0$$

or

$$\frac{(-1 + L_x) - 3}{\frac{1}{\sqrt{2}}} = \frac{(2 + L_y) - 2}{0} = \frac{(-1 + L_z) - 1}{\frac{1}{\sqrt{2}}} = \frac{(-1 + 3) - 1}{\frac{1}{\sqrt{2}}} = \frac{1}{\frac{1}{\sqrt{2}}} = C.$$

Hence, $L_x = 5$, and $L_y = 0$. Therefore, the translational submatrix is $\mathbf{L} = [5, 0, 3]^T$.

Thus, the transformation matrix $\mathbf{T}$ for relative orientation and location of the mating part with respect to the base part is obtained as

$$\mathbf{T} = \begin{bmatrix} R_{1x} & R_{1y} & R_{1z} & L_x \\ R_{2x} & R_{2y} & R_{2z} & L_y \\ R_{3x} & R_{3y} & R_{3z} & L_z \\ 0 & 0 & 0 & 1 \end{bmatrix} = \begin{bmatrix} -1 & 0 & 0 & 5 \\ 0 & 1 & 0 & 0 \\ 0 & 0 & -1 & 3 \\ 0 & 0 & 0 & 1 \end{bmatrix}.$$

One of the potential pitfalls in using the method discussed above for representing an assembly is that the rotation matrix derived from Eqn (4.8b) may not be orthogonal. To carry out a valid rotation, the rotation matrix must satisfy the two basic properties of orthogonality: $\mathbf{R}\mathbf{R}^T = \mathbf{I}$ and $|\mathbf{R}| = 1$.

We use the following example to illustrate the pitfall.

EXAMPLE 4.2

Consider the same assembly as in Example 4.1 except that the length of the base part is increased from 5 to 6, as shown in the figure on the next page. We only show side views with direction vectors to get to the points of this example.

It is apparent that the two mating constraints are in conflict in terms of determining an orientation of the mating part that satisfies both constraints. You may either impose a coincident-aligned on the top face of the base part and the bottom face of the mating part (shown in the lower left figure of next page), in which the vectors $\mathbf{n}_2^b$ and $\mathbf{n}_2^m$ align, or apply a concentric mate to the inner surface of the hole in the base part and the outer surface of the cylinder in the mating part (shown in the lower right figure of next page), in which the vectors $\mathbf{n}_1^b$ and $\mathbf{n}_1^m$ align. It is impossible to orient the mating part so that both sets of the normal vectors align simultaneously.

With this understanding, we will proceed with computing the rotation matrix $\mathbf{R}$ following the steps discussed and then point out the pitfall.

Continued

EXAMPLE 4.2—CONT'D

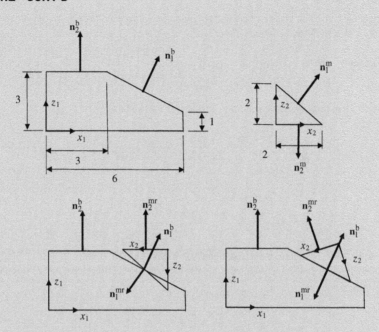

We first assume that both constraints are satisfied; hence, $\mathbf{n}_1^{mr} = -\mathbf{n}_1^b$ and $\mathbf{n}_2^{mr} = \mathbf{n}_2^b$. The table below lists the geometric data of the vectors and reference points.

	Base Part	Mating Part	Mating Part After Assembly
Concentric	$\mathbf{n}_1^b = \begin{bmatrix} \dfrac{2}{\sqrt{13}} \\ 0 \\ \dfrac{3}{\sqrt{13}} \end{bmatrix}$	$\mathbf{n}_1^m = \begin{bmatrix} \dfrac{1}{\sqrt{2}} \\ 0 \\ \dfrac{1}{\sqrt{2}} \end{bmatrix}$	$\mathbf{n}_1^{mr} = -\mathbf{n}_1^b = \begin{bmatrix} -\dfrac{2}{\sqrt{13}} \\ 0 \\ -\dfrac{3}{\sqrt{13}} \end{bmatrix}$
Coincident	$\mathbf{n}_2^b = \begin{bmatrix} 0 \\ 0 \\ 1 \end{bmatrix}$	$\mathbf{n}_2^m = \begin{bmatrix} 0 \\ 0 \\ -1 \end{bmatrix}$	$\mathbf{n}_2^{mr} = \mathbf{n}_2^b = \begin{bmatrix} 0 \\ 0 \\ 1 \end{bmatrix}$

With the two direction vectors $\mathbf{n}_1^m$ and $\mathbf{n}_2^m$ given, the third direction vector can be computed as

$$\mathbf{n}_3^m = \mathbf{n}_1^m \times \mathbf{n}_2^m = \begin{bmatrix} \dfrac{1}{\sqrt{2}} \\ 0 \\ \dfrac{1}{\sqrt{2}} \end{bmatrix} \times \begin{bmatrix} 0 \\ 0 \\ -1 \end{bmatrix} = \begin{bmatrix} 0 \\ \dfrac{1}{\sqrt{2}} \\ 0 \end{bmatrix}.$$

EXAMPLE 4.2—CONT'D

After normalizing it, we have $\mathbf{n}_3^m = \begin{bmatrix} 0 & 1 & 0 \end{bmatrix}^T$.

Similarly, the third vector after assembly is

$$\mathbf{n}_3^{mr} = \mathbf{n}_1^{mr} \times \mathbf{n}_2^{mr} = \begin{bmatrix} -\dfrac{2}{\sqrt{13}} \\ 0 \\ -\dfrac{3}{\sqrt{13}} \end{bmatrix} \times \begin{bmatrix} 0 \\ 0 \\ 1 \end{bmatrix} = \begin{bmatrix} 0 \\ \dfrac{2}{\sqrt{13}} \\ 0 \end{bmatrix}.$$

After normalizing it, we have $\mathbf{n}_3^{mr} = \begin{bmatrix} 0 & 1 & 0 \end{bmatrix}^T$.

Then the rotation matrix $\mathbf{R}$ can be calculated using Eqn (4.8) as

$$\mathbf{R} = \begin{bmatrix} \mathbf{n}_1^{mr} & \mathbf{n}_2^{mr} & \mathbf{n}_3^{mr} \end{bmatrix} \begin{bmatrix} \mathbf{n}_1^{m} & \mathbf{n}_2^{m} & \mathbf{n}_3^{m} \end{bmatrix}^{-1} = \begin{bmatrix} -\dfrac{2}{\sqrt{13}} & 0 & 0 \\ 0 & 0 & 1 \\ -\dfrac{3}{\sqrt{13}} & 1 & 0 \end{bmatrix} \begin{bmatrix} \dfrac{1}{\sqrt{2}} & 0 & 0 \\ 0 & 0 & 1 \\ \dfrac{1}{\sqrt{2}} & -1 & 0 \end{bmatrix}^{-1} = \begin{bmatrix} -\dfrac{2\sqrt{2}}{\sqrt{13}} & 0 & 0 \\ 0 & 1 & 0 \\ 1-\dfrac{3\sqrt{2}}{\sqrt{13}} & 0 & -1 \end{bmatrix}.$$

If we rotate the two direction vectors of the mating part before assembly, we have

$$\mathbf{n}_1^{mr} = \mathbf{R}\mathbf{n}_1^{m} = \begin{bmatrix} -\dfrac{2\sqrt{2}}{\sqrt{13}} & 0 & 0 \\ 0 & 1 & 0 \\ 1-\dfrac{3\sqrt{2}}{\sqrt{13}} & 0 & -1 \end{bmatrix} \begin{bmatrix} \dfrac{1}{\sqrt{2}} \\ 0 \\ \dfrac{1}{\sqrt{2}} \end{bmatrix} = \begin{bmatrix} -\dfrac{2}{\sqrt{13}} \\ 0 \\ -\dfrac{3}{\sqrt{13}} \end{bmatrix}$$

which is $-\mathbf{n}_1^b$, as it should be, and

$$\mathbf{n}_2^{mr} = \mathbf{R}\mathbf{n}_2^{m} = \begin{bmatrix} -\dfrac{2\sqrt{2}}{\sqrt{13}} & 0 & 0 \\ 0 & 1 & 0 \\ 1-\dfrac{3\sqrt{2}}{\sqrt{13}} & 0 & -1 \end{bmatrix} \begin{bmatrix} 0 \\ 0 \\ -1 \end{bmatrix} = \begin{bmatrix} 0 \\ 0 \\ 1 \end{bmatrix}$$

which is $\mathbf{n}_2^b$. Everything seems to be working fine mathematically. However, we know at the beginning that such a transformation is impossible physically. What is the problem? Let us take a look at the transformation matrix $\mathbf{R}$. First,

$$\mathbf{R}\mathbf{R}^T = \begin{bmatrix} -\dfrac{2\sqrt{2}}{\sqrt{13}} & 0 & 0 \\ 0 & 1 & 0 \\ 1-\dfrac{3\sqrt{2}}{\sqrt{13}} & 0 & -1 \end{bmatrix} \begin{bmatrix} -\dfrac{2\sqrt{2}}{\sqrt{13}} & 0 & 1-\dfrac{3\sqrt{2}}{\sqrt{13}} \\ 0 & 1 & 0 \\ 0 & 0 & -1 \end{bmatrix} = \begin{bmatrix} \dfrac{8}{13} & 0 & -\dfrac{2\sqrt{2}}{\sqrt{13}}+\dfrac{12}{13} \\ 0 & 1 & 0 \\ -\dfrac{2\sqrt{2}}{\sqrt{13}}+\dfrac{12}{13} & 0 & \dfrac{6\sqrt{2}}{\sqrt{13}}+\dfrac{44}{13} \end{bmatrix} \neq \mathbf{I}$$

Continued

EXAMPLE 4.2–CONT'D

and

$$|\mathbf{R}| = \begin{bmatrix} -\dfrac{2\sqrt{2}}{\sqrt{13}} & 0 & 0 \\ 0 & 1 & 0 \\ 1 - \dfrac{3\sqrt{2}}{\sqrt{13}} & 0 & -1 \end{bmatrix} = 0.7845 \neq 1.$$

Therefore, the rotation submatrix **R** is not orthogonal.

The problem with a nonorthogonal rotation matrix is that it does not perform the rotation correctly. For example, if we rotate the vectors that represent respectively axes x_2 and z_2 using the submatrix **R**, we have

$$\mathbf{n}_{x_2}^{mr} = \mathbf{Rn}_{x_2}^{m} = \begin{bmatrix} -\dfrac{2\sqrt{2}}{\sqrt{13}} & 0 & 0 \\ 0 & 1 & 0 \\ 1 - \dfrac{3\sqrt{2}}{\sqrt{13}} & 0 & -1 \end{bmatrix} \begin{bmatrix} 1 \\ 0 \\ 0 \end{bmatrix} = \begin{bmatrix} -\dfrac{2\sqrt{2}}{\sqrt{13}} \\ 0 \\ 1 - \dfrac{3\sqrt{2}}{\sqrt{13}} \end{bmatrix}$$

and

$$\mathbf{n}_{z_2}^{mr} = \mathbf{Rn}_{z_2}^{m} = \begin{bmatrix} -\dfrac{2\sqrt{2}}{\sqrt{13}} & 0 & 0 \\ 0 & 1 & 0 \\ 1 - \dfrac{3\sqrt{2}}{\sqrt{13}} & 0 & -1 \end{bmatrix} \begin{bmatrix} 0 \\ 0 \\ 1 \end{bmatrix} = \begin{bmatrix} 0 \\ 0 \\ -1 \end{bmatrix}.$$

These two vectors are nonperpendicular:

$$\mathbf{n}_{x_2}^{mr} \cdot \mathbf{n}_{z_2}^{mr} = \begin{bmatrix} -\dfrac{2\sqrt{2}}{\sqrt{13}} & 0 & 1 - \dfrac{3\sqrt{2}}{\sqrt{13}} \end{bmatrix} \begin{bmatrix} 0 \\ 0 \\ -1 \end{bmatrix} = -1 + \dfrac{3\sqrt{2}}{\sqrt{13}} \neq 0$$

which is wrong, to say the least.

This orthogonality properties of a rotation matrix can be easily verified by computer. When the properties are not satisfied, the mating constraints are in conflict, and CAD prompts an error message.

In practice, a product assembly must be properly parameterized so that a desired design intent is captured without invoking any conflict between mating constraints. In this example, there are two obvious possibilities. One intent is keeping the slope of the mating surfaces as 45°, as shown in Figure 4.17b. In this case, the width of the top edge of the base part must be related to that of the bottom edge, such as $d_{top}^{b} = d_{bottom}^{b} - 2$, where d_{top}^{b} and d_{bottom}^{b} are the widths of the top and bottom edges of the base part, respectively. The second intent, as illustrated in Figure 4.17b, allows the slope of the mating surfaces to vary. In this case, the widths of the top and bottom edges of the base part must be related to the bottom edge of the mating part, such as $d_{bottom}^{m} = d_{bottom}^{b} - d_{top}^{b}$, where d_{bottom}^{m} is the width of the bottom edge of the mating part.

The assembly modeling approach discussed is capable of supporting both cases, as long as there is no conflict between the mating constraints. For the first intent shown in Figure 4.17a, the

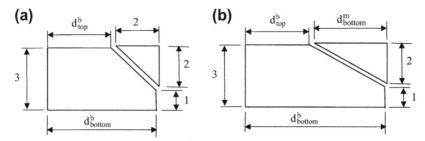

FIGURE 4.17 Illustration of Design Intents. (a) Slope of the Mating Surface Kept at 45°, and (b) Slope of the Mating Surface Varying.

transformation matrix can be calculated following the same steps shown in Example 4.1, in which points P_1^b is relocated according to the dimension d_{bottom}^b. The resulting transformation matrix is identical to that of Example 4.1, except for L_x, which is determined by the dimension d_{bottom}^b.

As for the second intent shown in Figure 4.17b, vectors n_1^b and n_1^m as well as points P_1^b and P_1^m must be calculated, based on the change of dimension d_{bottom}^b. This is left as an exercise.

4.3.2 Degree of freedom analysis

In the mating constraint method, designers specify the relative positions of parts by interactively defining spatial relationships between the geometric features of mating parts. These mating constraints are applied to the same type of mating features, such as a pair of planar faces. Each of the geometry mating constraints has a pair of direction vectors, called principal vectors, which characterize the mating geometric entities. For example, the principal vectors are two outbound unit vectors that are normal to the mating planes for the coincident-mate and coincident-aligned constraints, and two unit vectors parallel to the mating axis direction for the concentric constraint. The principal vectors are antialigned for coincident-mate, whereas they are aligned for coincident-aligned.

Using these mating geometric features and principal vectors, we can determine the remaining DOFs for a pair of mating parts. For example, the two-part assembly shown in Figure 4.13 is assembled using concentric and coincident-aligned constraints as discussed. After imposing the concentric constraint between the hole in the base part and the cylindrical surface of the mating part, the two unit vectors parallel to the mating axis direction align, creating one common principal vector. At this point, the mating part is allowed to rotate along the direction of the principal vector, as shown in Figure 4.13c. After imposing the coincident-aligned constraint between the top face of the base part and bottom face of the mating part, the two outbound unit vectors normal to the faces for the coincident-aligned constraints point in the same principal vector direction. These two principal vectors are not in parallel and are called independent. Imposing both mating constraints eliminates the rotation degree of freedom of the mating part. In general, one independent principal vector (IPV) allows one rotational DOF, and two or more IPVs eliminate all rotational DOF, as summarized in Table 4.7.

Note that for the bearing and crank example shown in Figure 4.3, there are two mating constraints imposed, concentric between the hole of the bearing and the cylindrical surface of the lower shaft of the crank, and a coincident-mate between the two opposite faces of the two components. These two

Table 4.7 Rotational DOF Analysis

Number of Independent Principal Vectors	RDOF
0	R3
1	R1
2 or more	R0

Table 4.8 Translational DOF Analysis

Intersection Mating Geometry	TDOF
Plane	T2
Line	T1
Point	T0

mating constraints form two principal vectors that are parallel, counting as one IPV, therefore allowing one rotational DOF according to Table 4.7.

Using the intersection mating geometry (IMG) of mating components, we can also compute the translational DOF as shown in Table 4.8. For example, a coincident-aligned constraint yields a face IMG (Figure 4.18a), therefore allowing translational movement in two directions on the face. A line

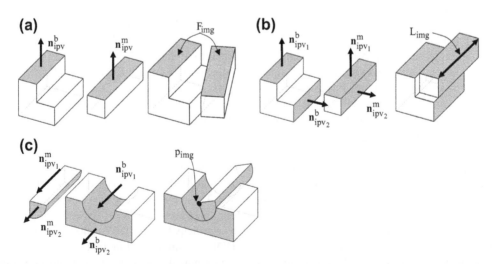

FIGURE 4.18 Examples of DOF Analysis. (a) One IPV and a Face IMG, (b) Two IPVs and a Line IMG, and (c) One IPV and a Point IMG.

IMG illustrated in Figure 4.18b as an example allows one translational DOF along the line, which is formed by intersecting two faces normal to the two respective IPV: $\mathbf{n}_{ipv1}$ and $\mathbf{n}_{ipv2}$ in Figure 4.18b. Figure 4.18c illustrates a case of point IMG, in which no translational movement is allowed. The point IMG is identified by intersecting the IPV ($\mathbf{n}_{ipv1}$ in Figure 4.18c) and the plane normal to the other IPV, $\mathbf{n}_{ipv2}$.

4.4 Kinematic modeling technique

The approach discussed in Section 4.3 sequentially positions the mating part with respect to the base part such that the given mating constraints are satisfied. The method is powerful and general. It is also capable of regenerating the assembly model after a design change is made as long as the mating constraints after the change are not in conflict. On top of that, the major advantage of the approach is that it solves the equations sequentially without dealing with a system of constraint equations simultaneously, as do many other methods proposed in the literature (e.g., Lee and Andrews, 1985).

However, this method requires that an assembly does not contain any closed-loop or under-constrained states, thus requiring that the mating part be positioned with respect to the base part whose position is determined. Such a limitation prevents the approach from dealing with two issues that are commonly encountered in product design involving assemblies. First, when a design change takes place, such as changing the dimensions of the crank and connecting rod for the slider-crank mechanism shown in Figure 4.19, the location and orientation of the individual parts are to be determined. Only when the location and orientation of the individual parts are determined can the method discussed in Section 4.3 be employed to calculate the transformation matrices for individual parts in the assembly. The second issue is that if the assembly is under-constrained, when a part in the assembly is moved, parts in the assembly must be repositioned according to how the parts are assembled. This is illustrated in Figure 4.20 using the slider-crank example.

These two common issues are discussed in this section. We discuss how to extract the kinematic information from mating constraints, construct a kinematic model, and carry out kinematic analysis that determines the location and orientation of individual parts in the assembly. There are several methods proposed for converting the mating constraints to kinematic joints. These include a mating

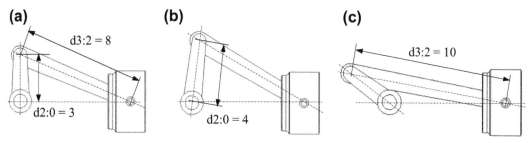

(a)   **(b)**   **(c)**

d3:2 = 8   d3:2 = 10

d2:0 = 3   d2:0 = 4

FIGURE 4.19 Design Changes in the Slider-Crank Assembly. (a) Dimensions d2:0 and d3:2, (b) d2:0 Changed to 4, and (c) d3:2 Changed to 10. All Assume a Stationary Slider.

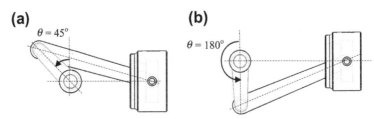

FIGURE 4.20 Crank Rotated in the Slider-Crank Assembly. (a) $\theta = 45°$, and (b) $\theta = 180°$.

relation-based method (e.g., Kim and Lee, 1989; Kim and Wu, 1990) and a contact condition based method (e.g., Sinha et al., 2002). There are also many methods developed for kinematic analysis of mechanisms (Dawari and Sen, 2007), including approaches based on configuration space (e.g., Joskowicz, 1990; Lozano-Pérez, 1983; Kim et al., 2003), screw theory (e.g., Adams et al., 1999), ports (e.g., Singh and Bettig, 2004), and features (e.g., Eng et al., 1999).

Before getting into the discussion, a few basic terminologies are mentioned. An assembly may be thought of as a set of rigid bodies connected by joints. A rigid body can be a single part or a subassembly, in which no relative motion is allowed between parts. These bodies are called links. An assembly of links and joints creates a kinematic chain, in which links are interconnected in a way to provide a desired output motion in response to an input motion. A mechanism is a kinematic chain in which at least one link has been grounded or attached to the frame of reference.

In this section, we first introduce the method proposed by Kim and Lee (1989) that maps mating constraints to kinematic joints, in which the joint information is automatically extracted from the mating relations for each link. Then, in Section 4.4.2, we discuss the Denavit–Hartenberg (D–H) representation, which is commonly employed to represent kinematic models in robotics applications. We then discuss, in Section 4.4.3, how to construct joint coordinate systems using mating constraints from the CAD assembly to construct a kinematic model. We discuss both open and closed-loop systems.

4.4.1 Mapping mating constraints to kinematic joints

As discussed in Section 4.2.2, joint constraints impose certain restrictions on the way the components can be assembled, and also on the way they move relative to one another. Each of the joint constraints is related to the rigid-body motion of a mating part and has DOFs associated with it.

As discussed in Section 4.3.2, counting the number of IPVs and types of intersections of the IMG, we are able to determine the remaining DOFs for a pair of mating parts. More precisely, one IPV allows one rotational DOF, and two or more IPVs eliminate all rotational DOFs, as summarized in Table 4.7. Moreover, as shown in Table 4.8, a face IMG allows two translational DOFs (Figure 4.18a), a line IMG allows one translational DOF (Figure 4.18b), and a point IMG allows no translational movement (Figure 4.18c).

Also, by reviewing the description of kinematic joints discussed in Section 4.2.2, an example of DOF analysis in Figure 4.18b shows a prismatic joint before and after applying two coincident-aligned constraints. The base and mating parts of the prismatic joint take two planar faces, respectively, as

IMGs. Each component has two IPVs and the line IMG; therefore, the joint has zero rotational DOFs and one translational DOF after assembly.

In addition to the case of Figure 4.18b, in which two coincident-aligned constraints are applied, cases such as two coincident-mates, one mate and one align, and one concentric and one mate (or align), as shown in Figure 4.21a, map to a prismatic joint.

The example shown in Figure 4.18c presents a revolute joint before and after applying one concentric and one coincident-aligned (or coincident-mate) constraints. Each component has one IPV. The point IMG is determined by interesting the axis (concentric) and the mate plane (coincident-aligned or coincident-mate), as illustrated in Figure 4.21b. As a result, the mating part is allowed to rotate along the axis (or IPV), resulting in a revolute joint.

With the discussion above, not only the number of DOFs can be determined by counting the number of IPV and checking the type of IMG between two mating parts, but also the type of joint between the two components can be determined. The mapping between mating constraints and kinematic joints is provided in Table 4.9. Figures 4.21c–e illustrate the mapping between mating constraints and kinematic joints of planar, cylindrical, and spherical joints, respectively.

4.4.2 D–H representation

After converting mating constraints to kinematic joints, the next step is to construct a kinematic model mathematically. We discuss a modeling approach that is commonly employed in robotics applications.

In robotics applications, most joints are associated with one actuator, either translational or rotational. Therefore, it is commonly assumed that all joints have only a single degree of freedom (Craig, 1989). Note that the assumption does not involve any real loss of generality because joints with multiple DOFs, such as a spherical joint (three rotational DOFs), can always be thought of as a succession of single DOF joints with zero-length links in between. This point is further illustrated in Example 4.4.

With the assumption that each joint has a single DOF, the action of each joint can be described by a single real number—that is, the angle of rotation in the case of a revolute joint or the displacement in the case of a prismatic joint. The objective of the kinematic analysis is to determine the cumulative effects of the entire set of joint variables.

An open loop mechanism with n joints has $n + 1$ links, as illustrated in Figure 4.22a schematically, because each joint connects two links. We number the joints from 1 to n, and we number the links from 0 to n, starting from the ground link 0. By this convention, joint i connects link $i − 1$ to link i. We consider the location of joint i to be fixed with respect to link $i − 1$. For example, Joint 2 is fixed to Link 1 in Figure 4.22a. When joint i is actuated, link i moves. Again, link 0 (the first link) is fixed and does not move when the joints are actuated.

Note that in general link n is not connected back to the base link 0, which is called open-loop. Figure 4.22b shows a closed-loop system, in which link n connects back to link 0 (or any link between 0 and $n − 2$).

With the ith joint, we associate a joint variable, denoted by q_i. In the case of a revolute joint, q_i is the angle of rotation; in the case of a prismatic joint, q_i is the joint displacement.

To construct a kinematic model, we rigidly attach a coordinate system to each link at the joint. In particular, we attach x_i–y_i–z_i to link i with an origin O_i at joint $i + 1$. We call this coordinate system C_i, defined by its origin O_i with three axes x_i–y_i–z_i. This means that whatever motion the joint imposes, the

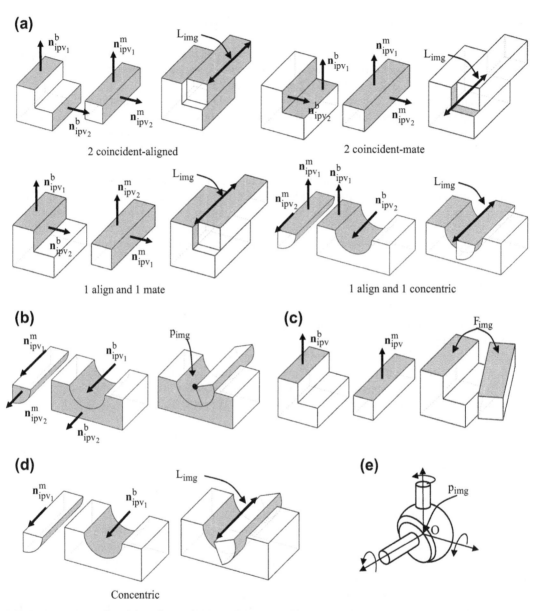

FIGURE 4.21 Illustration of Mapping between Mating Constraints and Kinematic Joints. (a) Prismatic, (b) Revolute, (c) Planar, (d) Cylindrical, and (e) Spherical Joints.

Table 4.9 The Mapping between Mating Constraints and Kinematic Joints

Joint	DOF	Number of IPVs	IMGs	Mating Constraints
Prismatic	T1R0	2	Line	Two coincident-mates, 2 coincident, coincident-mate and coincident-aligned, or coincident-mate (or align) and concentric
Revolute	T0R1	1	Point	Coincident-mate and concentric, or coincident-aligned and concentric
Planar	T2R1	1	Plane	Coincident-mate, or coincident-aligned
Cylindrical	T1R1	1	Line	Concentric
Spherical	T0R3	0	Point	Point coincident

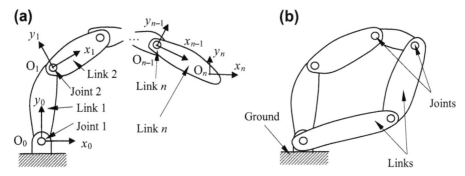

FIGURE 4.22 Schematic Representation of Kinematic Mechanism. (a) Open-loop and (b) Closed-loop.

location of each point on link i is constant when expressed in the ith coordinate frame. Furthermore, when joint i is actuated, link i and its attached frame C_i experience a resulting motion. The frame C_0, which is attached to the ground link, is referred to as the inertial frame. Note that O_n of link n (or end link), in which the axes x_n–y_n–z_n of the coordinate system C_n are attached, is usually located at a point of interest in design. In robotics applications, C_n is called the end-effector.

The transformation matrix similar to that of Section 4.3 can be employed to express the location and orientation of individual links. For example, matrix $\mathbf{T}_i^{i-1}$ defines the location and orientation of link i with respect to link $i-1$ or relating coordinate system of C_i with respect to C_{i-1}. More specifically, the matrix transforms a given vector $\mathbf{v}_i = [v_{ix}, v_{iy}, v_{iz}]^T$ in the ith link back to the coordinate system of the $(i-1)$th link $\mathbf{v}_{i-1} = [v_{i-1x}, v_{i-1y}, v_{i-1z}]^T$ in a homogeneous coordinate system:

$$\begin{bmatrix} \mathbf{v}_{i-1} \\ 1 \end{bmatrix} = \mathbf{T}_i^{i-1} \begin{bmatrix} \mathbf{v}_i \\ 1 \end{bmatrix} \tag{4.12}$$

in which $\mathbf{T}_i^{i-1}$ is a homogeneous transformation matrix like that of Section 4.3. The assumption that all joints are either revolute or prismatic implies that $\mathbf{T}_i^{i-1}$ is a function of only a single joint variable, namely q_i; that is,

$$\mathbf{T}_i^{i-1} = \mathbf{T}_i^{i-1}(q_i) \tag{4.12a}$$

in which

$$\mathbf{T}_i^{i-1} = \begin{bmatrix} \mathbf{R}_i^{i-1} & \mathbf{L}_i^{i-1} \\ 0 & 1 \end{bmatrix}. \tag{4.12b}$$

Now the homogeneous transformation matrix that expresses the location and orientation of C_j with respect to C_i is denoted by $\mathbf{T}_j^i$, which can be written as

$$\mathbf{T}_j^i = \mathbf{T}_{i+1}^i \mathbf{T}_{i+2}^{i+1} \dots \mathbf{T}_{j-1}^{j-2} \mathbf{T}_j^{j-1}, \text{ assuming } i < j. \tag{4.13}$$

Note that the following equations are valid because the homogeneous transformation matrices are orthogonal:

$$\mathbf{T}_j^i = \mathbf{I}, \text{ if } i = j, \text{ and} \tag{4.14a}$$

$$\mathbf{T}_j^i = \left(\mathbf{T}_i^j\right)^{-1}. \tag{4.14b}$$

By plugging Eqn (4.12b) into Eqn (4.13), we have

$$\mathbf{T}_j^i = \begin{bmatrix} \mathbf{R}_j^i & \mathbf{L}_j^i \\ 0 & 1 \end{bmatrix} \tag{4.15a}$$

where

$$\mathbf{R}_j^i = \mathbf{R}_{i+1}^i \mathbf{R}_{i+2}^{i+1} \dots \mathbf{R}_j^{j-1} \tag{4.15b}$$

which represents the orientation of the coordinate system C_j relative to coordinate system C_i, and

$$\mathbf{L}_j^i = \mathbf{L}_{j-1}^i + \mathbf{R}_{j-1}^i \mathbf{L}_j^{j-1} \tag{4.15c}$$

denoting the location of the coordinate system C_j relative to coordinate system C_i.

By the manner in which we have rigidly attached the various coordinate systems to the corresponding links, it follows that the position of any point on the end link (Link n), when expressed in coordinate system C_n, is a constant independent of the configuration of the mechanism. Then the location and orientation of the end link in the inertial frame are given by

$$\mathbf{T}_n^0 = \mathbf{T}_1^0(q_1)\mathbf{T}_2^1(q_2)\dots\mathbf{T}_n^{n-1}(q_n). \tag{4.16a}$$

Note that in general link n is not connected back to the base link 0; therefore, Eqn (4.16) represents an open-loop kinematic system. For a closed-loop system, link n usually connects back to link 0. In this case, Eqn (4.16a) becomes

$$\mathbf{I} = \mathbf{T}_1^0(q_1)\mathbf{T}_2^1(q_2)\dots\mathbf{T}_n^{n-1}(q_n)\mathbf{T}_0^n(q_0) \tag{4.16b}$$

in which the matrix $\mathbf{T}_0^n = (\mathbf{T}_n^0)^{-1}$ is multiplied from the right on both sides of Eqn (4.16a).

It is possible to simplify the transformation matrices by introducing conventions to represent a joint mathematically. In robotics applications, a commonly used convention for selecting frames of reference is the D–H convention (Denavit and Hartenberg, 1955; Hartenberg and Denavit, 1965). Following this convention, a considerable amount of streamlining and simplification in the mathematical representation of the kinematic model can be achieved. In this convention, each homogeneous transformation matrix $\mathbf{T}_i^{i-1}$ is represented as a product of four basic transformation matrices; that is,

$$\mathbf{T}_i^{i-1} = \mathbf{R}_z(\theta_i)\mathbf{T}_z(d_i)\mathbf{T}_x(a_i)\mathbf{R}_x(\alpha_i) \tag{4.17a}$$

where

$$\mathbf{R}_z(\theta_i) = \begin{bmatrix} c\theta_i & -s\theta_i & 0 & 0 \\ s\theta_i & c\theta_i & 0 & 0 \\ 0 & 0 & 1 & 0 \\ 0 & 0 & 0 & 1 \end{bmatrix} \tag{4.17b}$$

$$\mathbf{T}_z(d_i) = \begin{bmatrix} 1 & 0 & 0 & 0 \\ 0 & 1 & 0 & 0 \\ 0 & 0 & 1 & d_i \\ 0 & 0 & 0 & 1 \end{bmatrix} \tag{4.17c}$$

$$\mathbf{T}_x(a_i) = \begin{bmatrix} 1 & 0 & 0 & a_i \\ 0 & 1 & 0 & 0 \\ 0 & 0 & 1 & 0 \\ 0 & 0 & 0 & 1 \end{bmatrix} \tag{4.17d}$$

$$\mathbf{R}_x(\alpha_i) = \begin{bmatrix} 1 & 0 & 0 & 0 \\ 0 & c\alpha_i & -s\alpha_i & 0 \\ 0 & s\alpha_i & c\alpha_i & 0 \\ 0 & 0 & 0 & 1 \end{bmatrix}. \tag{4.17e}$$

Note that the four parameters θ_i, d_i, a_i, and α_i, illustrated in Figure 4.23, are associated with link i and joint i, and are generally named joint angle (θ_i), link offset (d_i), link length (a_i), and link twist (α_i), respectively. Also, in Eqns (4.17b and e), the short-hand notations, such as $c\theta_i = \cos\theta_i$, $s\alpha_i = \sin\alpha_i$, are employed.

As an example, the system shown in Figure 4.23 consists of four links (0, 1, 2, and 3) and three joints. Note that Joints 1 and 2 are revolute, and Joint 3 is a prismatic joint. θ_1 and θ_2 are the joint angles of the revolute joints 1 and 2, respectively. a_1 and a_2 are the link lengths of Links 1 and 2, respectively. α_1 and α_2 are the link twists of Links 1 and 2, respectively. d_2 is the offset between coordinate systems C_1 and C_2 (not necessarily a joint offset in this case). d_3 is the joint offset of the prismatic joint (Joint 3). Note that in this system, link lengths a_1 and a_2, offset d_2, and link twists α_1 and α_2 are constant.

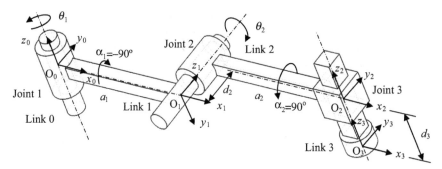

FIGURE 4.23 Illustration of Parameters: Joint Angle, Link Offset, Link Length, and Link Twist.

Plugging Eqns (4.17b–e) into Eqn (4.17a), we have

$$
\mathbf{T}_i^{i-1} = \begin{bmatrix} c\theta_i & -s\theta_i c\alpha_i & s\theta_i s\alpha_i & a_i c\theta_i \\ s\theta_i & c\theta_i c\alpha_i & -c\theta_i s\alpha_i & a_i s\theta_i \\ 0 & s\alpha_i & c\alpha_i & d_i \\ 0 & 0 & 0 & 1 \end{bmatrix} = \begin{bmatrix} \mathbf{R}_i^{i-1} & \mathbf{L}_i^{i-1} \\ 0 & 1 \end{bmatrix} \tag{4.18a}
$$

where

$$
\mathbf{R}_i^{i-1} = \begin{bmatrix} c\theta_i & -s\theta_i c\alpha_i & s\theta_i s\alpha_i \\ s\theta_i & c\theta_i c\alpha_i & -c\theta_i s\alpha_i \\ 0 & s\alpha_i & c\alpha_i \end{bmatrix} \tag{4.18b}
$$

and

$$
\mathbf{L}_i^{i-1} = \begin{bmatrix} a_i c\theta_i \\ a_i s\theta_i \\ d_i \end{bmatrix} \tag{4.18c}
$$

in which $\mathbf{R}_i^{i-1}$ and $\mathbf{L}_i^{i-1}$ are the rotation and translation matrices, respectively. In other words, $\mathbf{R}_i^{i-1}$ and $\mathbf{L}_i^{i-1}$ orient and locate Link i with respect to Link $i-1$.

Because the matrix $\mathbf{T}_i^{i-1}(q_i)$ is a function of a single variable q_i, as defined in Eqn (4.11), three of the four parameters in each individual transformation matrices are constant. Only one parameter is allowed to vary. It is apparent that for a revolute joint, joint angle θ_i is the variable; for a prismatic joint, link offset d_i is the only variable. For the system shown in Figure 4.23, joint angles θ_1 and θ_2 and the joint offset d_3 are variables.

One important note to make is that although the choices of coordinate systems are not unique, they have to be chosen carefully. If the coordinate systems chosen satisfy the following two conditions, then there exist unique numbers θ_i, d_i, a_i, and α_i such that Eqn (4.17a) can be determined (Spoong et al., 2005):

(1) The axis x_i is perpendicular to the axis z_{i-1}, and
(2) The axis x_i intersects the axis z_{i-1}.

Not that the choice of the origin of the coordinate system is less restrictive in general.

The coordinate systems defined for the links of the example shown in Figure 4.23 satisfy the conditions. A more important convention in choosing coordinate systems is to assign z_i to be the axis of actuation for joint $i + 1$. For example, the axis z_1 in Figure 4.23 is assigned at the actuation direction (axis of rotation) of the revolute joint, Joint 2.

To completely defined a coordinate system, we need the origin and axis x; then, the remaining y-axis can be determined by the right-hand rule. We will use examples to illustrate the construction of coordinate systems momentarily. For the time being, we assume coordinate systems for a given kinematic model have been created, and we discuss the measurement of the four parameters: joint angle (θ_i), link offset (d_i), link length (a_i), and link twist (α_i).

- Joint angle θ_i is the required rotation of x_{i-1}-axis about the z_{i-1}-axis to become parallel to the x_i-axis. For example, θ_1 in Figure 4.23 is the rotation angle of x_0 about the z_0-axis to become parallel to x_1-axis.
- Joint distance d_i is the distance between the x_{i-1} and x_i axes along the z_{i-1}-axis. Joint distance is also called link offset. For example, d_3 in Figure 4.23 is the distance between x_2 and x_3 about the z_2-axis.
- Link twist α_i is the required rotation of the z_{i-1}-axis about the x_i-axis to become parallel to the z_i-axis. For example in Figure 4.23, α_1 is the required rotation of the z_0-axis about the x_1-axis to become parallel to the z_1-axis, which is $-90°$ in this case.
- Link length a_i is the distance between z_{i-1} and z_i axes along the x_i-axis. Note that a_i is the kinematic length of link i. For example, in Figure 4.23, a_1 is the distance between z_0 and z_1 axes along the x_1-axis.

Apparently, the location of the origin of a coordinate system could affect link offset d_i and link length a_i.

Once the link parameters are identified, a table that lists link parameters for the system can be created. For example, the table for the system shown in Figure 4.23 can be created as below, with which transformation matrices can be written readily using Eqn (4.18a).

Note that link parameters a_1, a_2, α_1, α_2, and d_2 in Table 4.10 are constant.

The following example provides further illustration on the calculation of the transformation matrices. In Examples 4.3 and 4.4, we assigned coordinate systems that satisfy the two conditions mentioned above. In later examples, we illustrate the rules of specifying origin and x-axis of individual coordinate systems for both open- and closed-loop assemblies.

Table 4.10 List of Link Parameters for the System Shown in Figure 4.23

Link	θ_i	d_i	a_i	α_i
1	θ_1	0	a_1	$\alpha_1 = -90°$
2	θ_2	d_2	a_2	$\alpha_2 = 90°$
3	0	d_3	0	0

EXAMPLE 4.3

Consider a planar two-bar system shown below, which consists of two revolute joints and three links.

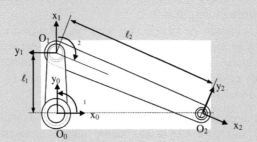

Note that joint axes z_i, $i = 0$, 2, are normal to the page, and all coordinate systems assigned satisfy the two conditions stated above. The base frame C_0 is fixed to Link 0 as shown. Calculate the transformation matrix for Link 2 with respect to Link 0; that is, $\mathbf{T}_2^0$.

Solutions

We first create a table that lists link parameters for the system as shown below.

Link	θ_i	d_i	a_i	α_i
1	θ_1	0	$a_1 = \ell_1$	0
2	θ_2	0	$a_2 = \ell_2$	0

From Eqn (4.18a), we have

$$\mathbf{T}_1^0 = \begin{bmatrix} c\theta_1 & -s\theta_1 & 0 & \ell_1 c\theta_1 \\ s\theta_1 & c\theta_1 & 0 & \ell_1 s\theta_1 \\ 0 & 0 & 1 & 0 \\ 0 & 0 & 0 & 1 \end{bmatrix}, \text{ and } \mathbf{T}_2^1 = \begin{bmatrix} c\theta_2 & -s\theta_2 & 0 & \ell_2 c\theta_2 \\ s\theta_2 & c\theta_2 & 0 & \ell_2 s\theta_2 \\ 0 & 0 & 1 & 0 \\ 0 & 0 & 0 & 1 \end{bmatrix}.$$

Therefore, from Eqn (4.16a), we have

$$\mathbf{T}_2^0 = \mathbf{T}_1^0 \mathbf{T}_2^1 = \begin{bmatrix} c\theta_{12} & -s\theta_{12} & 0 & \ell_1 c\theta_1 + \ell_2 c\theta_{12} \\ s\theta_{12} & c\theta_{12} & 0 & \ell_1 s\theta_1 + \ell_2 s\theta_{12} \\ 0 & 0 & 1 & 0 \\ 0 & 0 & 0 & 1 \end{bmatrix}$$

where $c\theta_{12} = \cos(\theta_1 + \theta_2)$, and $s\theta_{12} = \sin(\theta_1 + \theta_2)$.

Note that the first two entries of the last column of $\mathbf{T}_2^0$ are the x and y components of the origin O_2 referring to the base frame; that is,

$$O_{2x} = \ell_1 \cos \theta_1 + \ell_2 \cos(\theta_1 + \theta_2)$$

$$O_{2y} = \ell_1 \sin \theta_1 + \ell_2 \sin(\theta_1 + \theta_2).$$

The rotation part of $\mathbf{T}_2^0$ defines the orientation of the coordinate system C_2 relative to the base frame.

We mentioned earlier that a joint of multiple DOFs, such as a spherical joint of three rotational DOFs, can always be thought of as a succession of single DOF joints with zero-length links in between. We derive the transformation matrix for a spherical joint in the following example.

EXAMPLE 4.4

A spherical joint shown below (left) is defined by three joint axes, z_0, z_1, and z_2, which intersect at point O. Physically, the spherical joint connects Link A to Link O (base part). This spherical joint can be thought of as a succession of revolute joints with zero-length links in between, as illustrated in the figure below (right).

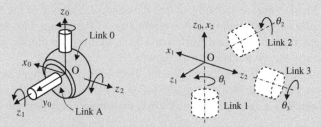

Derive a transformation matrix for the spherical joint.

Solutions

We first define the coordinate systems for the links, as shown above (right), satisfying the two conditions. Based on the coordinate systems, we have three joint angles, θ_1, θ_2, and θ_3, and two twist angles, α_1 and α_2, which are nonzero. Note that $\alpha_1 = -90°$ and $\alpha_2 = 90°$, according to the way that angles are measured as stated earlier.
We create a table that lists link parameters as below.

Link	θ_i	d_i	a_i	α_i
1	θ_1	0	0	$-90°$
2	θ_2	0	0	$90°$
3	θ_3	0	0	0

From Eqn (4.18a), we have

$$\mathbf{T}_1^0 = \begin{bmatrix} c\theta_1 & 0 & -s\theta_1 & 0 \\ s\theta_1 & 0 & c\theta_1 & 0 \\ 0 & -1 & 0 & 0 \\ 0 & 0 & 0 & 1 \end{bmatrix}, \mathbf{T}_2^1 = \begin{bmatrix} c\theta_2 & 0 & s\theta_2 & 0 \\ s\theta_2 & 0 & -c\theta_2 & 0 \\ 0 & 1 & 0 & 0 \\ 0 & 0 & 0 & 1 \end{bmatrix}, \text{ and } \mathbf{T}_3^2 = \begin{bmatrix} c\theta_3 & -s\theta_3 & 0 & 0 \\ s\theta_3 & c\theta_3 & 0 & 0 \\ 0 & 0 & 1 & 0 \\ 0 & 0 & 0 & 1 \end{bmatrix}$$

Therefore, from Eqn (4.16a), we have

$$\mathbf{T}_3^0 = \mathbf{T}_1^0 \mathbf{T}_2^1 \mathbf{T}_3^2 = \begin{bmatrix} c\theta_1 c\theta_2 c\theta_3 - s\theta_1 s\theta_3 & -c\theta_1 c\theta_2 s\theta_3 - s\theta_1 c\theta_3 & c\theta_1 s\theta_2 & 0 \\ s\theta_1 c\theta_2 c\theta_3 + c\theta_1 s\theta_3 & -s\theta_1 c\theta_2 s\theta_3 + c\theta_1 c\theta_3 & s\theta_1 s\theta_2 & 0 \\ -s\theta_2 c\theta_3 & s\theta_2 s\theta_3 & c\theta_2 & 0 \\ 0 & 0 & 0 & 1 \end{bmatrix}.$$

Note that $\mathbf{T}_3^0 = \mathbf{T}_A^0$, which transforms the rotation of Link A back to the base link O.

So far in the examples we discussed, all coordinate systems assigned satisfy the two important conditions. For an assembly model created by using kinematic joints in CAD, these coordinate systems can be created systematically. We use the slider-crank example shown in Figure 4.11 to illustrate the details. We first discuss open-loop system, in which we remove the prismatic joint between the slider and the ground. Then we discuss closed-loop system by resuming the prismatic joint.

4.4.2.1 Open-loop system

First, as discussed earlier, the z-axis of a joint aligns with the joint actuation direction. For a revolute joint, the z-axis aligns with the axis of rotation. Hence, the z-axes for all the three revolute joints are determined and illustrated in Figure 4.24a. Note that the positive direction of the z-axis is determined by Pro/ENGINEER internally, depending on the orientation of the datum axes selected for the individual joints. Users may flip the positive direction of a joint. In this example, the joint directions were adjusted to be pointing in the same direction as shown.

Next, the origin of individual coordinate systems associated with joints can be assigned at the datum points that were employed for defining the joints or at the intersection of joint axis and the mating faces. For example, for the first pin joint Pin1, the origin O_0 is located where datum points

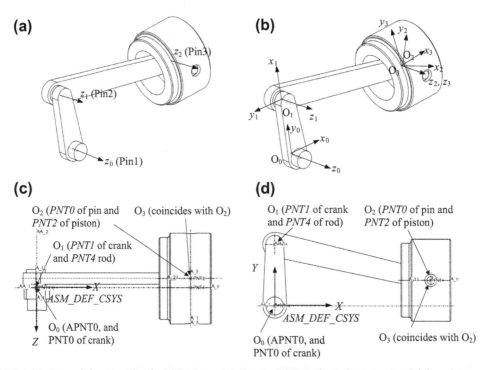

FIGURE 4.24 Determining the Coordinate Systems: (a) Z-axes, (b) Coordinate Systems, (c) Origins of the Coordinate Systems (top view), and (d) Origins of the Coordinate Systems (front view).

PNT0 (crank) and APNT0 coincide, as illustrated in Figures 4.24c and d. Similarly, O_1 (PNT1 of crank and PNT4 of rod), and O_2 (PNT0 of pin and PNT2 of piston) can be located. For O_3, because there is no datum point involved, O_3 can be located at any point along z_3; for example, locating O_3 to coincide with O_2, as shown in Figure 4.24b.

Now, we set the base coordinate system C_0. Because the origin O_0 and axis z_0 are determined, all we need is to determine axis x_0; axis y_0 can then be determined by the right-hand rule. Choosing axis x_0 is arbitrary for an open-loop system. For convenience, we may choose it to align with that of the global coordinate system (ASM_DEF_CSYS shown in Figures 4.24c and d as long as the x-axis of the global coordinate system is not parallel to the axis z_0. The x_0 axis is determined as shown in Figure 4.24b; hence, the y_0 axis and the coordinate system C_0 are determined.

After setting the base coordinate system C_0, we are now ready to assign x-axes for the remaining coordinate systems. There are three possible cases we considered. We discuss only Case A for the time being, which is relevant to the current open-loop example. If the axes z_i and z_{i-1} are parallel (such as z_1 and z_0, and z_2 and z_1 shown in Figure 4.24a), the axis x_i is chosen to be directed from O_i toward z_{i-1}, or as the opposite of this vector. In this example, we choose the latter: x_1 is chosen to be directed from O_1 toward the z_0-axis, but in the opposite direction, as shown in Figure 4.24b. Similarly, x_2 can be determined the same way. This is Case A, which is all we need for this open-loop example. More cases are discussed in the next example, closed-loop.

The coordinate system C_3 is added with its origin coinciding with that of C_2 and rotates θ_3 angle along the axis z_2, so that x_3-axis is parallel to x_0 for the time being, as shown in Figure 4.24b. Again, the coordinate system C_3 is called the end-effector in robotics applications.

Note that the approach of determining coordinate systems discussed above is systematic and general, which satisfy the two conditions mentioned above and can be implemented into computer. Once the individual coordinate systems are determined, the transformation matrices of the kinematic model can be created following the same approach discussed earlier.

We illustrate the calculation of the transformation matrices for the open-loop system in the following example.

EXAMPLE 4.5

Calculate the transformation matrices for the slider-crank mechanism shown in Figure 4.24b. The top and front views of the mechanism are sketched below with coordinate systems shown.

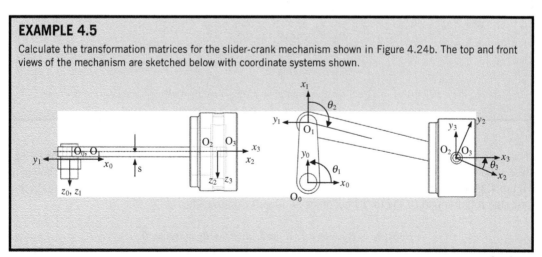

Continued

EXAMPLE 4.5—CONT'D

Solutions

We first create a table that lists link parameters for the system as below.

Link	θ_i	d_i	a_i	α_i
Crank (1)	θ_1	0	$a_1 = \ell_1$	0
Rod (2)	θ_2	$d_2 = -s$	$a_2 = \ell_2$	0
Slider (3)	θ_3	0	0	0

Note that although the rod does not involve any prismatic joint, its origin O_2 is offset a $-s$ amount (constant) from O_1 along the z_1-axis.

From Eqn (4.18a), we have

$$\mathbf{T}_1^0 = \begin{bmatrix} c\theta_1 & -s\theta_1 & 0 & \ell_1 c\theta_1 \\ s\theta_1 & c\theta_1 & 0 & \ell_1 s\theta_1 \\ 0 & 0 & 1 & 0 \\ 0 & 0 & 0 & 1 \end{bmatrix}, \mathbf{T}_2^1 = \begin{bmatrix} c\theta_2 & -s\theta_2 & 0 & \ell_2 c\theta_2 \\ s\theta_2 & c\theta_2 & 0 & \ell_2 s\theta_2 \\ 0 & 0 & 1 & -s \\ 0 & 0 & 0 & 1 \end{bmatrix}, \text{ and } \mathbf{T}_3^2 = \begin{bmatrix} c\theta_3 & -s\theta_3 & 0 & 0 \\ s\theta_3 & c\theta_3 & 0 & 0 \\ 0 & 0 & 1 & 0 \\ 0 & 0 & 0 & 1 \end{bmatrix}.$$

Therefore, from Eqn (4.16a), we have

$$\mathbf{T}_3^0 = \mathbf{T}_1^0 \mathbf{T}_2^1 \mathbf{T}_3^2 = \begin{bmatrix} c\theta_{123} & -s\theta_{123} & 0 & \ell_1 c\theta_1 + \ell_2 c\theta_{12} \\ s\theta_{123} & c\theta_{123} & 0 & \ell_1 s\theta_1 + \ell_2 s\theta_{12} \\ 0 & 0 & 1 & -s \\ 0 & 0 & 0 & 1 \end{bmatrix}$$

where $c\theta_{123} = \cos(\theta_1 + \theta_2 + \theta_3)$, and $s\theta_{123} = \sin(\theta_1 + \theta_2 + \theta_3)$.

Note that the first three entries of the last column of $\mathbf{T}_3^0$ are the x-, y-, and z-components of the origin O_3 referring to the base frame; that is,

$$O_{3x} = \ell_1 \cos\theta_1 + \ell_2 \cos(\theta_1 + \theta_2)$$

$$O_{3y} = \ell_1 \sin\theta_1 + \ell_2 \sin(\theta_1 + \theta_2)$$

$$O_{3z} = -s.$$

In addition, the rotation part of $\mathbf{T}_3^0$ defines the orientation of the coordinate system C_3 relative to the base frame C_0.

If we add a parallel mating constraint (called the align-oriented constraint in Pro/ENGINEER) between the horizontal plane (DTM3) of the piston and the datum plane ASM_TOP, as shown in Figure 4.25, the piston is allowed to translate on the x_0–y_0 plane, but it is not allowed to rotate. In this case, the rotation part of the of $\mathbf{T}_3^0$ becomes an identify matrix; that is,

$$\cos(\theta_1 + \theta_2 + \theta_3) = 1, \text{ and } \sin(\theta_1 + \theta_2 + \theta_3) = 0.$$

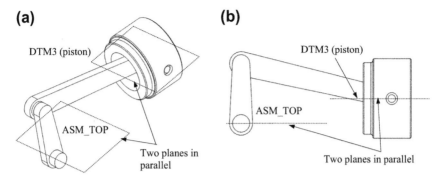

FIGURE 4.25 Parallel Mating Constraint Added Between Two Planes: (a) Iso-View, and (b) Front View.

Hence

$$\theta_1 + \theta_2 + \theta_3 = 0 (\text{or } 180°).$$

In this case, the system is no longer an open-loop.

In the following example, we illustrate the usage of the transformation matrices, in particular, to calculate the joint parameters in order to determine the configuration of the assembly.

EXAMPLE 4.6

We continue with Example 4.5 and assume that $\theta_1 = 90°$, $\ell_1 = 3$, and $\ell_2 = 8$. The vertical distance between the piston and the inertia frame is $h = 1$, as shown in the figure below. Determine the configuration of the assembly by calculating parameters θ_2, θ_3, and the distance between the piston and the base frame s_0, as shown in the figure below.

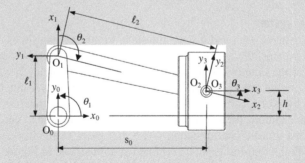

Solutions

The vertical position of the piston is given as $h = 1$, then

$$O_{3y} = \ell_1 \sin \theta_1 + \ell_2 \sin(\theta_1 + \theta_2) = \ell_1 + \ell_2 \sin(90 + \theta_2) = h.$$

Continued

EXAMPLE 4.6—CONT'D

The angle θ_2 can be solved as

$$\theta_2 = \sin^{-1}\left(\frac{h-\ell_1}{\ell_2}\right) - 90° = \sin^{-1}\left(\frac{1-3}{8}\right) - 90° = -105° \text{ or } 105°$$

giving two possible configurations, $\theta_2 = -105°$ shown above, and $\theta_2 = 105°$, where the piston is positioned to the left of the crank (see figure below).

Then the x-position of the slider can be found as

$$s_0 = O_{3x} = \ell_1 \cos\theta_1 + \ell_2 \cos(\theta_1 + \theta_2) = \ell_2 \cos(\theta_1 + \theta_2) = 8\cos(90° \pm 105°) = \pm 7.73.$$

Note that $s_0 = 7.73$, indicating the configuration shown above, and the configuration of $s_0 = -7.73$ is shown below. If the parallel mating constraint is present, the angle θ_3 can be calculated as
$$\theta_3 = -(\theta_1 + \theta_2) = -(90° \pm 105°) = 15° \text{ (see figure on previous page) or } -195° \text{ (see figure below).}$$

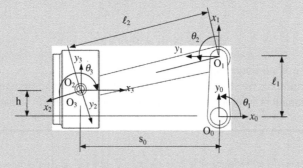

4.4.2.2 Closed-loop system

Now, we resume the prismatic joint and discuss the slider-crank mechanism as a closed-loop system.

For a closed-loop system, the x-axis of the base coordinate system C_0 cannot be determined arbitrary in general. For the time being, we assume a closed-loop system, in which the last link n is connected back to the ground link 0. The x_0-axis must be determined as if link 0 is connected to link n—that is, determined by axes z_0 and z_n following the same rule as any other joints as discussed above.

For the slider-crank example, the slider is connected back to the ground via a prismatic joint, and the z-axis aligns with its translational direction. With the z-axes of the three revolute joints shown before, the z-axes for all the four joints are determined and illustrated in Figure 4.26a. The origins of coordinate systems associated with joints are identical to those of the open-loop example, except for O_3. O_3 is assigned to datum point PNT4 because the axis A_6 that defines the translational direction of the prismatic joint passes PNT4. Note that in this case, O_3 is offset s from O_2 along z_2-axis.

Next, we set the x-axis for the coordinate systems C_1 and C_2 as before (Case A). However, for coordinate system C_3, the axis z_3 is not in parallel with z_2; instead, they intersect. For cases where z_i intersects z_{i-1} (Case B), x_i is chosen normal to the plane formed by z_i and z_{i-1} (with positive chosen

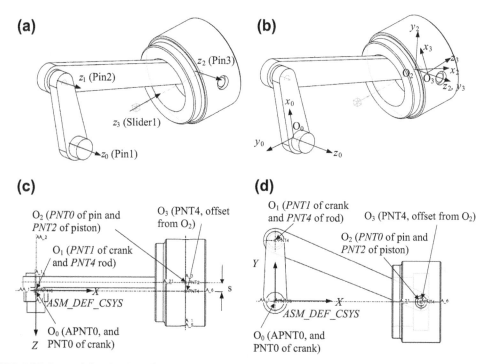

FIGURE 4.26 Determining the Coordinate Systems: (a) Z-axes, (b) Coordinate Systems, (c) Origins of the Coordinate Systems (top view), and Origins of the Coordinate Systems (front view).

arbitrarily). Hence, x_3 is determined by pointing (for example) upward, and y_3 is also determined by the right-hand rule as shown in Figure 4.26b.

Now, the slider connects back to the ground, and we must determine the x-axis of the coordinate system C_0. Because axis z_3 intersects z_0, we have Case B. Hence, x_0 is determined pointing (for example) upward, and y_0 is also determined by the right-hand rule as shown in Figure 4.26b.

Note that in both Cases A and B, z_i and z_{i-1} are coplanar. If z_i and z_{i-1} are not coplanar (Case C), for example, axes z_1 and z_0 shown in Figure 4.23, then there exists a line segment perpendicular to both z_i and z_{i-1} such that it connects both axes and it has a minimum length. The line containing this common normal to z_i and z_{i-1} defines x_i, and the axis y_i is determined to form a right-hand frame.

Note that the approach of determining coordinate systems discussed above is systematic and general, which satisfy the two conditions mentioned above and can be implemented into the computer. Once the coordinate systems of individual coordinate systems are determined, the transformation matrices of the kinematic model can be created following the same approach discussed earlier. Calculation of the transformation matrices for the slider-crank mechanism is illustrated in the following example.

EXAMPLE 4.7

Calculate the transformation matrices for the slider-crank mechanism shown in Figure 4.26b. The top and front views of the mechanism are sketched below with coordinate systems shown.

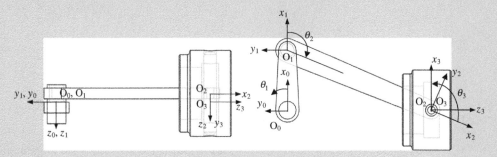

Solutions

We first create a table that lists link parameters for the system as below.

Link	θ_i	d_i	a_i	α_i
Crank (1)	θ_1	0	$a_1 = \ell_1$	0
Rod (2)	θ_2	$d_2 = -s$	$a_2 = \ell_2$	0
Slider (3)	θ_3	$d_3 = s$	0	90°
Ground (0)	0	d_0	0	−90°

Note that in this example, θ_1, θ_2, θ_3, and d_0 are variables.

From Eqn (4.18a), we have

$$\mathbf{T}_1^0 = \begin{bmatrix} c\theta_1 & -s\theta_1 & 0 & \ell_1 c\theta_1 \\ s\theta_1 & c\theta_1 & 0 & \ell_1 s\theta_1 \\ 0 & 0 & 1 & 0 \\ 0 & 0 & 0 & 1 \end{bmatrix}, \mathbf{T}_2^1 = \begin{bmatrix} c\theta_2 & -s\theta_2 & 0 & \ell_2 c\theta_2 \\ s\theta_2 & c\theta_2 & 0 & \ell_2 s\theta_2 \\ 0 & 0 & 1 & -s \\ 0 & 0 & 0 & 1 \end{bmatrix}, \mathbf{T}_3^2 = \begin{bmatrix} c\theta_3 & 0 & s\theta_3 & 0 \\ s\theta_3 & 0 & -c\theta_3 & 0 \\ 0 & 1 & 0 & s \\ 0 & 0 & 0 & 1 \end{bmatrix}, \text{ and}$$

$$\mathbf{T}_3^0 = \begin{bmatrix} c\theta_3 & 0 & s\theta_3 & 0 \\ s\theta_3 & 0 & -c\theta_3 & 0 \\ 0 & 1 & 0 & d_0 \\ 0 & 0 & 0 & 1 \end{bmatrix}$$

Therefore, from Eqn (4.16b), we have

$$\mathbf{I} = \mathbf{T}_1^0 \mathbf{T}_2^1 \mathbf{T}_3^2 \mathbf{T}_0^3 = \begin{bmatrix} c\theta_{123} & -s\theta_{123} & 0 & \ell_1 c\theta_1 + \ell_2 c\theta_{12} + d_0 s\theta_{123} \\ s\theta_{123} & c\theta_{123} & 0 & \ell_1 s\theta_1 + \ell_2 s\theta_{12} - d_0 c\theta_{123} \\ 0 & 0 & 1 & 0 \\ 0 & 0 & 0 & 1 \end{bmatrix} = \begin{bmatrix} 1 & 0 & 0 & 0 \\ 0 & 1 & 0 & 0 \\ 0 & 0 & 1 & 0 \\ 0 & 0 & 0 & 1 \end{bmatrix}.$$

Hence,

$$c\theta_{123} = \cos(\theta_1 + \theta_2 + \theta_3) = 1, \text{ and } s\theta_{123} = \sin(\theta_1 + \theta_2 + \theta_3) = 0,$$

$$\ell_1 \cos\theta_1 + \ell_2 \cos(\theta_1 + \theta_2) = 0$$

$$\ell_1 \sin\theta_1 + \ell_2 \sin(\theta_1 + \theta_2) - d_0 = 0.$$

EXAMPLE 4.7—CONT'D

Assuming $\theta_1 = 0°$, $\ell_1 = 3$, and $\ell_1 = 8$; we solve θ_2, θ_3 and d_0 from the above equations as follows:

$$\theta_2 = \cos^{-1}\left(\frac{-\ell_1 \cos \theta_1}{\ell_2}\right) - \theta_1 = \cos^{-1}\left(\frac{-3}{8}\right) - 0 = 112.0° \text{ or } -112.0°$$

which again gives two possible configurations, $\theta_2 = -112°$ shown above, and $\theta_2 = 112°$ where the piston is positioned to the left of the crank (see figure below).

Now the location of the piston can be found as follows:

$$d_0 = \ell_1 \sin \theta_1 + \ell_2 \sin(\theta_1 + \theta_2) = \ell_1 \sin \theta_1 + \ell_2 \sin(\theta_1 + \theta_2) = 3(0) + 8 \sin(\pm 112.0) = \pm 7.416.$$

Note that $d_0 = -7.416$, indicating the configuration shown above, and the configuration of $d_0 = 7.416$ is shown below. The sign of the link parameter d_0 is determined by the positive direction of the z_3 axis.

The angle θ_3 can be calculated as

$\theta_3 = -(\theta_1 + \theta_2) = -(\pm 112°) = 112°$ (see figure on previous page) or $-112°$ (figure below).

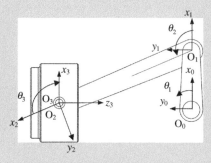

Now, let us go over the scenarios discussed at the beginning of this section. First, we change the dimensions of the crank and connecting rod for the slider-crank mechanism, as shown in Figure 4.19, and determine the location and orientation of the individual parts. We only take one set of link parameter values that result in a configuration with piston on the right to simplify the discussion.

EXAMPLE 4.8

Change the lengths of the crank and rod to 4 and 10, respectively; and calculate the parameters that determine the location and orientation of individual parts. First for Part A, we assume that the angle θ_1 is $\theta_1 = 0$. Then, in Part B, we assume the distance d_0 is a constant $d_0 = 7.416$.

Solutions

For Part A, we assume $\theta_1 = 0°$, $\ell_1 = 4$, and $\ell_2 = 10$; we solve θ_2, θ_3, and d_0 from the above equations as follows:

$$\ell_1 \cos \theta_1 + \ell_2 \cos(\theta_1 + \theta_2) = 0$$

$$\ell_1 \sin \theta_1 + \ell_2 \sin(\theta_1 + \theta_2) - d_0 = 0.$$

Continued

EXAMPLE 4.8—CONT'D

Solve for θ_2 from the first equation,

$$\theta_2 = \cos^{-1}\left(\frac{-\ell_1 \cos\theta_1}{\ell_2}\right) - \theta_1 = \cos^{-1}\left(\frac{-4}{10}\right) - 0 = \pm 114°.$$

We take only one value $\theta_2 = -114°$ for discussion, indicating the configuration where the piston is on the right of the crank, as shown below.

With this, we solve for d_0

$$d_0 = \ell_1 \sin\theta_1 + \ell_2 \sin(\theta_1 + \theta_2) = 4(0) + 10\sin(-114) = -9.165.$$

The figures below, left and right, show the assembly before and after changes, respectively.

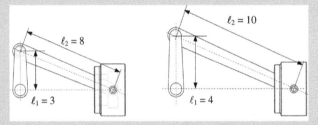

Now, for Part B, we assume the distance d_0 is a constant; that is, $d_0 = -7.416$. Note that we add a negative sign to the d_0 value in order to keep it consistent with the given configuration. With $\ell_1 = 4$ and $\ell_2 = 10$, we solve θ_1, θ_2, and θ_3 as follows.

$$\ell_1 \cos\theta_1 + \ell_2 \cos(\theta_1 + \theta_2) = 4\cos\theta_1 + 10\cos(\theta_1 + \theta_2) = 0$$

$$\ell_1 \sin\theta_1 + \ell_2 \sin(\theta_1 + \theta_2) - d_0 = 4\sin\theta_1 + 10\sin(\theta_1 + \theta_2) + 7.416 = 0.$$

The above nonlinear coupled equations can be solved using, for example, Matlab. For example, the Matlab script shown below (in italic) generates two sets of results.

$[q1, q2] = solve('4*\cos(q1) + 10*\cos(q_1 + q_2) = 0', '4*\sin(q_1) + 10*\sin(q_1 + q_2) = -7.416', 'Real', true).$

They are:

$$\text{Set 1: } \theta_1 = 0.51078 = 29.3° \text{ and } \theta_2 = -2.4380 = -139.7°$$

$$\text{Set 2: } \theta_1 = 2.6308 = 150.7° \text{(or} -29.3°\text{) and } \theta_2 = 2.4380 = 139.7°$$

which results in two configurations shown on the next page. The figure on the left shows the mechanism before design changes. The one in the middle indicates the configuration of the solutions of Set 1. The one on the right results from Set 2.

EXAMPLE 4.8—CONT'D

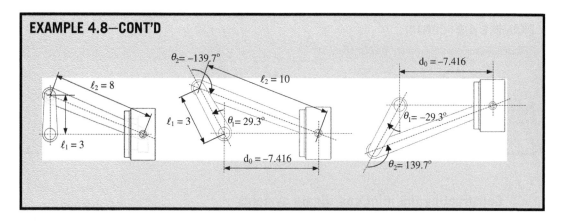

Now, we rotate the crank and reposition the parts in the assembly of the slider-crank example. This is illustrated in the next example. Again, we are only taking one set of link parameter values, which result in the configuration of the piston on the right, to simplify the discussion.

EXAMPLE 4.9

Change the angle θ_1 to 45° (Part A) and then 180° (Part B) as shown below, and calculate the parameters that determine the location and orientation of individual parts.

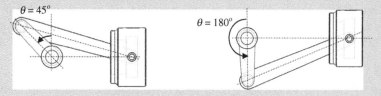

Solutions

We solve θ_2 and d_0 as follows, assuming $\ell_1 = 3$ and $\ell_2 = 8$, and $\theta_1 = 45°$ (Part A).

$$\ell_1 \cos \theta_1 + \ell_2 \cos(\theta_1 + \theta_2) = 0$$

$$\ell_1 \sin \theta_1 + \ell_2 \sin(\theta_1 + \theta_2) - d_0 = 0$$

Solve for θ_2, with $\theta_1 = 45°$,

$$\theta_2 = \cos^{-1}\left(\frac{-\ell_1 \cos \theta_1}{\ell_2}\right) - \theta_1 = \cos^{-1}\left(\frac{-3 \cos 45°}{8}\right) - 45° = -150.4° \text{ and } 60.4°$$

which indicates two respective configurations of the mechanism. We take only one value $\theta_2 = -150.4°$ for discussion, indicating the configuration where the piston is on the right of the crank.

Now, we solve for d_0:

$$d_0 = \ell_1 \sin \theta_1 + \ell_2 \sin(\theta_1 + \theta_2) = 3 \sin(45) + 8 \sin(45 - 150.4) = -5.59$$

which shows the configuration in the figure above (left).

Continued

EXAMPLE 4.9—CONT'D

Now, we solve for θ_2, with $\theta_1 = 180°$ (Part B).

$$\theta_2 = \cos^{-1}\left(\frac{-\ell_1 \cos \theta_1}{\ell_2}\right) - \theta_1 = \cos^{-1}\left(\frac{-3 \cos 180°}{8}\right) - 180° = \pm 112°$$

Again, the angles indicate two respective configurations of the mechanism. Like before, we take only one value, $\theta_2 = 112°$, for discussion. Now, we solve for d_0:

$$d_0 = \ell_1 \sin \theta_1 + \ell_2 \sin(\theta_1 + \theta_2) = 3 \sin(180) + 8 \sin(180 + 112) = -7.416.$$

4.4.3 Constructing the joint coordinate systems

The discussion presented in Section 4.4.2 assumes that the kinematic joints have been well defined in the assembly. This assumption is true if designers use CAD software, such as Pro/ENGINEER, and define kinematic joints using geometric entities such as datum axis, datum points, and so on. These datum entities can be used to determine the z-axis and origin of individual coordinate systems, construct transformation matrices, and solve for the location and orientation for individual links.

However, in most CAD systems, designers use mating constraints, instead of kinematic joints, to create assemblies. How do we construct transformation matrices and solve these equations for the location and orientation of individual parts? The missing link is the z-axis and the origin of the co-ordinate systems. If the information can be extracted from mating constraints, the same approach discussed in Section 4.4.2 can readily take over the remaining steps in positioning individual components in the assembly. Can the required information be extracted from mating constraints? The answer is yes. In this subsection, we introduce a method proposed by (Kim et al., 2001 and Kim.et.al., 2004).

In Section 4.3, we learned that the type of joints embedded in the CAD assembly can be determined by counting the number of IPVs and the type of IMG revealed in the mating constraints between the two mating parts. In this section, we discuss how to extract information from mating constraints and joint types in order to determine the z-axis and origin of the coordinate systems.

First, for a prismatic joint (Figure 4.27a) that was formed by for example two coincident-aligned constraints, the direction of the joint axis is determined by the direction in which the joint moves. For example, this joint axis can be determined by

$$\mathbf{n}_t = \mathbf{n}_{ipv_1}^{mr} \times \mathbf{n}_{ipv_2}^{mr} \quad \text{or} \quad \mathbf{n}_t = \mathbf{n}_{ipv_1}^{b} \times \mathbf{n}_{ipv_2}^{b}. \tag{4.19}$$

The origin of the coordinate system can be determined, for example, by intersecting a line formed by $\mathbf{P}_1^{ma}$ and $\mathbf{n}_{ipv_1}^{mr}$—that is, $L(\mathbf{P}_1^{ma}, \mathbf{n}_{ipv_1}^{mr})$—and a plane that is normal to $\mathbf{n}_{ipv_1}^{mr}$ and passes point $\mathbf{P}_2^{ma}$—that is, $P(\mathbf{P}_2^{ma}, \mathbf{n}_{ipv_1}^{mr})$, as shown in Figure 4.27a. The z-axis of the prismatic joint is aligned with the moving axis and starts from the origin O. A similar arrangement can be made for other combinations of mating constraints that yield a prismatic joint.

For a revolute joint, the origin of the coordinate system O can be set at point of P_{img}, and the z-axis aligns with the moving axis, say $\mathbf{n}_{ipv_1}^{mr}$, as shown in Figure 4.27b.

For a planar joint, the origin can be set at $\mathbf{P}_1^{ma}$, and the z-axis can be any vector on the plane, as shown in Figure 4.27c.

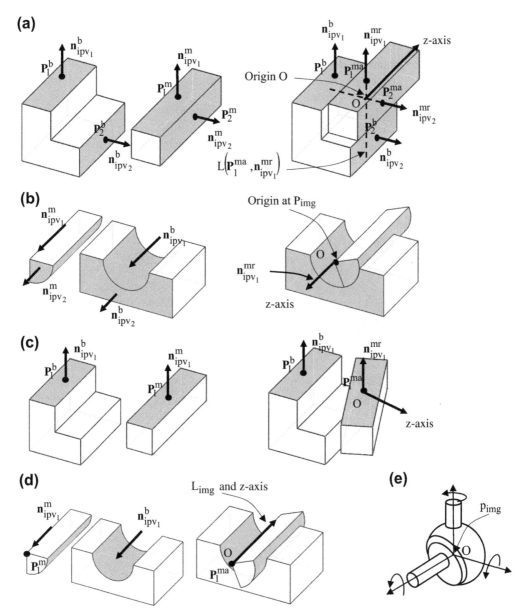

FIGURE 4.27 Determination of the Origin and z-axis of a Kinematic Joint. (a) Prismatic Joint, (b) Revolute Joint, (c) Planar Joint, (d) Cylindrical Joint, and (e) Spherical Joint.

For a cylindrical joint, the origin can be set at $\mathbf{P}_1^{ma}$, and the z-axis aligns with $\mathbf{n}_{ipv_1}^{mr}$, as shown in Figure 4.27d.

For a spherical joint, the origin can be set at P_{img} (Figure 4.27e), and the z-axis can be arbitrarily chosen, for example, to align with the z-axis of the global coordinate system.

Note that the origin can also be placed at the origin of the local coordinate system of the base or mating part assigned by the CAD system.

In the following, we use the same slider-crank example to illustrate the steps of identifying z-axis and coordinate systems of joints from mating constraints. We first assume an open-loop system by removing the two coincident-aligned constraints: Coincident3 between the piston and rod, and Coincident4 between the piston and bearing shown in Figure 4.7c. Note that if we add a parallel constraint between piston and bearing (Plane3@piston and Plane2@bearing), as shown in Figure 4.28a, the mechanism is like that of example shown in Figure 4.25 and is no-longer open-loop. We assume the assembly is underconstrained by suppressing the mating constraint Coincident2 between Plane3 of the crank and Plane3 of the bearing, as shown in Figure 4.7a.

In this example, the bearing is fixed to the inertial frame and is considered as a ground link, as shown in Figure 4.28a. The initial configuration of the crank and rod is shown in Figure 4.28b. We assume the same mating constraints for this example, as shown in Figure 4.7. We first illustrate the steps of determining IPV and IMG of each mating constraints, and the corresponding joints they represent. Then, we determine the z-axis and origin of the joint coordinate systems for each joints. A kinematic model, represented in the D–H convention, can be constructed and the transformation matrices that position and orient the links can be computed using approach discussed in Section 4.4.2.

The first two mates (*Concentric1* and *Coincident1*) assemble the crank to the fixed bearing, as shown in Figure 4.29a. According to Figure 4.27b, a revolute joint is extracted with the rotation axis z_0 pointing along a direction that aligns with $\mathbf{n}_{ipv_1}^{mr}$. Moreover, the origin of the coordinate system is located at the P_{img}. As a result, the origin O_0 is determined at the center of the hole of the bearing on the mating surfaces, as shown in Figure 4.29b. The x_0 and y_0 axes are chosen conveniently, such as to align with the WCS, to form a right-hand frame.

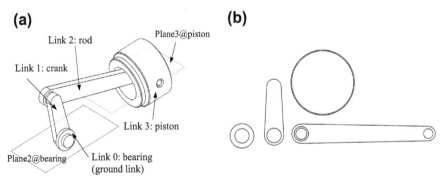

FIGURE 4.28 The Open-Loop Slider-Crank Example. (a) The Assembly in Iso-view, and (b) The Default Configuration of the Crank, Rod, and Piston (front view).

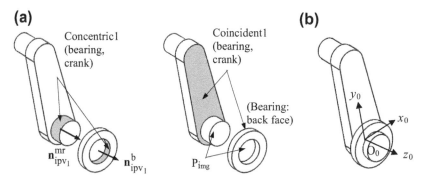

FIGURE 4.29 Joint Origin and z-axis for the Revolute Joint between Bearing and Crank. (a) Mating Constraints, IPV and IMG, and (b) Coordinate System for the Revolute Joint.

The next two mates (*Concentric2* and *Coincident3*) assemble the rod to the crank, as shown in Figure 4.30a. Again, a revolute joint is extracted, with the rotation axis z_1 pointing along a direction that aligns with $\mathbf{n}_{ipv_1}^{mr}$. Similarly, the origin O_1 is located at P_{img}, which is determined at the intersection of the axis of the upper shaft and the back face of the crank (i.e., on the mating surfaces between the rod and the crank), as shown in Figure 4.30b. The coordinate system C_1 can be determined following

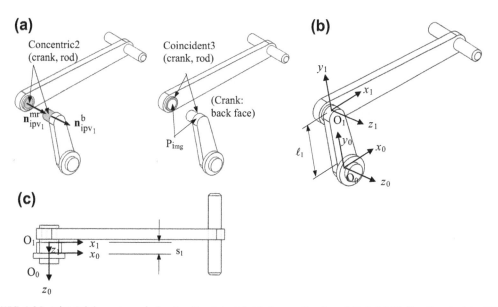

FIGURE 4.30 Joint Origin and z-axis for the Revolute Joint between Crank and Rod. (a) Mating Constraints, IPV and IMG, (b) Coordinate System of the Revolute Joint, and (c) Top View Showing the Offset s_1 between the Origins.

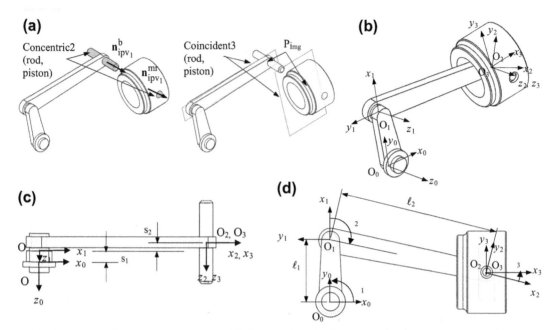

FIGURE 4.31 Coordinate Systems C_2 and C_3. (a) Mating Constraints, IPV and IMG, (b) Coordinate Systems in Iso-view, (c) Coordinate Systems in Top-view with Offsets s_1 and s_2, and (d) Coordinate Systems in Front-view.

the approach discussed earlier. The coordinate system C_1 aligns with C_0, except that it is offset along the x_0 direction by the amount that equals the length of the crank ℓ_1 and along the z_0-direction by an amount s_1, as shown in Figure 4.30c. Similar to the crank, the rod is allowed to rotate with respect to the crank.

Now, we assemble the piston to the rod by adding Concentric2 and Coincident3 constraints, as shown in Figure 4.31a. These two mating constraints create a revolute joint with the rotation axis z_2 pointing along a direction that aligns with $\mathbf{n}_{\text{ipv}_1}^{\text{mr}}$. Similarly, the origin O_1 is located at P_{img}, which is determined at the center of the circle at the midplane of the pin (on the mating surfaces between the rod and the piston), as shown in Figure 4.31b. The coordinate system C_2 aligns with C_1, except that it is offset along the x_1 direction by the amount that equals the length of the rod ℓ_2 and along the z_1-direction by an amount s_2, as shown in Figure 4.31c. The piston is allowed to rotate with respect to the rod.

Next, we add a coordinate system C_3 to the piston, as the end-effector, with its origin coinciding with that of C_2 and z-axis aligning with that of C_2. The coordinate system C_3 rotates a θ_3 angle along the z_3-axis, as shown in Figure 4.31d. Note that in Figure 4.31d, the position of the piston is lowered to simply better show the coordinate systems C_2 and C_3 as well as the rotation angle θ_3.

As a result, the table of link parameters for this open-loop system is created, as shown in Table 4.11. The transformation matrix for the mechanism can be constructed similar to that of Example 4.5, except that in the current example, we have $d_2 = -s_1$ and $d_3 = -s_2$ (instead of $d_2 = -s$ and $d_3 = 0$).

Table 4.11 Link Parameters for the Open-Loop System of the Slider-Crank Mechanism

Link	θ_i	d_i	a_i	α_i
Crank (1)	θ_1	0	$a_1 = l_1$	0
Rod (2)	θ_2	$d_2 = -s_1$	$a_2 = l_2$	0
Slider (3)	θ_3	$d_3 = -s_2$	0	0

Now, we discuss the closed-loop system. A prismatic joint is extracted by the mating constraints Coincident4 and Coincident5, in which the translational axis z_3 aligns with L_{img} formed by intersecting Front Plane@rod and Plane3@piston (or Plane2@piston and Plane3@bearing), as shown in Figure 4.32a. For convenience, we pick the z_3-axis pointing to the right as positive, and place the origin of the coordinate system at the same location as O_2, as shown in Figure 4.32b. Because the axis z_3 is not in parallel with z_2, instead, they intersect; x_3 is determined pointing upward, and y_3 is also determined by the right-hand rule.

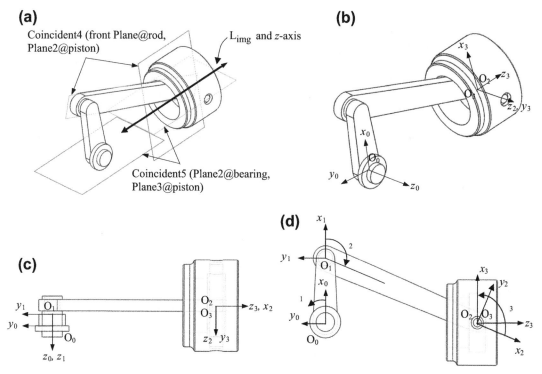

FIGURE 4.32 Determining the Coordinate Systems. (a) Z-axes, (b) Coordinate Systems, (c) Origins of the Coordinate Systems (top view), and (d) Origins of the Coordinate Systems (front view).

Table 4.12 Link Parameters for the Closed-Loop System of the Slider-Crank Mechanism

Link	θ_i	d_i	a_i	α_i
Crank (1)	θ_1	0	$a_1 = l_1$	0
Rod (2)	θ_2	$d_2 = -s_1$	$a_2 = l_2$	0
Slider (3)	θ_3	$d_3 = -s_2$	0	90°
Ground (0)	0	d_0	0	-90°

Now, the slider connects back to the ground. We must determine the x-axis of the coordinate system C_0. Just like that of the example shown in Figure 4.26, x_0 is determined to be pointing upward, and y_0 is also determined by the right-hand rule as shown in Figure 4.32b. The coordinate systems determined are shown in the top and front views in Figure 4.32c and d, respectively. Table 4.12 lists the link parameters for this closed-loop system. The transformation matrix for the mechanism can be constructed similar to that of Example 4.7, except that in the current system, we have $d_2 = -s_1$ and $d_3 = -s_2$, as shown in Table 4.12.

4.5 Case study and tutorial example

In this section, a case study and a tutorial example are presented. The case study presents briefly the applications of virtual reality technology to engineering design. The purpose of the case study is to showcase some of the interesting applications of CAD assembly to support engineering design. A single-piston engine is included as the tutorial example. Step-by-step instructions for creating the assembly model of the single-piston engine are given in Projects S1 and P1. Model files are available for download on this book's companion website (http://booksite.elsevier.com/9780123985132).

4.5.1 Case study: virtual reality

Virtual reality is the term used to describe a three-dimensional, computer-generated environment that can be explored and interacted with by a person. That person becomes part of this virtual world or is immersed within this environment and whilst there, is able to manipulate objects or perform a series of actions. One of the major development in virtual reality is CAVE (CAVE Automatic Virtual Environment), in which the person is fully immersed within it. CAVE takes the form of a cube-like space in which images are displayed by a series of projectors. Some systems enable the person to experience additional sensory input, such as sound or video, which contributes to the overall experience. A main feature of the CAVE system is interaction. The combination of interaction and total immersion is known as telepresence, in which a person can literally lose themselves within the virtual environment. Interaction takes place using a variety of input devices, such as a joystick, a wand or, more commonly, a haptics device (e.g., data glove). This enables the person to interact with objects, for example, by pulling, twisting, or gripping by means of touch. The ability to do this is known as haptics. An example of such a system is shown in Figure 4.33, in which a person wearing 3D eyeglasses and holding a virtual-reality controller steps into a room-sized, computer-generated version of a bathroom and

FIGURE 4.33 An Example of CAVE System. (Figure courtesy of www.news.wisc.edu/21313.)

interacts with medicine cabinet items and other moveable objects to simulate how patients self-administer health care at home (Taylor, 2012).

Virtual reality engineering includes the use of 3D modeling tools and visualization techniques as part of the design process. This technology enables engineers to view their project in 3D and gain a greater understanding of how it works. Plus they can spot flaws or potential risks before implementation. This also allows the design team to observe their project within a safe environment and make changes as necessary. What is important is the ability of virtual reality to depict fine-grained details of an engineering product to maintain the illusion. This means high-end graphics, video with a fast refresh rate, and realistic sound and movement. Automotive companies, such as Ford, uses CAVE (Figure 4.34a) to evaluate many aspects of the product design, including visibility, instrument reach, ergonomics, and roominess before building a physical prototype (Engine Technology International, 2012). GM also uses CAVE to interact with the layout of the interior (Figure 4.34b), in

FIGURE 4.34 Applications of CAVE for Engineering Design. (a) Design Evaluation of Vehicle Assembly at Ford (courtesy of www.enginetechnologyinternational.com/news.php?NewsID=41760), and (b) Automotive Interior Design Evaluation at GM (courtesy of www.parents.com/blogs/dadabase/2011/11/03/nostalgia/rise-of-the-dadmobile-the-chevy-traverse/).

FIGURE 4.35 Examples of Employing Virtual Reality for Other Applications. (a) Military Training (courtesy of www. vrs.org.uk/virtual-reality-military/index.html), and (b) The First Public Virtusphere (courtesy of http://www.popsci. com/gadgets/article/2010-06/human-sized-hamster-ball-lets-you-play-virtual-worlds).

which a designer wearing virtual reality glasses enters a small three-walled room where the proposed interior design on the vehicle is projected. The designer is immersed into the virtual interior of a vehicle that has not been built physically.

Virtual reality has been adopted by the military (this includes all three services—army, navy, and air force), where it is used for mainly training purposes. This is particularly useful for training soldiers for combat situations or other dangerous settings where they have to learn how to react in an appropriate manner. For example, a parachuting simulation can help train soldiers without flying them to 15,000 ft in the sky (Figure 4.35a). A virtual reality simulation enables them to do so but without the risk of death or a serious injury. They can re-enact a particular scenario, such as engagement with an enemy in an environment in which they experience this, but without the real-world risks. Virtual reality has also been quickly adopted by gaming industry. For example, the Excalibur Hotel and Casino in Las Vegas installed the first public Virtusphere, a human-sized hamster ball that lets players move through virtual worlds by walking, running, or crawling inside it (Duffy, 2010).

4.5.2 Tutorial example: a single-piston engine

The engine example consists of four major components: case, propeller, connecting rod, and piston, as shown in Figure 4.36. In both SolidWorks and Pro/ENGINEER, the assembly of the example is organized as three subassemblies (*case_asm*, *propeller_asm*, and *connectingrod_asm*) and one part (*piston*). The *case_asm* is fixed. The *propeller_asm* is assembled to *case_asm* using concentric and coincident-mate constraints, as shown in Figure 4.37a. The propeller is free to rotate along the *x*-direction. The *connectingrod_asm* is assembled to the propeller (at the crankshaft) using concentric and coincident-mate, as shown in Figure 4.37b. The connecting rod is free to rotate relative to the propeller (at the crankshaft) along the *x*-direction. Finally, the piston is assembled to the connecting rod (at pin) using a concentric mate, as shown in Figure 4.37c. The piston is also assembled to the case using another concentric mate. This mate restricts the piston movement along the *y*-direction, which in turn restricts the top end of the connection rod to move vertically.

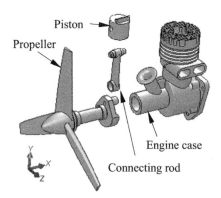

FIGURE 4.36 The Single-Piston Engine Example.

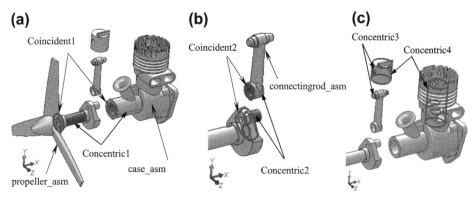

FIGURE 4.37 Assembly Mating Constraints Defined for the Engine Example. (a) Mates between Case and Propeller, (b) Mates between Propeller and Rod, and (c) Mates between Case and Rod.

4.6 Summary

In this chapter, we discussed assembly modeling that supports CAD to represent an assembly. Topics such as mating constraints, DOFs, and fully constrained vs underconstrained assemblies were discussed. We presented methods that support design changes and kinematic analysis in CAD assembly, which are the two most common activities encountered in assembly modeling using CAD. In addition to theoretical discussion, we included virtual reality as a case study that illustrated the application of CAD assembly for practical engineering designs. In addition, a single-piston engine assembly was introduced as a tutorial example.

After going over this chapter, we hope you have a fine understanding of the behind-the-scenes-operations when you work on CAD assembly models. By this time, you should know how CAD determines the location and orientation of individual parts that constitute the assembly. In addition, how CAD handles design changes and supports kinematic analysis when you drag a component should be clear. You should know what is fully constrained and what is underconstrained while you are creating an assembly. When you encounter an error message, such as constraints in conflict, you should know precisely the internal algorithm that makes this call. We hope this chapter has been useful to you in obtaining a general understanding of the methods employed for assembly modeling in CAD, becoming familiar with the behind-the-scenes operations in CAD, and most importantly, making you more confident as a designer in creating and handling CAD assemblies in support of product development.

Questions and exercises

1. Calculate the transformation matrix for the example shown in Figure 4.17b, assuming that the dimension d^b_{bottom} is changed to 6.
2. Calculate the transformation matrix for the example shown below. The base part (left) is a $2 \times 2 \times 2$ cube with the top right corner removed and a blind hole of diameter 0.5 drilled. The position of the hole is also shown. The mating part (shown below, center) is a tetrahedral block with a short cylinder, which simply complements the cut-out portion of the base part. Two mating constraints, concentric and coincident-aligned, are employed to create an assembly shown below (right).

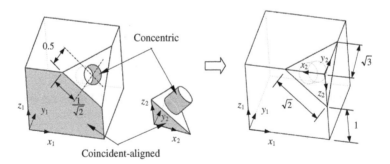

3. Define coordinate systems and calculate the transformation matrix for the single-piston engine mechanism shown on the next page (left). There are four joints, a pin joint (*Pin1*) between the propeller and case, the second pin joint (*Pin2*) between the connecting rod and the crankshaft (propeller), a third pin between the piston and the piston pin (mounted on the connecting rod), and a slider joint between the piston and the case. Kinematically, the system is a planar four-bar linkage shown on the next page (right), consisting of four links: crank, rod, slider, and ground. Note that the lengths of the rod and crank are 2.25 and 0.58333, respectively.

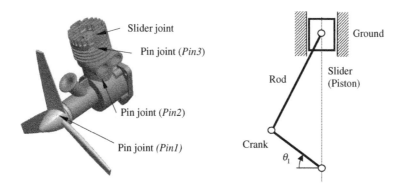

4. Continue with Problem 3 and answer the following questions by formulating and solving the equations involved.
   a. If $\theta_1 = 0°$, calculate all link parameters that determine the configurations of the system.
   b. Change the lengths of the crank and rod to 1 and 3, respectively; and calculate the parameters that determine the location and orientation of individual parts. First (Part A), we assume that the angle θ_1 is $\theta_1 = 0$. Then, in Part B, we assume the piston is stationary.
   c. Change the angle θ_1 to 45° (Part A) and then 180° (Part B), and calculate the parameters that determine the configurations of the system.
5. Conduct a case study in the application of virtual reality (or CAVE) technology for engineering design that was not included in Section 4.5. In your one-page report, please include the following:
   a. Name of the company or organization
   b. Source of the information (article, paper, magazine, website, YouTube, etc.)
   c. What is the nature of the application? What kind of equipment is employed to support such an application? What is the value added to the company or organization by employing the virtual reality technology?

References

Adams, J.D., Gerbino, S., Whitney, D.E., July 1999. Application of screw theory to motion analysis of assemblies of rigid parts. In: Proceedings of the 1999 IEEE International Symposium on Assembly and Task Planning Porto, Portugal, pp. 75–80.

Chang, K.H., 2010. Motion Simulation and Mechanism Design Using SolidWorks Motion 2011. SDC Publications.

Craig, J.J., 1989. Introduction to Robotics: Mechanics and Control, second ed. Addison-Wesley.

Dawari, A., Sen, D., December 12–13, 2007. Relation of Part Positions and Kinematic Freedom: a Survey, 13th National Conference on Mechanisms and Machines (NaCoMM07), IISc, Bangalore, India.

Denavit, J., Hartenberg, R.S., 1955. A kinematic notation for lower-pair mechanisms based on matrices. Trans. ASME J. Appl. Mech. 23, 215–221.

Duffy, J., June 23, 2010. Human-Sized Hamster Ball Lets You Play in Virtual Worlds. Popsci. http://www.popsci.com/gadgets/article/2010-06/human-sized-hamster-ball-lets-you-play-virtual-worlds.

Eng, T.-H., Ling, Z.-K., Olson, W., McLeanb, C., 1999. Feature based assembly modeling and sequence generation. Comput. Ind. Eng. 36, 17–33.

Ford Starts Work on Virtual Assembly Lines, August 09, 2012. Engine Technology International. www.enginetechnologyinternational.com/news.php?NewsID=41760.

Erdman, A.G., Sandor, G.N., Kota, S., 2001. Mechanism Design: Analysis and Synthesis, fourth ed. Prentice Hall.

Hartenberg, R.S., Denavit, J., 1965. Kinematic Synthesis of Linkages. MCGraw-Hill Series in Mechanical Engineering. McGraw-Hill, New York, p. 435.

Joskowicz, L., 1990. Mechanism comparison and classification for design. Artif. Intell. Eng. Des. 2 (1), 149–166.

Kim, S.H., Lee, K., 1989. An assembly modeling system for dynamic and kinematic analysis. Comput. Aided Des. 21 (1).

Kim, M.G., Wu, C.-H., November 1990. A formal part mating model for generating compliance control strategies of assembly operations. Los Angeles, CA, USA. In: IEEE International Conference on Systems, Man and Cybernetics, pp. 611–616.

Kim, J., Kim, K., Choi, K., Lee, J.Y., 2000. Solving 3D geometric constraints for assembly modelling. Int. J. Adv. Manuf. Technol. 16, 843–849.

Kim, J., Kim, K., Lee, J.Y., 2001. Solving 3D Geometric Constraints for Closed-Loop Assemblies, The 5th International Conference on Engineering Design & Automation, pp. 368–373, Las Vegas.

Kim, K.-J., Sacks, E., Joskowicz, L., 2003. Kinematic analysis of spatial fixed-axis higher pairs using configuration spaces. Comput. Aided Des. 35, 279–291.

Kim, J.S., Kim, K.S., Lee, J.Y., Jung, H.B., 2004. Solving 3D geometric constraints for closed loop assemblies. Int. J. Adv. Manuf. Technol. 23, 755–761.

Kim, J.S., Kim, K.S., Lee, J.Y., Jeong, J.H., 2005. "Generation of assembly models from kinematic constraints." Int. J. Adv. Manuf. Technol. 26, 131–137.

Lee, K., Andrews, G., 1985. Inference of the positions of components in an assembly part: 2. Comput. Aided Des. 17 (1), 20–24.

Lombard, M., March 2013. SolidWorks 2013 Bible. Wiley.

Lozano-Pérez, T., 1983. Spatial planning: a configuration space approach. IEEE Trans. Comput. 32 (2), 108–120.

Reyes, A., 2013. Beginner's Guide to SolidWorks 2013-Level I. SDC Publication.

Shih, R.H., 2013. Parametric Modeling with Pro/ENGINEER Wildfire 5.0. SDC Publication.

Singh, P., Bettig, B., 2004. Port-compatibility and connectability based assembly design. J. Comput. Inf. Sci. Eng. 4.

Sinha, R., Gupta, S.K., Paredis, C.J.J., Khosla, P.K., 2002. Extracting articulation models from CAD models of parts with curved surfaces. J. Mech. Design 124 (1), 106–114.

Spoong, M.W., Hutchinson, S., Vidyasagar, M., 2005. Robot Modeling and Control. Wiley.

Taylor, C., December 4, 2012. Virtually Healthy: 'CAVE' Lets Researchers Experience Patients' Behavior. University of Wisconsin-Madison News. www.news.wisc.edu/21313.

Toogood, R., Zecher, J., 2012. Creo Parametric 1.0 Tutorial and MultiMedia DVD. SDC Publication.

Whitney, D.E., 2004. Mechanical Assemblies Their Design, Manufacture, and Role in Product Development. Oxford University Press, New York.

Design Parameterization

5

CHAPTER OUTLINE

Design changes are frequently encountered in the product development process. The complexity of the design change is multiplied when the product design involves large-scale assemblies with multiple engineering disciplines. Very often, a simple change in one part may propagate to its neighboring parts, therefore affecting the entire product assembly. Both parts and assembly must be regenerated (or rebuilt) for a valid product model. At the same time, the regenerated product model must satisfy the geometric design requirements and meet the designer's expectations.

When a product is being developed in a virtual environment, the design changes are often implemented first by altering the geometry of the product represented in solid models using a computer-aided design (CAD) tool. If the product solid model is not parameterized properly, the changes in geometry often lead to invalid parts or assembly. At the part level, the changes may yield a solid model with invalid solid features if it is not properly parameterized. In this case, the entire product assembly is in vain. Even when individual parts of the product are regenerated correctly, parts may still penetrate to their neighboring parts or leave excessive gaps between them, if the solid model is not properly parameterized at the assembly level.

In this chapter, the fundamental principles of design parameterization for parts and assembly will be discussed. A set of guidelines will be presented for designers to parameterize solid models in order to capture design intents more effectively. These guidelines, which are provided at both part and assembly levels, support designers in successfully conducting product design in the e-Design environment.

A number of simple examples are included to explain concepts and methods. A slider-crank mechanism and its crankshaft are employed to illustrate and demonstrate the practicality of guidelines developed for both Pro/ENGINEER and SolidWorks. In addition, a single-piston airplane engine and a high-mobility multipurpose wheeled-vehicle (HMMWV) suspension are presented as case studies to demonstrate the parameterization method for practical applications. Note that in this chapter, parts and assembly are created in respective CAD tools. Issues of solid model translations between CAD systems will be addressed in Chapter 6. CAD models of the examples employed in this chapter can be found on the book's companion site. More detailed instructions for bringing up these models and steps for carrying out studies described in this chapter can be found in Projects P1 and S1 of the book.

Overall, the objectives of the chapter are to (1) introduce fundamental principles of design parameterization for designers to capture design intents at both part and assembly levels, and (2) offer practicing guidelines and illustrations for designers to facilitate the construction of parametric solid models. Note that the guidelines provided in this chapter may be used to support design practice for your project team and can be extended as needed.

5.1 Introduction

After intensive research and development in recent decades, the feature-based parametric modeling technique has become a reality (Lee, 1999; Zeid, 1991). This technique has been widely adopted in the mainstream CAD tools, such as *Pro/ENGINEER*, *SolidWorks*, *SolidEdge*, *Unigraphics*, *CATIA*, and even *Mechanical Desktop* of *AutoCAD*. With such a technique, designers are able to create parts through solid features and assemble parts or subassemblies for a complete product digital mockup in the CAD environment. In addition, the designer is able to define design variables by relating dimensions of the part features and create assembly mating constraints between parts to parameterize the product model through the parametric modeling technique. With the parameterized product model, the designer can make a design change simply by changing geometric dimension values and asking the CAD software to automatically regenerate the parts that are affected by the change, and hence the entire assembly.

For example, the bore diameter of an engine case is defined as the design variable, as shown in Figure 5.1a. When the diameter is changed from 1.2 in. to 1.6 in., the engine case is regenerated first by properly updating its solid features that are affected by the change. As shown in Figure 5.1b, the engine case becomes wider and the distance between the two exhaust manifolds is larger, just to name a few. At the same time, the change propagates to other parts in the assembly, including the piston, piston pin, cylinder head, cylinder sleeve, cylinder fins, and crankshaft, as illustrated in Figure 5.1b. More important, the parts stay intact, maintaining adequate assembly mating constraints, and the

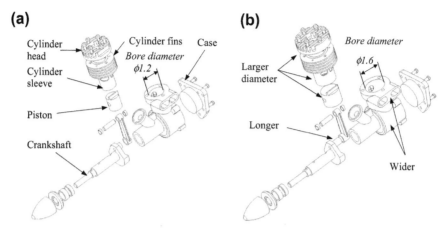

(a)

Cylinder head
Cylinder fins   Case
Bore diameter
$\phi 1.2$
Cylinder sleeve

Piston

Crankshaft

(b)

Bore diameter
$\phi 1.6$

Larger diameter

Longer

Wider

FIGURE 5.1 An Exploded View of a Single-piston Engine with a Bore Diameter of 1.2 in. (a) and a Bore Diameter of 1.6 in. (b).

change does not induce interference nor leave excessive gaps between parts. With such parametric models, designers are given tremendous freedom to explore design alternatives efficiently and accurately. In addition, this parametric technology supports the cross-functional team in conducting parametric studies and designing trade-offs in the e-Design environment (Chang et al., 1999). More about parametric study and design trade-off methods are discussed in the fourth book of the Computer-Aided Engineering Design series: Design Theory and Methods Using CAD/CAE.

We start in Section 4.2 by introducing design intents in product solid models. With the understanding of design intents, we discuss the two design axioms in Section 4.3 that form the basis of the design parameterization methods. In Sections 4.4 and 4.5, we offer guidelines for design parameterization at part and assembly levels, respectively. Section 4.6 includes two case studies, an airplane engine and an HMMWV suspension, which demonstrate the application of parameterization method and guidelines to practical examples.

5.2 Design intents

In a broader scope, design intent (DI) is a realization of design requirements (DR) in the shape of the product solid model. In the context of the e-Design paradigm, design intent is defined as the geometric shape of parts and/or configuration of the product assembly that the designer desires to attain while changing dimension values of the product solid model in CAD for better design alternatives.

In practice, design intents are derived from design requirements, satisfying physical requirements through the geometric shape of the product solid model. These design intents must be implemented in CAD for the purpose of exploring design alternatives that reveal better product performance and meet the design requirements.

While exploring design alternatives, changes are realized in CAD by modifying geometric dimension values and regenerating (or rebuilding) the product solid models automatically. In order to capture the DI, the product solid model must be properly constructed and parameterized.

Design intent for a single part can be captured by properly creating individual solid features and carefully relating dimensions within or between features so that when a dimension value is changed, the solid features affected by the change can be regenerated or rebuilt successfully. The geometric dimensions that can be changed independently to capture design intents are called *design variables* (DVs). The relationships between DR, DI, and DV are shown in Figure 5.2a. Figure 5.2b depicts an uncoupled design, in which one design requirement is realized in two DIs. DI1 is implemented in CAD in such a way that one design variable DV1 (an independent dimension) affects the DI. On the other hand, DI2 is captured by two design variables, DV2 and DV3. A change in DV1 does not affect DI2, and changes in DV2 or DV3 do not affect DI1. Figure 5.2c illustrates a coupled design intent, in which a change in DV2 affects both DIs.

In general, an uncoupled design is much more desirable than coupled one. In some cases, a coupled design intent may be decoupled by adding more DIs and/or DVs, which will be illustrated later in Section 5.3. The design requirements depicted in Figures 5.2b and c are decoupled. It is possible that DRs may be coupled, in which one DI may affect both DRs. Note that in this chapter, we focus more on capturing DIs and less on DRs, except for the slider-crank example to be discussed in Section 5.3.1.

To illustrate more, we use a block with a hole, which is shown in Figure 5.3, as an example. This part consists of a base extrusion block and a through hole. The design intent derived from the design requirement is to keep the hole right at the center of the block while varying its size. It is apparent that the DI is uncoupled because it is the only DI being considered.

To capture the intent for this simple example, relations between dimensions must be created between the hole's center point and the size of the block. As shown in Figure 5.3, the relations are $d1 = d2/2$, and $d3 = d4/2$, where d2 (block width) and d4 (block height) are design variables.

Before defining the relations, a rectangular profile was first created in the sketch before extruding the base block feature. In the rectangular profile, several sketch relations were imposed so that the size of the block can be determined by the width and height dimensions only (i.e., d2 and d4). The sketch relations specify that the top and bottom edges of the rectangle are horizontal, the left and right edges are vertical, and one of the corner points is anchored to the coordinate system of the sketch plane. As a result, the sketch relations and dimensions fully constrain the profile in the sketch. Once the base block is extruded, the center hole can be created as, for example, an extruded cut feature. Details about the sketch and sketch relations were discussed in Chapter 3.

Note that it is important to also capture the upper and lower limits of a design variable. In general, these limits must ensure a design that is always physically meaningful. In addition, the limits must

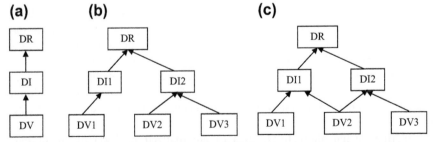

FIGURE 5.2 Illustration of the Relationships between Design Requirement (DR), Design Intent (DI), and Design Variables (DV). (a) Relationship between DR, DI, and DV. (b) Uncoupled Design Intent. (c) Coupled Design Intent.

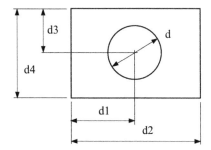

FIGURE 5.3 The Example of a Block with a Center Hole.

ensure that a valid solid model can be regenerated in CAD. For the example shown in Figure 5.3, the lower limits of the width (d2) and height (d4) must be greater than the hole diameter 'd' for a physically meaningful design. However, if the concern is simply about the solid model regeneration, both variables should at least be greater than zero.

At the assembly level, DI is captured by defining adequate mating constraints and relating dimensions across parts so that a change in dimension value can be propagated to all parts affected. The parts affected must be regenerated successfully; at the same time, they must maintain proper positions and orientations with respect to one another without violating any mating constraints nor revealing part penetration or excessive gaps between them. Moreover, the regenerated solid model must meet the designer's expectations. The bracket assembly shown in Figure 5.4 illustrates these points.

The bracket assembly consists of four parts: bracket, bushing, shaft, and arm. Mating constraints, such as concentric and surface mate, are defined to assemble the parts. In addition, one relation is defined to relate the shaft diameter and inner diameter of the hole in the bushing to ensure that the assembly is properly retained when the shaft size is changed. The relation is defined as

$$\phi d1:2 = \phi d3:4 + 0.01 \tag{5.1}$$

(a)   **(b)**

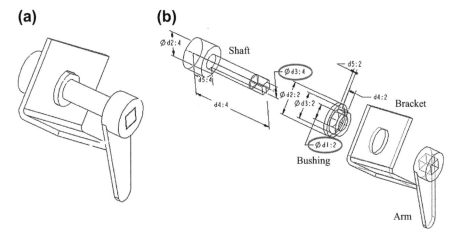

FIGURE 5.4 The Bracket Assembly Example. (a) Unexploded View. (b) Exploded View.

where ϕd1:2 is the diameter of the hole in the bushing, ϕd3:4 is the outer diameter of the shaft, and 0.01 is the prescribed clearance between them. Note that in Eqn (5.1), ϕd1:2 becomes a dependent dimension and ϕd3:4 stays independent.

5.3 Design axioms

As mentioned earlier, a set of guidelines for design parameterization will be presented. These guidelines were developed following two important axioms from Suh (1990):

Axiom 1: The Independence Axiom (maintains the independence of design intent)
Axiom 2: The Information Axiom (minimizes the information content of the design intent)

Axiom 1 implies that changing the DV values has an effect only on the referent DI. In other words, it is desirable to create uncoupled DIs whenever possible. Because DIs are derived from design objectives, Axiom 1 is often exercised to address coupled design objectives in practice.

Axiom 2 states that the amount of information (usually number of DVs) that is available to the designer for making design changes must be minimized for each DI.

5.3.1 Independence axiom

Generally speaking, the independence axiom is easier to comply with than that of the information axiom. Often the challenge lies in deriving uncoupled design intents from (sometimes) coupled design objectives or potentially conflict design requirements. It may not always be possible to create uncoupled DI. When coupled DIs are unavoidable, additional DIs (sometimes DVs) may have to be added in order to resolve or alleviate the conflict. In this case, the DIs are referred to as *decoupled*. A decoupled DI is less desirable, it is simply an unavoidable compromise.

An uncoupled design is always superior to a coupled or decoupled one. This is because that the DIs in the uncoupled design can be attended much easier, especially for complex design problem with large-scale assembly, because the effect of individual DV on the referent DI is completely separated. Moreover, an uncoupled design often carries less information for the designer to attend.

A simple example of an uncoupled design is shown in Figure 5.5a. The design of a plate with an orifice used to measure flow rates. The design requirement is simply positioning the hole inside the plate for a physically viable design. We assume two DIs in this case:

DI1: position of the orifice (i.e., d0 and d1), which keep the orifice completely inside the plate.
DI2: height of the plate (i.e., d2), which is sufficiently large to enclose the orifice.

The design in Figure 5.5a is uncoupled because perturbation in the value of d0 and d1 that defines the position of the hole has no effect on the height of the plate.

The same plate example may be created as a coupled design, where the height of the block is determined by the sum of the two DVs, d1 and d3, as shown in Figure 5.5b. Perturbing d1 not only alters the orifice position, but it also affects the height of the plate. For the two DIs to remain independent, they need to be referenced to a datum that is not a DV, such as the dimensions shown in Figure 5.5a, where the bottom edge of the plate serves as the datum. Both solid models in Figures 5.5a and b are valid. The solid model in Figure 5.5a provides the designer with a clearer perspective on how

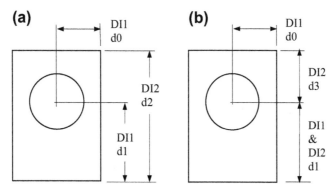

FIGURE 5.5 Illustration of Design Intents (DIs). (a) Uncoupled DIs. (b) Coupled DIs.

each DV affects its own DI. Although the solid model in Figure 5.5b is valid, it does not comply with the independence axiom; therefore, its use in the design process may be cumbersome.

A slider-crank example shown in Figure 5.6 is presented to further illustrate the issues involved in the design parameterization. This slider-crank mechanism consists of four parts: crankshaft, connecting rod, piston pin, and piston. Two design requirements are defined in this example:

DR1: Horizontal velocity of the piston increases 20% when the crankshaft is driven at the same angular velocity.

DR2: Weight of the mechanism reduces 5%.

It is first assumed that both DRs are realized by the stroke of the mechanism.

DI1: Stroke of the mechanism, which is realized by the lengths of the crankshaft and connecting rod. Moreover, the first design requirement can be realized, for example, by increasing the length of the crankshaft d2:0 (or the length of the connecting rod d3:2), as shown in Figure 5.6b. The dimension d2:0 becomes the DV of the first design intent. However, changing the DV also affects the second design requirement—the weight of the mechanism. In this case, these two design requirements are coupled.

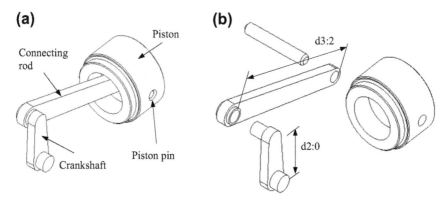

FIGURE 5.6 The Slider-crank Mechanism. (a) Unexploded View. (b) Exploded View.

In order to reduce the coupling effect, a second DI can be defined that, for instance, reduces the width of the connecting rod:

DI2: Width of the connecting rod.

This will help achieve the second design requirement or alleviate the effect of the change in d2:0 on the second design requirement, the weight. Adding the second DI helps to decouple the design requirement and, therefore, better comply with the independence axiom.

In particular, it is desirable for the designer to change only the value of the DV d2:0 while exploring design alternatives for the first DI. This is what Axiom 2 (information axiom) asks. The crankshaft must be properly parameterized (more to be discussed in Section 5.4.3) in order to capture the length design variable. When the DV is changed, the change must be propagated to the affected parts. The remaining parts must be kept unchanged, and the entire assembly must be maintained intact, as illustrated in Figure 5.7.

5.3.2 Information axiom

The second axiom, the information axiom, can be primarily addressed in the sketch of the solid feature. A bracket example created in Pro/ENGINEER is shown in Figure 5.8. The bracket profile consists of two horizontal and three vertical line segments, two perpendicular line segments, two quarter circular arcs, and two circles. By using the Intent Manager of Pro/ENGINEER (Toogood and Zecher, 2012), the profile is fully constrained with ten dimensions and a number of sketch relations, including vertical (V), horizontal (H), tangent (T), concentric ($\oplus$), and vertical alignment ($\mathbf{I}$) (see Figure 5.8a). The symbols of the geometric constraints that appear in Figure 5.8a were explained in Chapter 3.

Note that the information contents of the profile shown in Figure 5.8a are not minimized. Assume that a DI is to keep the profile symmetric with respect to its middle horizontal line. In order to capture the symmetry DI, entities must be related. For examples, the radii of the two circular arcs, the radii of the circles, and lengths of the two vertical line segments must be changed simultaneously. Keeping the sketch profile as it is and creating dimension relations to capture the DI is complex and unnecessary. A better option is to add sketch relations to properly parameterize the profile. While adding sketch relations, redundant or conflict dimensions will be removed by CAD automatically.

As shown in Figure 5.8b, two equal radii constraints (R_1–R_1 and R_2–R_2), three equal lengths constraints (L_1–L_1, L_2–L_2, and L_3–L_3), and a perpendicular constraint ($\perp$) are added. As a result, the symmetry DI is properly captured and the number of dimensions is reduced to five, as shown in

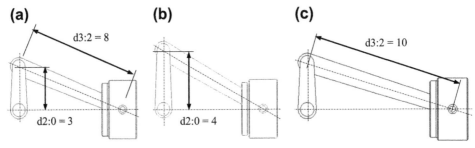

FIGURE 5.7 Changes of the Length of the Design Variables. (a) Design Variables d2:0 and d3:2. (b) d2:0 Changes to 4. (c) d3:2 Changes to 10.

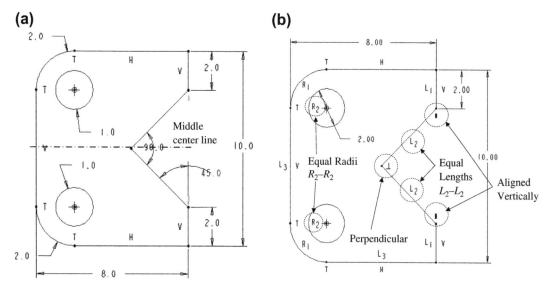

FIGURE 5.8 Minimization of Information Contents. (a) Information not Minimized. (b) Information Minimized.

Figure 5.8b. Changes in any of the dimensions yield a symmetric sketch with respective to its middle horizontal line. Note that the profile in Figure 5.8b is simpler and complies better with the information axiom.

The profile shown in Figure 5.8b can also be created in SolidWorks because similar sketching capabilities and sketch relations are available in SolidWorks sketch mode. Typical sketch relations in both Pro/ENGINEER and SolidWorks were summarized in Appendix A in Chapter 3.

In addition to the sketch profile, relations in Pro/ENGINEER or equations in SolidWorks can be added to relate dimensions between features. When new features are created by copying, mirroring, or patterning existing features, additional dimensions may be assigned as dependent to those of the original feature in order to reduce the information contents, if it is consistent with the DI to capture.

5.4 Design parameterization at the part level

Design parameterization must be carried out at both part and assembly levels. At the part level, sketch relations and dimensions must be defined to fully constrain the sketch profile of each solid feature and to capture the design intent. In addition, the geometry of the part will be regenerated following certain rules that were established when the part was created. In this section, the general modeling capabilities and the modeling procedure in CAD will be briefly reviewed. Guidelines for design parameterization at the part level will be presented, followed by examples.

5.4.1 Profile in sketch

A general solid modeling procedure in Pro/ENGINEER and SolidWorks was discussed in Chapter 3. The solid modeling procedure usually starts with defining datum features, such as datum planes, datum coordinate systems, and datum axes, which serve as the references to facilitate solid feature creations. One of the datum planes is usually chosen to sketch a two-dimensional profile that is protruded to create the first (or base) solid feature.

A build plan that describes the design intents, solid features, and sketch profiles with relations and dimensions is highly recommended. As discussed in Chapter 3, a build plan is especially useful for the beginners to develop before creating any solid features.

In sketch, geometric entities, such as lines, arcs, and splines, are drawn as vectors for a single open or closed profile that can be protruded for a surface or solid feature or making a cut. A set of characteristic points is created for these vector entities. As discussed in Chapter 3, the profile is determined by the x- and y-positions of the characteristic points. In both Pro/ENGINEER and SolidWorks, sketch relations, such as a concentric of circular arcs or parallel of lines, are generated automatically when these entities are drawn. The designer may define additional dimensions and sketch relations that fully constrain the profile if needed. Note that it is necessary to fully constrain the sketch profile in order to avoid unexpected errors while conducting design changes.

In general, the range of design variables is critical. When a profile dimension in a solid model is changed, the solid feature may become physically invalid. For a complex solid model, this problem may not be easy to detect. Determining the proper range of design variables that ensure valid solid features in advance is critically important for part parameterization. Note that finding the range for design changes is often conducted on a case-by-case basis. When more than one dimension is involved, the complexity of determining proper ranges for design changes is multiplied. Example 5.1 is presented to illustrate the point.

EXAMPLE 5.1

In the sketch profile shown in Figure 5.8b, 11 characteristic points are created; therefore, 22 equations must be generated to determine their locations. As shown in the figure, geometric relations, including equal radii, concentric, perpendicular, and alignment, are imposed. As a result, four dimensions can be changed independently. In this example, we will focus on changing the height dimension (current value: $H = 10.00$), as circled in the figure below. What will be the width W of the sketch if the height dimension H is increased from 10 to 12? What are the upper and lower limits of the height dimension H?

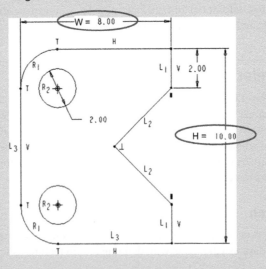

EXAMPLE 5.1—CONT'D

Solutions

The first thing to ask is how the other entities will change when we vary the height dimension H. Because the equal length constraint L_3 is imposed on the bottom and left edges, changing H will affect the arc radius R_1 because the total width of the profile (current value $W = 8$) is retained.

In addition, the circles are concentric with the circular arcs; when the radius R_1 is changed, these two circles will move accordingly. Will the circles be pulled toward or pushed away from the arcs when the height dimension value increases? This question can be answered by the following equations.

$$H = L_3 + 2R_1$$

$$W = L_3 + R_1$$

Hence, $R_1 = H - W$. Currently, H is 10 and W is 8; therefore, R_1 is 2. When H is increased from 10 to 12, R_1 becomes 4; that is, the circles are pushed away from the arcs, as shown in the figure below (Case A). Note that H cannot be set equal to W because it yields a zero-radius R_1, which is invalid and will not be accepted by CAD. When H is less than W, the radius will become negative; the profile will be regenerated as shown below (Case B), which is physically invalid. In addition, the height dimension cannot be too large. A large H value may cause the circles to run across the two 45° perpendicular edges, as shown in Case C. Therefore, the valid range of the height dimension is between 8 and 12.3, if W is kept unchanged.

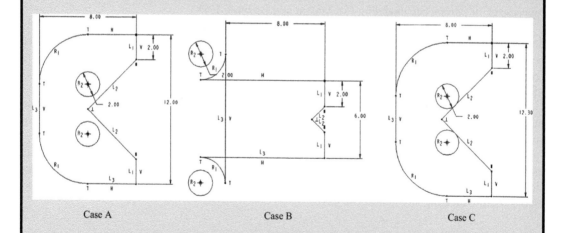

| Case A | Case B | Case C |

Note that the lower limit of H can be determined by $H = W + R_1$. Because W is 8 and R_1 is greater than 0, H must be greater than 8. How can the upper limit of the height H (which is 12.3) be determined? This is left as an exercise (hint: see Case C).

When the width dimension W is also considered for a change, the range of W depends on the current H value. If H is 12, the upper limit of W is 12. What will be the lower limit of W? (Again, this question is left as an exercise.) In general, defining the range of simultaneous changes for more than two design variables is not straightforward. Usually, all but one design variable is changeable.

5.4.2 Solid features in part

When the first feature is protruded, parametric surfaces (Zeid, 1991; Mortenson, 2006) that represent the boundary of the solid feature are generated by CAD. After that, the designer may create additional datum, sketches, and protrusion features using options such as extrusion, sweep, revolve, and blend, as discussed in Chapter 3. The designer may also cut the existing features; generate chamfers or rounds; or copy, mirror, and/or pattern the existing features to create additional features.

When additional features are created, a model tree or feature tree is generated by CAD following the feature creation sequence. Boolean operations are employed to union or subtract the features from the previous ones according to their definitions. At the same time, the intersection curves between boundary surfaces are calculated to evaluate the Boolean operations and display the features. This is essentially the constructive solid geometry (CSG) method.

Note that in general the intersecting curves are approximated by interpolating a number of intersection points using B-spline or nonuniform rational basis spline curves (Zeid, 1991). The evaluated geometry and topology of boundary faces, edges, and vertices are stored in the CAD database for display. This is the boundary representation (B-rep) method. When features are being created, the designers can define relations in Pro/ENGINEER (or equations in SolidWorks) to relate feature dimensions to capture DIs. In this process, independent and dependent dimensions will be created to define a one-way relation. The independent dimensions become DVs. This is so-called *unidirectional*, *procedural*, or *parametric* modeling (McMahon and Browne, 1998).

Once all the features are created and relations are established, the part solid model is completely defined. When a design change is conducted by changing the DV values, the solid model will be regenerated in Pro/ENGINEER (or rebuilt in SolidWorks) by updating features (both datum and solid features) following the model tree, one at a time. Pro/ENGINEER or SolidWorks carries out steps (discussed in Chapter 3) to update individual features.

Note that Pro/ENGINEER and SolidWorks employ the concepts of both CSG and B-rep for solid modeling. In general, CSG keeps the relationship between features, whereas B-rep stores topological and geometric data for display and computations.

If the DIs are not properly captured in features and relations, the regeneration may lead to an undesirable or an invalid solid model. It is strongly recommended, especially for journeyman designers, that the designer creates a model build plan (with details of features, dimensions, and relations) before creating any features in CAD.

5.4.3 Guidelines for design parameterization

Based on the previous discussions, a set of guidelines for part parameterization is established in Table 5.1. These guidelines are separately listed according to the two axioms and the steps in solid modeling. Note that these guidelines are by no means complete. Readers may add more guidelines to suit their needs. These guidelines are not entirely objective. Some may be modified. The crankshaft example of the slider-crank mechanism is employed as an example to illustrate some of these guidelines.

The crankshaft is created in both Pro/ENGINEER and SolidWorks in the following sequence: three default datum planes (DTM1, 2, 3 or Front, Top, Right), a default coordinate system (CS0), a base extrusion feature, the lower extrusion feature, and the upper extrusion feature, as illustrated in

Table 5.1 Guidelines for Part Parameterization

	Independence Axiom (1)	Information Axiom (2)
Datum (D)	D1a: A solid model should always start with the default datum features (i.e., three orthogonal datum planes) and the default coordinate system. D1b: Additional datum features should be referenced to the default datum features instead of geometric entities (e.g., an edge) of a solid feature whenever possible.	D2a: Never duplicate datum features.
Sketch (S)	S1a: A sketch should be created on a default datum plane if possible (instead of on a face of an existing solid feature) in order to minimize parent–child coupling between solid features. S1b: Dimensions on a sketch should be created by using datum features as references instead of geometric entities (e.g., an edge) of a solid feature whenever possible. S1c: A design variable (DV) should never be referenced to another geometric dimension (unless the design intent requires it).	S2a: One characteristic point of the sketch profile should be anchored to default datum features (e.g., intersection of two datum planes). S2b: A face and a geometric entity of an existing feature can be chosen as the sketch plane and the anchor point, respectively, only for the purpose of capturing a design intent (DI). S2c: Geometric entities on a sketch profile should be aligned to datum features or existing entities to minimize the number of dimensions. S2d: Sketch relations should be defined as much as possible to reduce the number of dimensions. S2e: The information contents should be minimized by using symmetry constraint. S2f: Relations between dimensions must be added not only to minimize the information content but also to capture the DI. It is desirable to define fewer relations by adding more sketch relations. S2g: Redundant and zero-valued dimensions should never be defined. S2h: Range of the design variables should be determined in advance.
Solid features (F)	F1a: A solid feature should be decoupled from existing solid features by referencing only to default datum features whenever possible.	F2a: Use attribute of the solid feature instead of addition dimension to define the feature; for example, a through hole should not be created with an extrusion cut with a depth dimension of larger value. F2b: The amount of information in solid features that have one or more planes of symmetry can be minimized using pattern, copy, and mirror.
Parts (P)	P1a: After the solid model is built, the designer should only have access to dimensions that form a DI (i.e., only to the DVs). P1b: Relate dimensions directly to the DV to avoid loop or chain relations.	P2a: Define relations between dimensions of different features to capture the design intent.

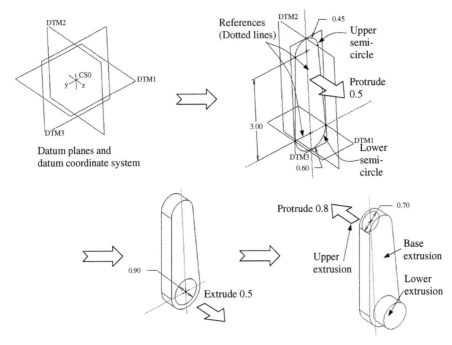

FIGURE 5.9 Feature Creation Steps of the Crankshaft.

Figure 5.9. Note that in both Pro/ENGINEER and SolidWorks, datum planes and coordinate systems are given for each part by default. They will be used as references for sketches and features (Guideline D1a). To simplify the presentation, only Pro/ENGINEER model will be discussed. The steps in SolidWorks are similar.

The base extrusion feature is created by sketching its profile on *DTM3* (Guideline D1a) and extruding 0.5 unit along the normal direction of *DTM3*. The sketch is drawn using two semicircles and two straight-line segments, with the dotted lines (representing *DTM1* and *DTM2*) shown in Figure 5.9 as the references (Guidelines S1b). With the center points of the semicircles aligned with the references and various sketch relations (see Figure 5.10), only three dimensions are needed to completely define the sketch: the radii of the semicircles and the vertical distance between the center points (Guideline S2a, S2c, and S2d). Because the minimal number of dimensions is employed for the sketch, the crank length design variable can be easily captured in the base feature.

There are six characteristic points generated in the profile, as shown in Figure 5.10. Hence, it requires 12 independent equations to uniquely determine the positions of the characteristic points. This profile consists of six sketch relations, as listed in Figure 5.10, and three dimensions. These 12 equations can be formulated by employing the sketch relations and dimensions, as shown in Figure 5.10. Note that in this case, these equations are linear; hence, they can be solved by matrix operations. When a design variable is changed in the sketch, the same set of equations is solved for the new positions, hence updating the profile.

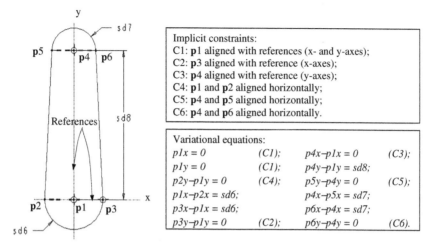

FIGURE 5.10 Variational Equations for the Sketch Profile.

As shown in Figure 5.9, the lower extrusion feature is created by drawing a circle of diameter 0.9 with its center point concentric with that of the lower semicircle of the base feature (Guideline S2b) and extruding 0.5. Similarly, the upper extrusion feature is created by drawing a circle of diameter 0.7 with its center point concentric with that of the upper semicircle of the base feature (Guideline S2b) and extruding 0.8 in the opposite direction.

By imposing the alignment and concentric rules, the crankshaft is properly parameterized, yet the number of dimensions in the crankshaft solid model is minimized. The change of crank length can be realized by simply modifying the dimension d2:0, as shown in Figure 5.11. The base extrusion feature is updated according to its sketch shown in Figure 5.10. The lower extrusion feature is unchanged because its center point is concentric with the lower semicircle of the base feature. The upper extrusion feature is pushed upward because its center point is concentric to the upper semicircle of the base

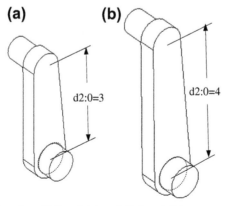

FIGURE 5.11 Design Intent Captured as d2:0 = 3 (a) and d2:0 = 4 (b).

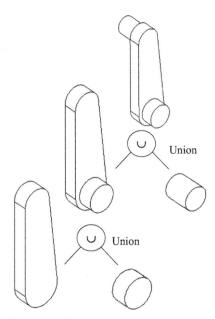

FIGURE 5.12 The Constructive Solid Geometry Tree.

feature and the center point moves up due to the references chosen. The change is propagated to features in the crankshaft through a model tree (or a CSG tree) established following the feature creation sequence and Boolean operations, as shown in Figure 5.12.

5.5 Design parameterization at the assembly level

Before reading this section, you are encouraged to review Chapter 4 for the mating constraints employed in Pro/ENGINEER and SolidWorks that support assembly modeling. With the understanding of the mating constraints, we discuss the guidelines for assembly developed following the two axioms. The slider-crank example is used to illustrate the assembly capabilities and design parameterization in both CAD tools.

5.5.1 Guidelines for design parameterization

Similar to the part level, the DIs in assembly can be uncoupled, coupled, and decoupled. The uncoupled design is again always superior to the others, according to the independent axiom. However, uncoupled DIs may not be always possible in practical applications. In general, it is required that the designer decouple the coupled DIs by adding DVs that alleviate the coupling effect, as discussed previously.

The information axiom at the assembly level can be exercised by adding relations or equations for dimensions across parts. For example, the diameter of a shaft must be related to that of the hole it inserts into to reduce the number of DVs and capture the DIs. Note that, at the assembly level, in

Table 5.2 Guidelines for Assembly Parameterization

	Independence Axiom (1)	Information Axiom (2)
Datum (D)	D1a: An assembly should always start with the default datum features (i.e., three orthogonal datum planes) and the default coordinate system. D1b: Additional datum features should be referenced to the default datum features whenever possible.	D2a: Never duplicate datum features.
Mating constraints (C)	C1a: Whenever possible, mate all components to one or two fixed components or references. Long chains of components take longer to solve and are more prone to mate errors. C1b: Do not create loops of mates. They lead to mate conflicts when you add subsequent mates. C1c: Drag components to test their available degrees of freedom and see if the design intent is captured.	C2a: Avoid redundant mates. Although SolidWorks allows some redundant mates, these mates take longer to solve and make the mating scheme harder to understand and diagnose if problems occur. C2b: Eliminate all degrees of freedom (dof), except the dof needed for kinematic analysis.
Assembly dimensions (A)	A1a: Define relations across parts to capture DI.	A2a: Minimize the number of dimensions and relations while assembling parts. The assembly options that require defining more dimensions and relations should only be used when the new dimension is a DV.

addition to complying with the two axioms, mating constraints and datum features must be properly defined to capture DIs.

A set of guidelines for assembly is listed in Table 5.2. Again, more guidelines may be added, and some of the guidelines stated in Table 5.2 may be subjective. The slider-crank assembly is employed to illustrate these guidelines in Section 5.5.2.

In addition to the guidelines stated above, the following are useful tips:

- Fix mate errors as soon as they occur. Adding mates never fixes earlier mate problems.
- If a component is causing problems, it is often easier to delete all its mates and recreate them instead of diagnosing each one.
- Whenever possible, fully define the position of each part in the assembly, unless you need the part to move to visualize the assembly motion.
- Assemblies with many available degrees of freedom (dof) take longer to solve, have less predictable behavior when you drag parts. Drag components to check their remaining degrees of freedom.

5.5.2 Slider-crank assembly in Pro/ENGINEER

At the assembly level, the intent is to orient the crankshaft vertically and align the piston and piston pin horizontally with the center point of the lower shaft of the crankshaft, as shown in Figures 5.13a and b.

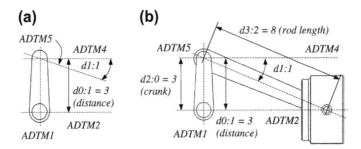

FIGURE 5.13 Parameter Relations. (a) Crankshaft. (b) Assembled Slider-crank Mechanism.

Three assembly datum planes and a datum coordinate system are given by default. The crankshaft is assembled by properly aligning its datum planes with the assembly datum planes for a vertical orientation, as shown in Figure 5.13a (Guideline D1a). In order to assemble the rod, two additional datum planes are created in the assembly. *ADTM4* is created by offsetting *ADTM2* 3 units upward, as shown in Figure 5.13b (Guideline D1b). The datum plane *ADTM5* is created by rotating *ADTM4* with an angle d1:1 = $\sin^{-1}(3/8)$. Note that *ADTM5* will be used to orient the rod (Guideline D1b). The rod is assembled to the crankshaft by three mating constraints: axis alignment, surface mate, and surface alignment, as shown in Figure 5.14.

In addition, the vertical position of *ADTM4* and the rotation angle of *ADTM5*, which determine the configuration of the assembly, will be related to the crankshaft and rod lengths through the following equations:

$$d0:1 = d2:0 \tag{5.2}$$

$$d1:1 = \sin^{-1}(d2:0/d3:2). \tag{5.3}$$

Note that Eqn (5.2) defines a relation that moves the datum plane *ADTM4* up or down according to the crank length (d2:0). Eqn (5.3) defines the trigonometric relation of angle d1:1 to the design variables d2:0 (crank length) and d3:2 (rod length). Dimension d1:1 actually rotates *ADTM5* according to the changes of d2:0 and d3:2. This is how the slider-crank mechanism is parameterized. These equations

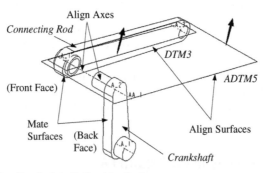

FIGURE 5.14 Assembly Mating Constraints Defined for Rod.

define two independent design variables (i.e., d2:0 and d3:2) by relating four dimensions in assembly. Therefore, the information contents of the first design intent are minimized (Guideline A2a). Details about this assembly are available in the tutorial lesson P1.3.

Note that the way that the slider-crank mechanism is parameterized, as discussed above, is not unique. They are presented for the purpose of illustrating some of the guidelines listed in Table 5.2. Other ways of parameterizing this mechanism exist. For example, instead of offsetting *ADTM2* for *ADTM4* with a dimension d0:1, a datum axis can be created by intersecting *ADTM1* and *ADTM3*, then *ADTM5* can be created by rotating *ADTM1* along the datum axis. By doing so, dimension d0:1 can be removed, and Eqn (5.2) is not necessary, thus further reducing the information contents.

5.5.3 Slider-crank assembly in SolidWorks

The slider-crank mechanism is assembled in SolidWorks in a slightly different way. Because one of the objectives in SolidWorks assembly is to conduct kinematics analysis of the mechanism, as illustrated in Figure 5.15a, a bearing part is introduced and is fixed in the assembly, as shown in Figure 5.15b. Moreover, no additional datum plane is needed to orient the rod because its orientation will be determined by SolidWorks when the crankshaft rotates.

The crankshaft is assembled to bearing using *Concentric* and *Coincident* constraints, leaving one rotational dof (please refer to Figure 4.7a in Chapter 4 for details). The connecting rod and piston pin are assembled in a similar way, also leaving one rotational dof for each part assembled (please see Figure 4.7b). The piston is assembled using one *Concentric* and two *Coincident* constraints, as shown in Figure 5.16. Note that the second *Coincident* constraint that coincides with *Plane3* of the piston and *Plane2* (horizontal plane) of the bearing confines the movement of the piston horizontally. When the length of the crankshaft or rod is changed, the assembly will be rebuilt, as shown in Figure 5.17, according to the trigonometric equation (see Figure 5.17d), with the distance d between the piston and the crankshaft fixed temporarily:

$$\alpha = \sin^{-1}\left(\frac{d2:0^2 + d^2 - d3:2^2}{2 \times d2:0 \times d}\right). \tag{5.4}$$

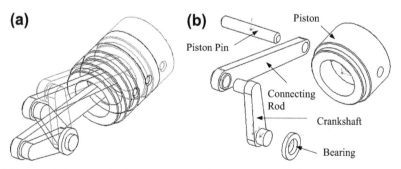

FIGURE 5.15 Slider-crank Assembly in SolidWorks. (a) Kinematic Analysis. (b) Exploded View.

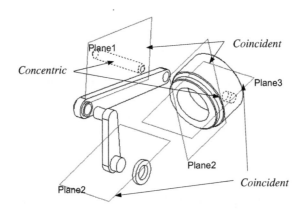

FIGURE 5.16 Mating Constraints Defined for the Piston.

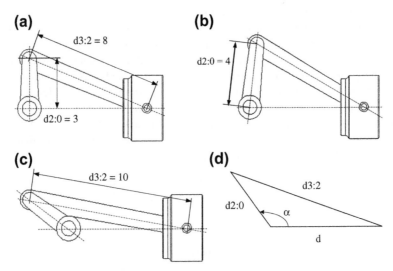

FIGURE 5.17 Change of Length Design Variables in SolidWorks. (a) Design Variables d2:0 and d3:2. (b) d2: 0 Changes to 4. (c) d3:2 Changes to 10. (d) Trigonometric Relation of the Assembly Dimensions.

5.6 Case studies

Two case studies are presented to demonstrate the parameterization method for practical applications, including a single-piston airplane engine and a HMMWV suspension. Parameterization at both the part and assembly levels are discussed.

5.6.1 Single-piston engine

Solid models of a single-piston airplane engine were created in Pro/ENGINEER, as shown in Figures 5.1 and 5.18. The models, consisting of 18 parts, were first created by a third party without adequate parameterization. Even though this original model is geometrically valid, it is not properly parameterized. As a result, any simple change will lead to invalid features, parts, and hence assembly.

Moreover, the amount of information the model carries is sometimes excessive or, in other cases, incomplete. A typical example of the use of excessive information is shown in Figure 5.19a, in which redundant dimensions are founded. For instances, d45 and d22 both specify the depth of the inner hole of the horizontal cylinder of the case, and ϕd44 and ϕd21 refer to the diameter of the same hole. In addition, as shown in Figure 5.19b, there are several dimensions with zero value as circled, which must be removed.

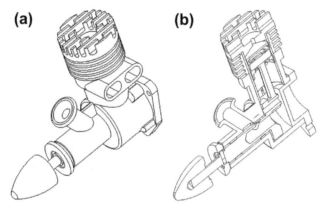

FIGURE 5.18 The Single-piston Engine. (a) Unexploded View. (b) Section View.

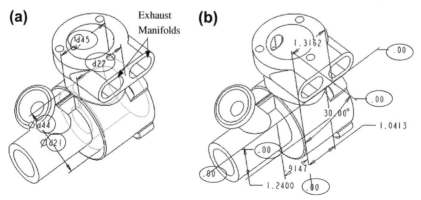

FIGURE 5.19 Excessive Information in the Original Model. (a) Redundant Dimensions. (b) Dimensions with Zero Value.

The DIs of the engine design are defined as:

DI1: The stroke length, composed of the crankshaft length d6:10 (10 represents crankshaft in engine assembly), as shown in Figure 5.20a.

DI2: The volume above the piston at top dead center position, composed of the bore diameter d46: 0, and the length of the connecting rod d0:14, as shown in Figures 5.20c and b, respectively.

Modifications to all 18 parts in the original model were carried out in order to comply with Axioms 1 and 2 at both the part and assembly levels. A typical example, the engine case, is presented to illustrate how the guidelines are followed at the part level. Note that changes in any of the three DVs will affect the engine case. The objective is to parameterize the case so that d46:0 can be changed independently, and changes in d6:10 and d0:14 will propagate to the engine case correctly.

5.6.1.1 Part level: engine case

Independence Axiom: Three default orthogonal datum planes are kept in the case solid model (Guideline D1a). Redundant datum planes that coincide with the three default datum planes are deleted (Guideline D2a). For example, *DTM1* and *DTM4* coincide with each other, and *DTM4* is deleted. Before deleting *DTM4*, all the features that were referenced to *DTM4* must be redefined by referring them to *DTM1* (Guideline D2b).

Also, a design variable should be referred to nonchanging features, such as datum planes (Guideline S1c). In the engine case, the bore is created as a hole with its axis intersected by *DTM1* and *DTM3*, as shown in Figure 5.21a. Consequently, the design variable d46:0 is not dependent on other dimensions, but refers to the nonchanging datum features.

The dimensions that are not DVs should not be available to the designer. However, they should be updated automatically via relations (Guideline P2a). Neither d21 nor d22 are DVs. The depth of the hole d22:0 is related to the bore diameter d46:0, and the diameter of the hole d21:0 depends on the length of the crankshaft d6:10. The relations and the modified solid model of the case are shown in Figure 5.22.

Information Axiom: Redundant and zero-value dimensions must be eliminated (Guideline S2g). In the engine case, both the depth and the diameter of the hole were redundantly defined in the original model, as shown in Figure 5.19a. Even though d44 and d21 have the exact same value, they belong to two different features. Dimension d44 is eliminated by aligning the circle defined by d44 to the circle

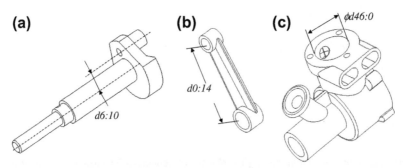

FIGURE 5.20 Design Variables Defined for the DIs. (a) Crankshaft. (b) Connecting Rod. (c) Engine Case.

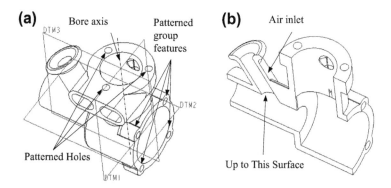

FIGURE 5.21 Engine Case. (a) Datum Planes. (b) Air Inlet (Section View).

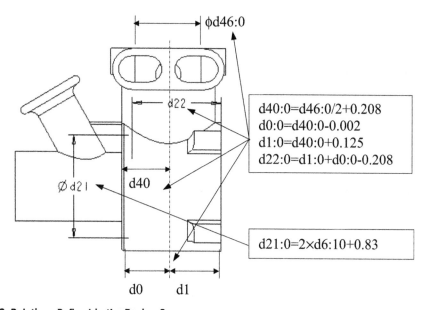

d40:0=d46:0/2+0.208
d0:0=d40:0-0.002
d1:0=d40:0+0.125
d22:0=d1:0+d0:0-0.208

d21:0=2×d6:10+0.83

FIGURE 5.22 Relations Defined in the Engine Case.

defined by d21 (Guideline S2c). A similar problem arises for dimensions d45 and d22. Dimension d45 is eliminated by resorting to alignment (Guideline S2c). In addition, all the zero-value dimensions are removed because they do not compose any of the DIs (Figure 5.19b).

The air inlet is defined as a through hole, instead of a hole with depth up-to-surface (Guideline F2b). Consequently, it will always be a through hole when a design change is committed. Also, copy and mirror are used in the creation of one of the exhaust manifolds (Guideline F2a), as shown in Figure 5.21a. It is desired that both manifolds maintain the same dimensions. In the original model, they were defined independently. In this case, one design change must be implemented twice, which is unnecessary and error-prone. The same guideline is applied to pattern the three holes on top of the case (see Figure 5.21a). In addition, the group of extrusion, round, and hole features, as shown in Figure 5.21a, is patterned for additional three instances. Furthermore, the dimensions are properly related (see Figure 5.22) to capture the DIs (Guideline P1a).

5.6.1.2 Assembly level: engine

The other 17 parts are also parameterized following the guidelines at part level for the two DIs. At the assembly level, guidelines are followed for assembling the 18 parts.

Independent Axiom: Default assembly datum features, including the three orthogonal datum planes and coordinate system, are used as references (Guideline D1a). Proper mating constraints are employed to assemble all 18 parts without looping (Guideline C1b).

Information Axiom: Only two rotational dof (between connecting rod and piston pin and between crankshaft and connecting rod) are kept for kinematic analysis. All the other dof are eliminated (Guideline C2b). Dimensions are related across the parts (e.g., d21:0 in Figure 5.22) to minimize the contents of the DIs (Guideline A2a). Moreover, relations are created for the length of the piston pin, diameter of the piston, and the bore diameter of the case. The DIs are properly captured at the assembly level, as illustrated in Figure 5.1.

5.6.2 HMMWV suspension

The HMMWV solid model, discussed in Chapter 1, was initially created in Pro/ENGINEER and then converted to SolidWorks, partially for testing the process of solid model conversion between CAD systems. There are more than 200 parts and assemblies (see Figure 5.23a). The suspension is modeled in detail (Figure 5.23b) such that the main characteristics of the vehicle performance can be captured accurately in motion simulation. A more detailed view on the front right suspension quarter is shown in Figure 5.24a.

A dynamic simulation model was created for dynamic simulation on a bumpy 100 ft × 100 ft terrain, as shown in see Figure 5.24b. The vehicle vibrates significantly towards the later stages of the simulation due to the bumpy road conditions. The overall design goal was to optimize the vehicle dynamic characteristics (Chang and Joo, 2006).

(a) **(b)**

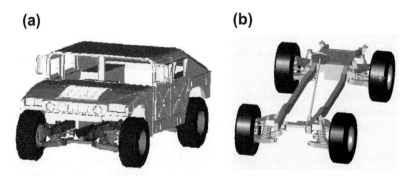

FIGURE 5.23 HMMWV CAD Model. (a) Vehicle Assembly. (b) Suspension Assembly.

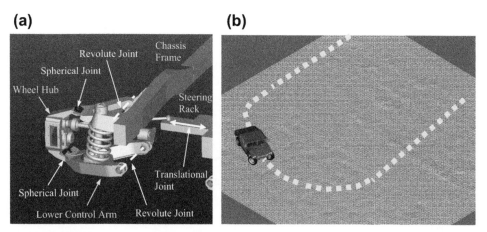

FIGURE 5.24 HMMWV Dynamic Simulation Model. (a) Front-right Suspension. (b) Vehicle Motion Simulation.

The vehicle track and wheelbase shown in Figure 5.25 are the two primary design variables defined for HMMWV suspension. In order to support HMMWV design optimization, the suspension assembly must be parameterized in CAD (in this case, SolidWorks). The design parameterization must be conducted at both part and assembly levels.

5.6.2.1 Track design variable

For the track design variable, two parts are involved: differential (Figure 5.26a) and steering rack (Figure 5.26b). The geometry of both parts is simple, and their width dimensions are to be related to capture the track design variable. The outer width of the differential d2@sketch1, as shown in Figure 5.26a, is chosen as an independent dimension. All the geometric features in the differential will be changed according to d2@sketch1 following the relations defined in Table 5.3. The relations show

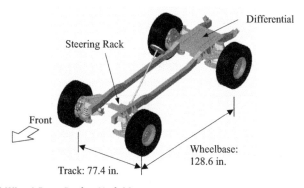

FIGURE 5.25 Track and Wheel Base Design Variables.

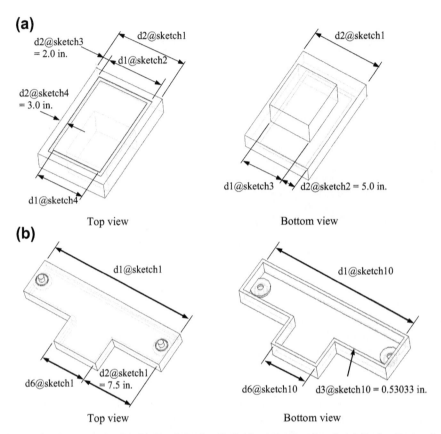

FIGURE 5.26 Design Parameterization for Track Design Variable at the Part Level. (a) Design Parameterization for the Differential. (b) Design Parameterization for the Steering Rack.

Table 5.3 Relations Defined for the Differential

Equations	Design Intents
d1@sketch2 = d2@sketch1−2×d2@sketch3	d2@sketch1 is independent, d2@sketch3 = 2.0, and is fixed.
d1@sketch3 = d2@sketch1−2×d2@sketch2	d2@sketch1 is independent, d2@sketch2 = 5.0, and is fixed.
d1@sketch4 = d2@sketch1−2×d2@sketch4	d2@sketch1 is independent, d2@sketch4 = 3.0, and is fixed.

that d1@sketch2, d1@sketch3, and d1@sketch4 will be changed according to d2@sketch1. In addition, d2@sketch3, d2@sketch2, and d2@sketch4 are fixed. Note that in the equations of Table 5.3, dimensions shown on the left hand side of the equal sign become dependent.

For the steering rack shown in Figure 5.26b, dimension d1@sketch1 is chosen as independent, and d6@sketch1 will be changed with the same amount as d1@sketch1, as defined by the first equation listed in Table 5.4. Dimensions d1@sketch10 and d6@sketch10 are related to d1@sketch1 and d6@sketch1 via the last two equations shown in Table 5.4, respectively, with a fixed wall thickness of d3@sketch10 = 0.53033 in.

At the assembly level, mating constraints are defined for the differential and both frame rails, as shown in Figure 5.27a. First, the side faces of the differential and frame are assembled using surface

Table 5.4 Relations Defined for the Steering Rack

Equations	Design Intents
d6@sketch1 = d1@sketch1 − 2 × d2@sketch1	d1@sketch1 is independent, d2@sketch1 = 7.5 in., and is fixed
d1@sketch10 = d1@sketch1 − 2 × d3@sketch10	d1@sketch1 is independent, wall thickness d3@sketch10 = 0.53033 in. fixed
d6@sketch10 = d6@sketch1 − 2 × d3@sketch10	Wall thickness d3@sketch10 = 0.53033 in. fixed.

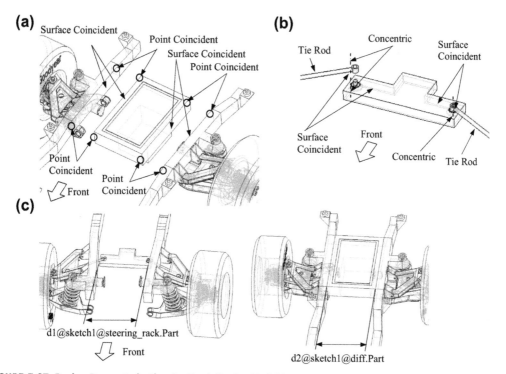

FIGURE 5.27 Design Parameterization for Track Design Variable at the Assembly Level. (a) Mating Constraints Defined between the Differential and Frame. (b) Mating Constraints Defined between Steering Rack and Tie-rods. (c) Relation between Widths of the Differential and Steering Rack.

coincident (mate) constraints. In addition to surface coincident constraints, point coincident constraints are added between the corner points of the differential and points on the top edge of the frame rails. The steering rack is assembled to the tie-rod on each side by using concentric (axis alignment) and surface coincident (mate) constraints, as shown in Figure 5.27b.

Next, the relationship between the width of the differential and width of the steering rack is defined at assembly level, as shown in Figure 5.27c. The relationship between dimensions d1@sketch1 in the steering rack and d2@sketch1 in the differential is defined as d1@sketch1@steering_rack.Part = d2@sketch1@diff.Part, so that widths of the steering rack and differential change simultaneously. Therefore, d2@sketch1@diff.Part represents the track design variable which is independent. Note that d1@sketch1@steering_rack.Part and d2@sketch1@diff.Part have the same numerical value.

5.6.2.2 Wheelbase design variable

Defining the wheelbase design variable is straightforward. It involves changing the length of the two center frame rails at the same time (Figure 5.28a). The center frame rails are assembled to the rear frame using surface coincident constraints as well as point coincident constraints at the end faces of the frame (Figure 5.28b). Similar constraints are defined for assembling the center frame rails to the front frame. A relation d1@sketch5@right_frame.Part = d1@sketch3@left_frame.Part is defined to capture the wheelbase design variable represented by d1@sketch3@left_frame.Part. As a result, when d1@sketch3@left_frame.Part is increased, the rear portion of the vehicle gets pushed backwards, and vice versa. Note that when the track or wheelbase design variable is changed, both mass properties and joint locations of the HMMWV vehicle model are altered, therefore varying the vehicle dynamic performance. In this case, both the track and wheelbase design variables are called global design variables.

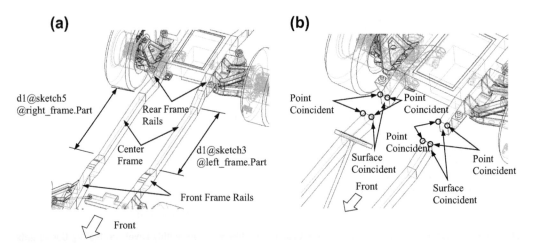

FIGURE 5.28 Design Parameterization for the Wheelbase Design Variable. (a) Relation between Two Center Frame Rails. (b) Mating Constraints Defined for the Center and Rear Frame.

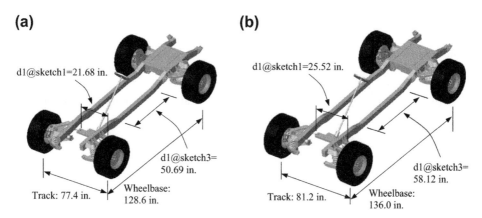

FIGURE 5.29 HMMWV Suspension at Initial Design (a) and Optimal Design (b).

5.6.2.3 Design change

The change of HMMWV suspension geometry due to the change of these two design variables is shown in Figure 5.29. The suspension is properly parameterized. The parameterized HMMWV model was employed to search for a design that optimizes the dynamic characteristics of the suspension (Chang and Joo, 2006).

5.7 Summary

In this chapter, solid modeling and assembly techniques implemented in Pro/ENGINEER and SolidWorks were discussed. Usually in e-Design, a design change is first realized by modifying geometric dimension values in CAD and automatically regenerating or rebuilding the solid models. To capture a product DI in CAD, the product solid model must be properly created and parameterized. With this understanding, a design parameterization method that supports the capture of design intents has been introduced.

Design intent for a single part can be captured by properly creating individual solid features and carefully relating dimensions within or among features. Consequently, when a dimension value is changed, the solid features affected by the change can be regenerated or rebuilt successfully. At the assembly level, DI is captured by defining adequate mating constraints and relating dimensions across parts so that a change in dimension value can be propagated to all parts affected. The parts affected must be regenerated successfully; at the same time; they must maintain proper positions and orientations with respect to one another without violating any assembly mating constraints nor revealing part penetration or excessive gaps. Moreover, the regenerated solid model must meet the designer's expectations.

Design parameterization guidelines based on the independent and information axioms have been introduced. These guidelines will facilitate the creation of parametric solid models that support design engineers in exploring design alternatives in the e-Design environment.

Questions and exercises

1. Show how the upper limit of the height dimension $H = 12.3$ in Example 5.1 is determined based on the geometry of the sketch profile. If H is set to 12, determine the upper and lower limits of the width design variable W.

2. Use the given three parts (link1, link2, and link3) to create an assembly like the one shown below.

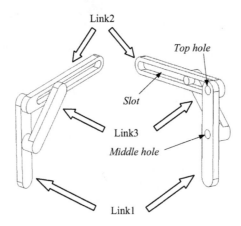

Required configuration:

- Link1 must be vertical;
- Link2 must be horizontal, and the shaft on link2 must insert to the top hole in link1.
- The two shafts of link3 must insert to the middle hole of link1 and slot of link2.

(i) Show/explain what mating constraints you employed for this assembly.
(ii) Define the vertical location of the middle hole of link1 as the design variable, as shown in the next figure, and find the following:

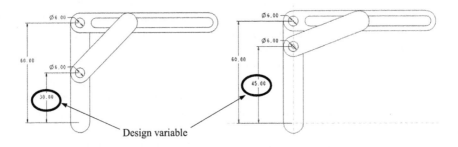

Upper bound of the design variable without interference (showing how you reached the answer):

_____

Lower bound of the design variable without interference (showing how you reached the answer):

_____

References

Chang, K.H., Joo, S.-H., July 2006. Design parameterization and tool integration for CAD-based mechanism optimization. Advances in Engineering Software 37, 779–796.

Chang, K.H., Silva, J., Bryant, I., December 1999. Concurrent design and manufacturing for mechanical systems. Concurrent Engineering Research and Applications (CERA) Journal 7 (4), 290–308.

Lee, K., 1999. Principles of CAD/CAM/CAE Systems. Addison Wesley Longman, Inc. ISBN: 0-201-38036-6.

McMahon, C., Browne, J., 1998. CADCAM, second ed. Addison-Wesley. ISBN: 0-201-17819-2.

Mortenson, M.E., 2006. Geometric Modeling, third ed. Industrial Press.

Suh, N.P., 1990. The Principles of Design, Oxford Series on Advanced Manufacturing. Oxford University Press, New York, NY.

Toogood, R., Zecher, J., 2012. Creo Parametric 1.0 Tutorial and MultiMedia DVD. SDC Publication.

Zeid, I., 1991. CAD/CAM Theory and Practice. McGraw-Hill, Inc. ISBN: 0-07-072857-7.

Product Data Management

CHAPTER OUTLINE

Product Design Modeling using CAD/CAE.

Product data management (PDM) is the technology and associated software systems that support the management of both engineering data and process information during the product development phase and beyond. Engineering data management involves organizing, structuring, storing, and tracking the product information created by a design team while conducting engineering and product development activities. PDM aims at providing product design teams with the right data and information at the right time for making proper design decisions. There are significant benefits offered by PDM, such as interdisciplinary collaborations, reduction of product development time, reduction of the complexity in information access, and improvement of project management.

PDM is an essential subject in e-Design and a broad topic in scope. To focus the discussion, this chapter is organized with an emphasis on engineering practice and aimed at providing practicing engineers and engineering students with a brief introduction to the topic, as well as a fine understanding of the use of PDM systems to support engineering design. We narrow our scope with more emphasis on product design and less on product life cycle management (PLM). PLM is sometimes interchangeable with PDM. However, PLM is a subject of substantially larger scale. In general, PDM focuses on managing product design data as it relates to the product development phase, whereas PLM centers on reengineering product development and manufacturing processes as they relate to product life cycles. PDM is a design-focused technology that increases efficiencies within existing product development processes by improving the management of product design data. PLM, on the other hand, is a strategic, process-centered approach that leverages PDM and other technologies to manage product life cycles, remake processes, and increase output. As a result, PLM aims at improving productivity across the enterprise rather than in a single department or a specific process (Dassault

Systèmes, 2010). For a complete discussion of PLM, readers are referred to textbooks such as Stark (2011). In this chapter, our discussion stays mainly within the scope of the e-Design paradigm introduced in Chapter 1.

In addition to the discussion of PDM, we include in this chapter a highly practical and important issue in PDM—product data exchange, or more specifically, solid model translations between heterogeneous or dissimilar computer-aided design (CAD) systems (also called the interoperability issue). Solid model translations between heterogeneous CAD systems are still an ongoing research topic. We introduce the means currently available, as well as their strengths and shortfalls. Among the possible approaches, solid feature recognition (FR) has been one of the most recent developments; it provides the best possible support to address the interoperability issues within the context of e-Design. We provide a brief discussion on the underlying technology and present examples in the tutorial lessons of Projects S1 and P1 for SolidWorks and Pro/ENGINEER, respectively.

Overall, the objectives of this chapter are: (1) to provide a brief overview on PDM that introduces students to this research area and helps practicing engineers gain an understanding of PDM, (2) to present an overview of PDM software systems so that readers may explore options for software selections when an opportunity comes, (3) to discuss product data exchange and help readers understand the interoperability issue and available means for addressing it, and (4) to offer tutorial lessons that support readers in properly handling the CAD model translation issues.

6.1 **Introduction**

As discussed in Chapter 1, the e-Design paradigm and tool environment supports a cross-functional team for product design and development. One of the key advantages of e-Design is the intensive product data and knowledge gained in the early design stage that support better design decision-making, thus breaking the Ullman's design paradox.

In general, the amount of product data generated during the design phase is substantial. The data have the characteristics of being tentative and iterative, and intermediate with heterogeneous formats and complex relationships. Moreover, the product data often evolve along the design cycle because product development often takes significant time, especially for a complex system. The design logics and tools may vary with the development of science and technology, which leads to the revisions of data, files, and parameters. The product design team is often geographically distributed, which adds to the complexity in the management and access of product data and information. Therefore, the efficient organization and management of the massive product data becomes essential in support of product development in general, particularly when using the e-Design paradigm.

PDM is the technology and associated software systems that support the management of both engineering data and process information during the product development phase and beyond. PDM involves organizing, structuring, storing, and tracking the product information created by the product design team as they carry out engineering and development activities. With the explosion of engineering knowledge and advancement in computer-aided software tools for engineering design, PDM becomes indispensable in product development and is essential in ensuring an effective and efficient product development process. Overall, PDM aims at providing the product design team with the right data and information at the right time, and more importantly in the right form, for carrying out engineering assignments and making proper design decisions.

In the early 1980s, many large corporations, often the original equipment manufacturers (OEMs), realized their efficiency was severely downgraded by paper-based systems. With no commercially available systems at that time, they had no choice but to develop their own data management solutions. In the late 1980s, a number of software companies started to realize the potential market of efficient data management systems and began to introduce the first generation of commercial PDM systems (Liu and Xu, 2001). The majority of those vendors at the time were already involved in the CAD/computer-aided engineering (CAE)/computer-aided manufacturing (CAM) software market, so PDM development was a natural extension of their products and service to existing customers. They focused on developing data management solutions and added PDM to their product lines (Hepplemann, 1998). Since the late 1990s, the focus has been shifted to the improvement of the product life cycles—that is, the PLM—with an aim at improving productivity across the enterprise rather than in a single department or a specific process. Also, the on-premises software from the early years has been gradually replaced with the new, alternative deployment-and-use model: the so-called cloud-based or Software as a Service (SaaS), which typically uses the Internet to remove the need for the user to install any software on premises. Such software offers benefits, such as running software remotely, which can result in considerable cost savings because of reduced staffing, maintenance, and other factors.

PDM systems are increasingly being used in industrial applications for long-term archival of product information, as well as to enhance collaboration and communication throughout the design process, support distributed design teams through advanced document sharing, track changes in product information, and control design documents (ranging from requirements information to CAD). The adoption of PDM systems has caused a change in how design processes are managed and how individual designers collaborate. Additionally, companies are pushing the limits of currently available PDM software, resulting in continual development and new application domains. For example, commercial software vendors have integrated PDM systems with other design support tools, included automated workflow management and suites of CAD/CAM/CAE tools, and refined and developed new functionality (Caldwell and Mocko, 2008).

Today, PLM is primarily used in the automotive and aerospace industries, as well as in the machinery industry (Lee et al., 2008; Abramovici and Sieg, 2002). For examples, GM credits PLM initiatives with decreasing time-to-market from 48 to 18 months (Tang and Qian, 2008). Automotive industry leaders such as Autoliv, Eaton, Honda, and Johnson Controls are driving success by using the MatrixONE solutions (Tang and Qian, 2008). Regarding the importance of PLM to the automotive industry, Reale and Burkett concluded that "the smarter the car, the more automakers need PLM" (Tang and Qian, 2008, p. 288). Among many successful stories, we include two of the most notable in Section 6.3.4 to illustrate a few insights.

Although PLM is meant to manage product information throughout the entire life cycle of a product, an international study revealed that the adoption of PLM is still mainly limited to product design (Abramovici and Sieg, 2002). Today, PLM is an active research topic, especially in supply chain integration and the integration of business process into the overall product life cycle development. Nevertheless, in practice, PDM and PLM are often interchangeable, particularly from a product development perspective.

In practice, product data is largely embedded in files. Many student teams may not have access to full-range PDM or PLM systems and must reply on ad-hoc approaches for file management. We include a number of commonly employed approaches to support file management in Section 6.2.1. We also include a case study in Section 6.6.1 to illustrate the use of SolidWorks Workgroup PDM, a mid-range PDM system, which is commonly available to students.

If you work with multiple CAD systems, you might need to translate solid models from one CAD system to another for numerous purposes. You may need to bring parts from CAD system A to CAD system B in order to conduct Finite Element Analysis (FEA), generate a machining toolpath, or perform other engineering activities. You may be given an assignment to bring in parts and sub-assemblies from other CAD systems to the major CAD software your company is using in order to generate a complete product model in the designated CAD system. In industry, the OEM integrates and communicates with their suppliers, during which they may encounter the issue that the CAD models created by suppliers may not be compatible with the major CAD system used by the OEM. CAD model translation (also called interoperability among CAD systems) is a practical and essential issue in product data integration for engineering design. The question is how to handle CAD model translations. What are the available options? How practical are these options? Therefore, in addition to discussing the PDM, we address this practical and important interoperability issue encountered in product data exchange—that is, CAD model translations between heterogeneous (or dissimilar) systems that employ different geometric kernels.

In this chapter, we address two issues that are very relevant to product design. First, we provide an overview on the practical means for file management and an introduction to PDM technology and systems. We start by discussing file management in Section 6.2. In Section 6.3, we present the fundamentals of PDM, including the product and process data that PDM manages, the functionalities of PDM, and the benefits and successful stories of PDM technology. In Section 6.4, we discuss commercially available PDM systems. Then, in Section 6.5, we shift focus to the second issue—the product data exchange. We discuss the viable options and available model translators, as well as their strengths and limitations. We include examples that outline the part and assembly model translations between Pro/ENGINEER and SolidWorks, as well as feature recognition (FR) in both Pro/ENGINEER and SolidWorks. In Section 6.6, we offer case studies that illustrate the practical use of PDM, including SolidWorks Workgroup PDM.

6.2 File management

During the product development phase, a large amount of data is generated. This includes product data and process data in the forms of files, documents, and diagrams, etc. Typical product-related data include CAD geometry, engineering drawing, specifications, project plans, part and assembly files, bill of materials (BOM), engineering simulation data, engineering change requests, and so forth, which are shared throughout the product development phase by the product development team and by the extended enterprise. The product-related data and information are stored in the forms of paper documents, digital files, and information extracted and stored in databases. In the context of e-Design, digital files are the most common and largest in quantity, including document files such as specifications, configuration, and purchase orders; product models, such as CAD drawings, part files, and assembly files; CAE analysis model and result files; and manufacturing related information, such as numerical control (NC) programs.

The management of information in modern engineering design projects is typically characterized by the following (Caldwell and Mocko, 2008):

- A large amount of digital product information, generated by different engineering software, is often stored in a variety of formats.
- Documents often go through several revisions during the product development phase, which may be initiated and completed by different designers and engineers of different disciplines. Access to documents must be controlled across members of the design team.
- Documents are highly interrelated, such that the changes in a single document may be propagated through several other documents.
- Data sharing and design collaboration take place between design team members in distributed locations.

Nevertheless, the most basic function of a PDM system is digital file management and sharing. In any team design projects, large or small in scale, file management and sharing among the team members is critical; it should be the first issue addressed at the onset of the project. PDM certainly offers excellent capabilities in support of file management. However, in academic environments, student teams may not have access to full-range PDM systems to support their design activities. Therefore, in this section, we first discuss file management without a PDM and then provide a short introduction to the file management aspect of PDM systems. In Section 6.6.1, we discuss SolidWorks Workgroup PDM, which is part of SolidWorks Premium and Professional package and is popular among engineering students.

6.2.1 Ad-hoc methods

Product information can be shared by using a variety of technologies, including email, web-based workspaces (i.e. Google Groups, Yahoo Groups), shared network drives with file management systems, and PDM systems, to name a few. In this subsection, common ad-hoc methods are discussed for sharing product data embedded in digital files. These methods may be useful to student teams for exchanging product information in collaborative design projects.

Ad-hoc methods of file management work reasonably well for self-contained files, such as Word documents and Excel spreadsheets. However, they break down quickly for more complex CAD file management. Some of the most common methods (Buchal, 2006) include email file attachments, peer-to-peer file sharing (e.g., Windows Messenger), removable media (CD-R or USB drive), FTP, shared network folders, web folders, and Microsoft SharePoint.

Email file attachments, peer-to-peer file sharing, and removable media are simple and widely employed. However, this type of management is difficult to maintain and does not fully support multiple users accessing data from distributed systems. The main issue is that distributing a file to multiple recipients immediately creates multiple instances of a file, with no mechanism for version control or reconciliation. In some cases, recipients may encounter problems in reviewing files. For example, if a CAD assembly file is sent, it may not be opened because the referenced component files may not be located properly due to incorrect drive letter and/or folder path.

Files may be shared using standard network communication and shared file folders, such as shared network folders, web folders, and Microsoft SharePoint. These approaches provide a simple way to

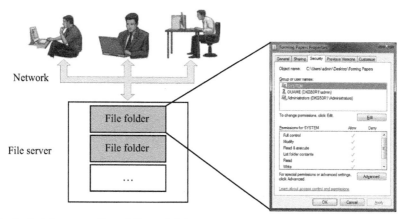

FIGURE 6.1 Controlled Access to Shared Network Drives.

share files within a team, either via a network or the web browser, while offering access control and basic check-in/check-out. For example, using network folders, if a user has a file open for editing, it becomes available as read-only to other team members. Additionally, it is possible to control file and folder using security properties in Windows operating systems (see Figure 6.1). Access is restricted to computers connected to a Windows LAN domain, so students with laptops or home computers may not access the shared folder. On the other hand, anyone with an Internet connection and user account can access web folders. Web folders provide access control and allow editing in place without downloading and uploading. Some software supports accessing to files via network, such as the Open from Web Folder dialog box in SolidWorks. The assembly files and related part files are all located on the web server, and they are accessed using a URL rather than a drive letter and network path. However, SolidWorks web folders maintain product structure, but offer no version management or check-in/check-out capabilities.

SharePoint is a web-based shared workspace with many collaboration tools, which provides access control, check-in/check-out, revision management, and many other collaboration capabilities. SharePoint is not designed to manage product structure, so files can only be opened or downloaded individually (Caldwell and Mocko, 2008). Figure 6.2 shows a view of a SharePoint document library containing SolidWorks part and assembly files (Buchal, 2006). It is worth noting that SharePoint document libraries can be accessed as web folders from SolidWorks, but SharePoint version management and check-in/check-out are not integrated with SolidWorks.

Another important issue encountered by a project team without access to a PDM system is viewing the product model. On many occasions, team members need to view CAD models without the ability to change them or to incorporate them into other models. Some viewers, developed by CAD vendors, can open and view CAD files in their respective native formats and more; for example, NX viewer views NX part and assembly, I-DEAS files, Parasolid, and JT files, whereas the SolidWorks viewer views SolidWorks part and assembly and JT files. Viewers allow designers to zoom, pan, and rotate models. Some viewers allow users to take sections with multiple section planes. Some support users to add notes, text with leaders, and dimensions. Most viewers incorporate markups and save them as JT format.

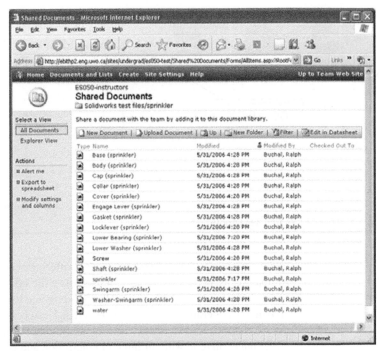

FIGURE 6.2 Document Library View in SharePoint (Buchal, 2006).

Many viewers support JT files, which is a three-dimensional (3D) data format developed by Siemens PLM Software (formerly UGS) and used for product visualization, collaboration, and CAD data exchange to some extent. JT viewer (Figure 6.3) is free for download and was recently extended to support iPad, iPhone and iPod Touch (JT2Go), which moves mobile engineering design one step forward.

Although ad-hoc approaches to file management support design teams in managing files and sharing design data to some extent, they are less desirable in practice. There is not a standardized revision system. Revisions made to documents are not structured. Changes between documents are not explicitly captured, thus losing the evolution and refinement between documents. There is not a means for controlling specific aspects of a file and/or contents within the file. In addition, standardized locations that are accessible by all designers may not exist.

In the next section, PDM systems are presented as a means for addressing the problems associated with information management using a traditional file-based approach. Please note that, in general, PDM has a lot more capabilities than just managing files.

6.2.2 PDM approach

PDM systems attempt to address file management issues by (1) structuring existing document meta-information, (2) adding meta-information, and (3) enforcing rules for creating, accessing, sharing, and modifying documents. Specifically, PDM syste ms provide greater control for enforcing effective

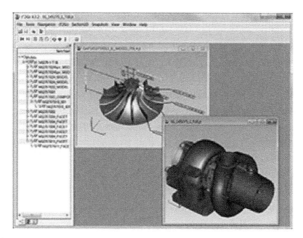

FIGURE 6.3 Computer-Aided Design Viewers. JT Viewer (www.plm.automation.siemens.com/en_us/products/teamcenter/lifecycle-visualization/jt2go).

document management practices through the use of revisions, locations, editors, owners, and much more information about the data stored within a file. Metadata, in a digital context, is the data used for describing the file or the content of a document. The meta-database in a PDM system controls the document's relation to other documents and the rules for how the system links information. The basics of how PDM systems work in managing files are illustrated in Figure 6.4.

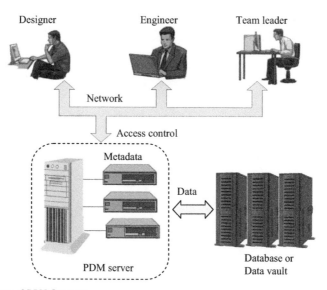

FIGURE 6.4 Architecture of PDM Systems.

PDM systems offer functionality at the server side and the client side. Typical PDM systems provide controlled access to document vaults using secured login. Individual designers have controlled access ability to check-in/check-out documents to ensure changes are not being concurrently made resulting in document conflicts. They also support workflow management and trigger for automated document tracking and notification to design team members. PDM systems provide a means for flagging changes for affected team members for document review, approval, and release. They provide storage of documents in a shared location with the ability to create specialized locations for projects and groups, and they support the query and retrieval of documents based on richer document descriptions. They also standardize revision schemes, which is essential for data management. In addition to enabling document sharing, PDM systems are typically integrated closely with CAD software as integrated add-ons or stand-alone applications. More about commercial PDM systems is discussed in Section 6.4.

6.3 Fundamentals of PDM

PDM forms the product information backbone for a company and its extended enterprise, which allows the cross-functional team to contribute throughout the product design and development phase. In addition to file management, PDM can be viewed as a data or information integration tool connecting different functional areas. It ensures that the right data and information are available to the right person at the right time—and more importantly, in the right form. In addition, PDM improves communication and collaboration between groups of diverse functions and engineering expertise in the enterprise. The area of application of a typical PDM system is shown in Figure 6.5 (CIMdata, 1998).

Although it is highly desirable that knowledge is accessible when a design decision is to be made, in reality, knowledge is not directly available but is obtained by interpretation of information deduced from analysis of data. In general, data is available to an organization in the form of observations, computational results, and factual quantities. Interpretation, abstraction, or association of this data leads to generation of information. Finally, knowledge is obtained by experiencing and learning from this information and putting it into action (Owen and Horváth, 2002). In fact, looking at engineering

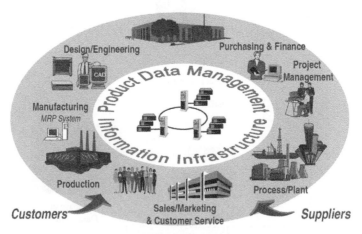

FIGURE 6.5 Application of a PDM System in an Organization (CIMdata, 1998).

design from a teleological point of view, it can be said that the primary function of engineering design research should be to transform empirical or rational knowledge into a form that can be used for practical deployment (Horváth, 2004). Design information extraction and knowledge management is an ongoing research topic. Interested readers are referred to excellent review articles as an introduction, such as Chandrasegaran et al. (2013).

Our goal in this section is to provide readers with a fundamental understanding of PDM from the context of e-Design. In Section 6.3.1, we discuss the product data model that describes the data and its structure to be managed in support of e-Design. In Section 6.3.2, we discuss the basic functions of a PDM system that supports the design team in managing the product data. In Sections 6.3.3 and 6.3.4, we present the benefits that PDM offers and the impact of PDM to industry, respectively.

6.3.1 Engineering data models

Engineering data models, which consist of product data models and process data models, provide the engineering team with a consistent and unified engineering data set that supports engineering activities for product development. A product data model is evolving throughout the product development process and beyond; for example, a revolution of CAD models of a high-mobility multipurpose wheeled-vehicle (HMMWV) throughout the design phase is shown in Figure 6.6. On the other hand, a process data model is relatively less involved.

6.3.1.1 Product data model

In general, the size and contents of the product data can be different from one product to another, depending on many factors, such as the nature of the product being developed, the design process employed, and the tools and technology adopted for the product development, among others.

From the data authoring perspective, product data can be categorized into three types: documents, files, and data or parameters. Documents are usually authored by engineers, such as product requirement and specifications, organizational structure of the product development team, major milestone and workflow, reports, guidelines, standards, and manuals. Documents can be in the form of Microsoft Word, PowerPoint, Excel, or Project files, or as Adobe PDFs. In addition, pictures and videos in numerous formats are created to support visual aid of the product in the design phase. In e-Design, a large amount of files are created by software tools. Geometric model files, including part and assembly files, are created by CAD and exported in other formats, such as IGES (IGES, 2001) or STL (en.wikipedia.org/wiki/STL_(file_format)), for numerous purposes. Simulation model files are

(a)   **(b)**   **(c)**

FIGURE 6.6 HMMWV CAD Model for Concept and Detailed Designs. (a) A 15-Part Concept Model, (b) An 18-Part Model for Intermediate Design, and (c) A Detailed Design Model with More Than 200 Parts (Chang et al., 1998).

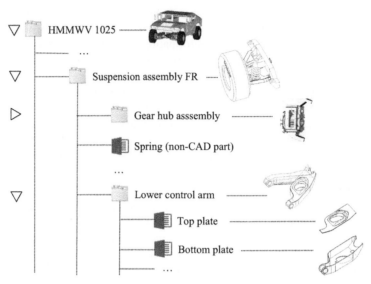

FIGURE 6.7 Example of Bill of Materials for an HMMWV.

created for engineering analysis, such as FEA, and the result files are generated by the respective FEA or simulation software tools. In addition, toolpath files and machining model are generated using CAM software. Data or parameters extracted from files or documents or entered by the product development team are stored in the database, such as the BOM of the product extracted from the CAD model of the product.

All these documents, files, and data must be structured and organized to support the needs of product design, including product data model and process data model. There are many different ways to structure and organize the data, as long as the data models are logical and facilitate the product development team in accessing product information in a timely manner. One of the most common ways to structure and organize product data is using the BOM s. For example, a BOM of the HMMWV is shown in Figure 6.7, in which the product is broken down into parts and subassemblies. Each entity is given a name (and identification number) and icon that links to more data and information.

Data can be linked by using, for example, HyperText Markup Language (HTML) for creating web pages and other information that can be displayed in a web browser. For example, the BOM shown in Figure 6.7 is displayed in a scrollable window (shown in Figure 6.8a with each entity name implemented as a hypertext). Once clicked, its associated data are displayed in the window below. For example, if Suspension Assembly FR is clicked, its geometry is brought into a viewer (left window) and its properties (e.g., mass properties) appear in the right window. Engineers may pan, rotate, and zoom in/out the model in the viewer to gain a better understanding of the model geometry and its constituent components. In addition, right-clicking an entity brings up its associated page for more information. For example, right clicking the HMMWV 1025 at the top of the BOM in Figure 6.8a brings up the product development page shown in Figure 6.8b. In this page, the introductory

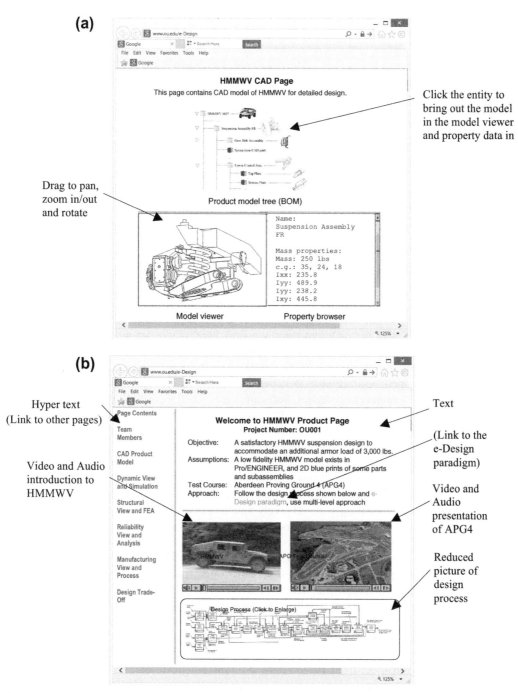

FIGURE 6.8 Sample Web-Browser Pages of PDM of a Product Data Model. (a) HMMWV CAD Model Page, and (b) HMMWV Product Page.

information about the design project, such as objective and assumption, are listed, together with videos offering visual aid to the HMMWV physical model and the test course. In addition, an overall design process is defined and shown in the bottom half of the page. Hypertext on the left connects to other pages, such as connecting back to the CAD model page shown in Figure 6.8a.

A typical PDM system supports such product and process data models in some way. Even without a commercial PDM system, implementing such web pages for support of PDM is not too difficult. For those who took courses such as engineering multimedia, creating data management pages like those of Figure 6.8 is generally straightforward. The model viewer on the lower left portion of the CAD model page (Figure 6.8a) can be implemented using JT Viewer (http://www.plm.automation.siemens.com/en_us/products/teamcenter/lifecycle-visualization/jt2go) or Virtual Reality Modeling Language (VRML; en.wikipedia.org/wiki/VRML) viewers, for example. You may publish your solid models (part or assembly) as JT or VRML models. These files can be stored on a website or emailed as an attachment. Anyone can view the JT or VRML files by using a free plug-in to their browser, such as Cosmo VRML Player for Windows; view3dscene for Windows, Macintosh, and Linux; or JT viewer for JT models.

The BOM shown in Figure 6.7 is the most basic product data model, which provides information that helps team members gain a first-level understanding of the product being developed at a given time. In e-Design, the product data model should support follow-up activities, including engineering modeling and analysis, manufacturing process and machining simulations, and design trade-offs. A BOM is just a start, and a BOM alone is not sufficient to support all design activities.

How do we proceed with the e-Design paradigm from here? There are at least three options: (1) using commercial CAD/CAE/CAM, (2) using commercial PDM, and (3) developing one's own tool and information integration infrastructure. The option of using commercial CAD/CAE/CAM suite is for those who have access and are able to depend on the engineering capabilities offered in the software for support of engineering design. Software suites, such as SolidWorks (with SolidWorks Motion, SolidWorks Simulation, and CAMWorks), Pro/ENGINEER (with Pro/MECHANICA Structure, Pro/Mechanism, and Pro/MFG), or CATIA (with FEA and CAM modules) may be used for this option. In this case, the design team may proceed from CAD to CAE and CAM in a straightforward fashion because the transition from CAD solid models to CAE simulation or CAM toolpath generation is seamless. There is no need to do anything extra when going forward to carry out CAE and CAM activities. Most PDM is taken care of by the commercial software. This option is ideal for a small project team who is dealing with relatively smaller-scale design projects, such as capstone design projects for senior engineering students. The design team will have to manually organize the product data beyond the analysis phase, such as conducting design changes across disciplines. One key condition for using the commercial CAD/CAE/CAM software is that the CAE and CAM capabilities offered must be able to support all engineering analysis requirements for the design problem at hand. If this is not the case (e.g., the design problem involves structural analysis of engine mounts made of rubber and none of the CAD/CAE/CAM suites offer FEA for rubber), then an FEA code that is capable of supporting the required analysis, such as ABAQUS (www.3ds.com/products-services/simulia/portfolio/abaqus/overview), must be employed. In that case, the CAD model of the engine mount must be imported into ABAQUS (or other modeling tools, such as PATRAN or Hypermesh) for mesh generation, and loads obtained from motion analysis must be converted into a format that can be incorporated to the FEA model in ABAQUS for analysis. This approach is suitable for a design problem that heavily involves engineering analysis.

The second option is using commercial PDM (or PLM) software, such as Windchill from Parametric Technologies Corporation (PTC) (www.ptc.com), to support product data integration without a fully integrated CAD/CAE/CAM software suite. In this case, the design team will have to manually handle simulation model generation based on the product data embedded in the CAD solid models. For example, in order to create a motion simulation model using, for instance, ADAMS (www.mscsoftware.com/product/adams) or DADS (www.lmsintl.com), the design team will have to first organize the product CAD model into subassemblies or parts as individual bodies and collect mass properties of individual bodies with their respective coordinate systems. Then, the team will use the mass properties provided by CAD models to create motion simulation model using ADAMS or DADS externally to the CAD software. Some software, such as ADAMS/Car (www.mscsoftware. com/product/adamscar), offers templates that facilitate the creation of a motion model. This approach is suitable for a design problem that is less involved with engineering analysis and more dependent on the design process and data integration. An example of such an application is provided in Section 6.6.2 as a case study, in which an integrated testbed for reverse engineering using Windchill is discussed.

The third option is developing a tool and information integration infrastructure to support product development, in which a customized software infrastructure is implemented to integrate CAD, CAE, and CAM software and provide design trade-off and product management capabilities to meet specific product development needs. This is certainly not a trivial task. However, this approach offers maximum flexibility for a cross-functional team in product development. An example of such an application is provided in Section 6.6.3, in which a software infrastructure of tool integration for e-Design developed a few years ago is presented.

6.3.1.2 Process data model

In general, process management in a PDM system supports the engineering team in defining, disseminating, coordinating, and tracking design activities. The design processes are often described and modeled in a flow chart. The modeled workflow is then executed and the PDM system manages the actual workflow, so that the right work is done at the right time with the right information by the right person (Lee et al., 2010). The PDM system gives notifications to control the activities. For example, a designed part that has not been approved yet by the management will not be manufactured.

There are two types of workflows: static and dynamic. Static workflows are fixed; once they are modeled and started, they have to be finished according to the model. Dynamic models can be modified easily because usually there is a visual flow chart that can be used in a drag-and-drop style. Once a dynamic model has been started, it can be changed if the process needs to be changed while the workflow is in progress (Qiu and Wong, 2007). Older PDM systems usually apply the schematic static workflow, whereas the newer systems use graphical workflow modeling. After modeling the workflow, permissions are assigned to different users, allowing them to approve, release, or modify the documents.

As discussed earlier, a PDM system also serves as a process data management tool. A logical way to organize process data is associating the data with the design process. In general, design process can be defined in different levels. One task, represented in one block in the process chart of a higher level, can be expanded into more detailed tasks in the lower-level process, as illustrated in the upper box of the web page shown in Figure 6.9. When an entity of a task is clicked, the task-associated data is

Click the entity to bring out the task information in the window below, including team organization (left) and task information (right)

Right click email address to bring up MS Outlook for sending emails

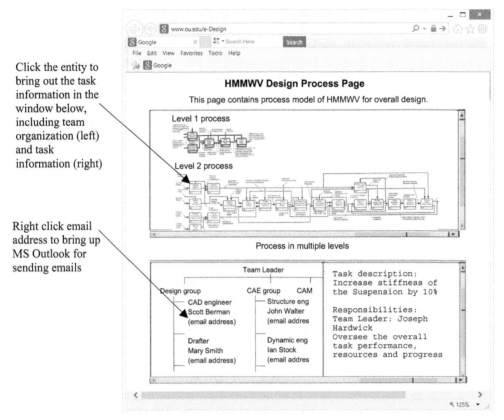

FIGURE 6.9 Sample Web-Browser Page of PDM for Process Data Model.

displayed in the window below. For example, if a top-level task is clicked, the organizational structure of the project team that is in charge of the task is displayed in a viewer (left window) and the task assignments, personnel, and individual assignment and responsibility are displayed in the right window. Right-clicking an email address below the name of the team member brings a Microsoft Outlook page for sending email to the person.

6.3.2 Basic functions of PDM systems

Although there is no clear consensus in industry and academia regarding the functionality of PDM systems, many articles refer to CIMdata (1998) and Crnkovic et al. (2003) for the common denominator. According to Crnkovic et al. (2003), the functionality of PDM systems can be grouped into two categories: user functions and utility functions. The user functions allow users to interact with the PDM system either as a user or as an author of information. The utility functions connect to the network infrastructure and support user functions by providing interfaces between different operating environments.

6.3.2.1 *User functions*
These user functions include data vault and document management, workflow and process management, as well as program management.

6.3.2.1.1 Data vault and document management
The core element in a PDM system, which is closely related to document management, is vault. Vault, from a logical point of view, is a single place where all documents are stored. Typically, it is a computer server (or group of servers) that physically stores the documents that are encompassed by operation of PDM. The documents in vault are accessible only by PDM users through PDM client functions. After the document is created on a computer, the operation "check-in" is performed, which transfers the document from the client computer to the vault. From this point on, the management of the document is seized by PDM and all further actions involving this document have to be performed by using the PDM client functions. If a team member wants to access this document, the operation "check-out" should be performed. Documents stored in the vault can be reviewed without checking out. An example of document processing using a PDM system is illustrated in Figure 6.10.

Because a PDM system usually involves many users, there will be situations when two or more persons want the same information at the same time. A PDM system controls the access to information and to what extent the information is available. Therefore, another important aspect of document management in PDM is version and status. In a typical PDM system, when a document is sent to the vault by check-in operation, it receives the status of "checked-in," meaning that the document is ready for next actions. The first person to access the current information, such as a Word document, checks it out and becomes the temporary owner of the information. If another person tries to access the same information, the information will either be blocked or made available as a read-only copy. This state is maintained until the first person checks the original information back in again. After a cycle of processing, the document status can be changed to "Released," indicating that document obtained a satisfactory status and is ready for general use. The history of changes, including dates, persons, etc., together with all versions that were created during such a cycle, are collected by the PDM system. An example of the cycle is illustrated in Figure 6.11.

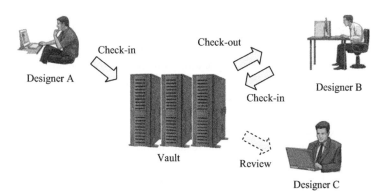

FIGURE 6.10 Example of Document Processing in a PDM System.

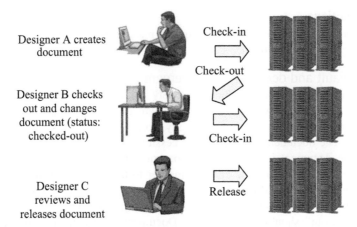

FIGURE 6.11 Document Status and Versions during Processing in a PDM System.

6.3.2.1.2 Workflow and process management

Workflow and process management is used to define and control the workflows and information flows, such as engineering change procedures and release procedures. In PDM workflow management, these processes can be modeled and managed, allowing automated distribution of the right information to the right users. Discussion of Figure 6.9 presents a practical implementation for work and process management.

6.3.2.1.3 Program and project management

Program management connects the product data with project data, thus allowing resource allocation and project tracking. In this way, the projects using a specific product or part can be found. Program and project management issues might involve planning and work performance control. Project leaders can overview the effort and performance of a project team and check if the progress is within the time schedule.

6.3.2.2 Utility functions

Underlying those user functions are the utility functions. They are connected to the network infrastructure, and insulate the end user from that. The utility functions provide interfaces between different operating environments and include communication and notification, data transport and translation, image services, administration, and application integration.

6.3.2.2.1 Communication and notification

Within an organization, different users are interested in different information. PDM systems can provide notification features that inform certain users when a specific event occurs (e.g., a task is finished or the project state has changed), in addition to initiating and organizing web meetings. The users who are to be informed about the event can be connected to roles and assignments of the individual team members. The notification is commonly sent as an email.

6.3.2.2.2 Data transport and translation

Sometimes, data need to be transported from one subgroup to another. On many occasions, data need to be translated from one engineering tool to another. These tools may not support the same data formats. By providing translation service, a PDM system can overcome this problem. The translation can be performed manually or automatically. More about product data exchange and CAD model translations is discussed in Section 6.5.

6.3.2.2.3 Image services

In the product development process, CAD and different engineering tools are used to model a product. For easier access to the product model and information, the PDM system provides tools or add-ons that, for instance, allow models stored in a CAD system to be viewed from the PDM system. Model viewers discussed in Section 6.2 represent a typical scenario of the image service function in PDM system.

6.3.2.2.4 Administration

Because PDM systems are quite extensive, they require a lot of administration. In addition to the usual administration tasks such as installation, configuration, maintenance, user authorization, role management, and data backup, the PDM administrator defines workflows, translations, and tailors the system for the company.

6.3.2.2.5 Tool integration

In PDM, it is important to collect all data in one location in order to avoid data inconsistency. The integration between engineering tools and the PDM system is important to enable this. In Section 6.3.1, three options were discussed to address the integration between CAD/CAE/CAM and the PDM system. In Section 6.6.3, we present a software infrastructure that supports tool integration for e-Design.

6.3.3 Benefits of PDM systems

In general, organizations that successfully implement PDM can achieve multiple advantages in terms of productivity and competitiveness. The benefits can be summarized as follows (CIMdata, 1998; Miller, 1998).

- *Collaboration between design team members.* PDM software provides a virtual workspace in which design team members can store and share documents related to projects. Additionally, interdisciplinary collaboration between designers, marketing, and manufacturing is supported. A PDM system can lead to collaborative development of new products, as well as improvements on existing products.
- *Reduced product development cycle time.* Due to the increased collaboration with all areas of an organization and its supply chain, as well as the easy access to product information, the product development time can be greatly reduced. This enables organizations to respond to the market with greater effectiveness and consistently provide their customers with new and initiative products. In addition, when properly implemented, PDM can simplify many day-to-day user operations by managing and automating routine tasks, such as searching for drawings, tracking approvals, and completing status reports. This improvement dramatically decreases the user's non-value-added time.

- *Workflow and project management.* Project management is made easier using a PDM system because all those involved in the project have access to the same information and can work with a common product model. A PDM system also allows project managers to track the progress of a project more effectively and therefore ensure that the work being carried out is correct, on schedule, and on target.
- *Improved life cycle design.* The information captured within PDM systems is increasing from solely a CAD focus to include several engineering domains, and it supports easy access to information on new product development. It allows manufacturing staff and production engineers to access design information at a much earlier stage of the product development, hence making it easier for problems to be identified earlier rather than later.
- *Supply chain collaboration.* PDM systems are considered to have a strong impact on supply chain relationships by linking subcontractors, vendors, consultants, partners, and customers and giving them access to the same information. PDM systems can also act as a data store for internally developed parts and external parts available from suppliers. By using a PDM system's database of existing parts, a designer can eliminate duplicate work and therefore considerably reduce development time and cost.

6.3.4 Impact to industry

As mentioned earlier, the automotive and aerospace industries are the biggest adopters of PLM. The high degree of penetration of PLM in the automotive and aerospace industries is due to the fact that their products have long life cycles, are very complex, and have nearly no possibility of physical prototyping (Liu and Xu, 2001). We briefly mention two notable success stories in using PDM or PLM for product or project development: Boeing 777 and Ford C3P; both are extracted mostly from Caldwell and Mocko (2008).

As reported by Caldwell and Mocko (2008), Boeing began using 3D solid modeling software and a PDM system during the design of the 777 jet aircraft. Previously, Boeing used two-dimensional (2D) modeling software to design its airplanes, which required many stages of design of parts and subsystems of the airplane in order to ensure that all components fit together properly. Boeing used three stages of mock-up before producing a final design; even with three rounds of mock-ups, the final design would have parts with mismatched geometries. For the design of the 777 jet aircraft, Boeing implemented a new 3D CAD system, which consisted of CATIA and Electronic Preassembly Integration (EPIC) on CATIA. These two programs helped Boeing eliminate mock-ups by allowing parts to be designed and assembled together in the computer. This allowed Boeing to ensure that part geometries would match up properly. By converting to a 3D CAD system with assemblies, Boeing was able to speed up its design process and eliminate many errors. Boeing also took advantage of the all-computer design by using a PDM system. Boeing stored all of its CATIA files on the world's largest (at the time) grouping of IBM mainframe computers in Bellevue, Washington. This allowed companies in Japan, the United States, and the United Kingdom who were working with Boeing to access the CAD files at any time. These suppliers, therefore, were aware of changes made to the design very soon after the changes were made. The implementation of this new system allowed Boeing to reduce engineering change requests by 90%, reduce cycle time for these requests by 50%, reduce material rework by 90%, and improve fuselage tolerances by 5000%.

Another well-known story is the C3P initiative at Ford Motor Co. As reported by Caldwell and Mocko (2008), Ford Motor Company implemented a PDM system worldwide in the 1990s. It was part of a new CAD initiative that Ford called C3P, an acronym for CAD/CAM/CAE/Product Information Management (PIM). PIM is a Ford-specific term for PDM. Although this project was called C3P, Ford's focus in this endeavor was on the PDM system. This project began in 1996; by mid-1998, 16 vehicle programs had already begun using the new system. Ford's plans were to use the PDM system worldwide, so that all of its operations and suppliers accessed CAD files that were stored in Dearborn, Michigan. Ford's C3P program was a $200 million deal with SDRC, which included both software and services. Before this deal, Ford had used other CAD and PDM systems, including a PDM system developed in-house. Ford experienced the obvious improvement, which was a faster time-to-market of its products. Engineering efficiency rose around 30–40% due to new solid modeling capabilities. Prototype costs decreased by 40–50%, saving hundreds of millions of dollars. Late changes were reduced by 50%, and programs were able to be completed in less than 2 years. Ford was able to extend the benefits of C3P beyond the design of the vehicle itself. They used computer programs to analyze the solid models in order to determine a vehicle's manufacturability within an existing plant. In one case, Ford was able to prevent a $60 million tooling modification that would have been required had the design not been analyzed ahead of time. Ford's C3P program was a success that continues today. Thus, by implementing a PDM system, Ford was able to reduce time-to-market of new vehicles, increase engineering efficiency, reduce prototype costs, and reduce late changes to parts.

6.4 **PDM Systems**

From the mid-1980s through the late-1990s, we saw the development of many capable PDM systems, including iMAN from UG Solutions, Metaphase from SDRC, Optegra and Windchill from PTC, MatrixOne, Pro/PDM and Pro/Intralink from PTC, ENOVIA from Dassault Systèmes, and Workgroup PDM and Enterprise PDM from SolidWorks. Since that time, we have seen both company and product consolidation. UG Solutions acquired SDRC, which was in turn acquired by Siemens. Their respective products were combined to create TeamCenter. ComputerVision was acquired by PTC and their collective products were integrated to produce Windchill. Dassault Systèmes acquired both MatrixOne and SolidWorks. MatrixOne lives on as Enovia.

There is currently a multitude of PDM products available on the market. Their popularity is also steadily increasing mainly due to quicker and easier implementation. Some are offered by CAD vendors who have reinvented themselves as PLM software companies. This is meant to indicate that they provide not just design and manufacturing software products, but services and solutions that integrate product development into an enterprise. In most cases, a tight integration exists primarily for the CAD system developed by the same software provider. PDM systems currently available on the market include, among others, AutoDesk ProductStream, ENOVIA Smarteam, PTC Windchill, Siemens UGS TeamCenter, and SolidWorks Enterprise PDM. Table 6.1 depicts some PDM systems and the primarily supported CAD system offered by the same vendor. In addition to the PDM (or PLM) offered by CAD vendors, there are number of popular PLM software developed by non-CAD vendors, such as SofTech, Arena, and so forth.

Table 6.1 Available PDM Systems and the Corresponding CAD Systems

PDM System	Corresponding CAD System
AutoDesk ProductStream	Autodesk Inventor
ENOVIA Smarteam	CATIA
PTC Windchill	Pro/Engineer
Siemens UGS TeamCenter	Siemens UGS NX
SolidWorks Enterprise PDM	SolidWorks

In general, systems offered by CAD vendors built their PDM with a strong connection to product data models created in CAD. Data sharing, such as BOM built upon the model tree of the CAD models, is a natural approach. In addition, they facilitate engineering collaboration by taking the advantage of the existing fully integrated CAD/CAE/CAM suite of existing software product line. However, such systems are usually less flexible in terms of data integration or exchange with other engineering tools. On the other hand, systems offered by non-CAD vendors are more flexible and more general. For example, Arena Cloud PLM was developed in support of general engineering products without tying it with any specific mechanical CAD software. However, such software often requires more effort in creating product data. It offers strong data management capabilities but relies on external engineering capabilities for product design.

6.4.1 Systems offered by CAD vendors

In this subsection, we briefly mention prominent commercial PDM systems offered by CAD vendors, including AutoDesk ProductStream, ENOVIA Smarteam, Windchill, and Enterprise PDM.

6.4.1.1 AutoDesk® ProductStream® of Autodesk Inventor

AutoDesk® ProductStream® is the major software module that supports a design team in organizing, managing, and automating key design and release management processes (AutoCAD, 2009). With this software, the design can be reviewed and approved before releasing it to manufacturing. The software stores and manages work-in-progress design data and related documents with data management tools for workgroups (see sample screen in Figure 6.12a). Team members can accelerate development cycles and increase their company's return on investment in design data by driving design reuse. In addition, AutoCAD offers Balloons and BOM, which use standards-based balloons and part lists (see Figure 6.12b), and automatically update the BOM to seamlessly track any changes, which helps to keep teams on schedule by reducing costly breaks in production due to incorrect part counts, identification, and ordering.

6.4.1.2 ENOVIA Smarteam of CATIA

Dassault Systèmes' solutions for PLM include two products categories: ENOVIA and Smarteam. ENOVIA solutions include PDM, intellectual property life cycle management, virtual product design, collaboration solutions, and configured digital mock-up. Smarteam for life cycle and PDM enhances and accelerates the proliferation of product knowledge and business processes across the enterprise and product value chain with tighter CAD integrations.

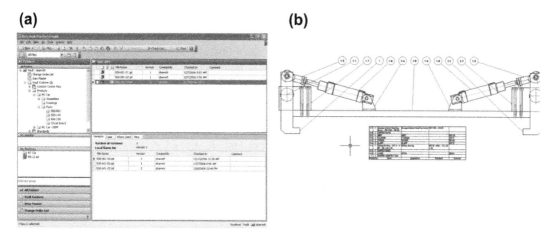

FIGURE 6.12 AutoDesk® ProductStream®. (a) Sample Screen Shot, and (b) BOM Shown in Standards-Based Balloons and Part Lists. Figure courtesy of http://203.53.66.237/Data/Attachments/AutoCAD%20Mechanical%202009%20Detail_HR.pdf (AutoCAD, 2009).

One of the key advantages of ENOVIA Smarteam is that the system provides collaborative offerings focused on product development processes supporting design, engineering, and enterprise activities (Dassault Systèmes, 2013). Design Collaboration enables dispersed design teams to work in collaboration in a single or Multi-CAD environment, to innovate new products and to reuse existing ones for faster time-to-market. Engineering Collaboration seamlessly combines cross-functional engineering-based activities throughout the product life cycle into a unified environment for effective data management and sharing among various organizational roles. Enterprise Collaboration provides a PLM solution throughout and across the extended enterprise, including the value chain. Supply Chain allows companies to leverage supply chain capabilities throughout the product life cycle and make their suppliers an integral part of product development.

6.4.1.3 Windchill by PTC

Windchill from PTC is considered to be one of the most advanced PDM systems available. It combines the power of client–server technologies with the implementation, manageability, and usability benefits of the web. It provides complete support for managing and communicating information about product structures and changes throughout the product life cycle. Windchill is built from scratch using all modern web technologies, and it is fundamentally based upon standard Internet, web, Java, and Oracle technologies at all levels of its architecture. Hence, Windchill claims their product to be "web-centric" as opposed to other "web-enabled" products.

6.4.1.4 TeamCenter by Siemens UGS NX

TeamCenter connects team members throughout the life cycle with a single source of product and process knowledge. TeamCenter's comprehensive portfolio of end-to-end PLM solutions gives users the flexibility to choose the right mix of solutions for product development needs.

TeamCenter's engineering process management solution allows users to integrate company's global engineering teams by bringing together the product designs from all sites within a single PDM system. Team members can capture, manage, and synchronize product design data, then facilitate engineering change, validation, and approval processes. TeamCenter supports NX, Pro/ENGINEER, CATIA V5, AutoCAD, Inventor, Solid Edge, and SolidWorks.

6.4.1.5 SolidWorks Enterprise PDM

SolidWorks Enterprise PDM (EPDM) provides an easy way for designers to collaborate on product designs without worrying about version control or data loss. It stores CAD models and supporting documents in an indexed central repository that tracks versions and automates workflows to eliminate wasteful repetition. SolidWorks Enterprise PDM simplifies the process of managing design changes while improving product reuse by integrating within SolidWorks, AutoCAD, Autodesk Inventor, Pro/ENGINEER, and Windows Explorer.

6.4.2 Systems offered by non-CAD vendors

In this subsection, we briefly mention commercial PDM systems that are not offered by CAD vendors, including SofTech ProductCenter® PLM (www.softech.com) and Arena Cloud PLM (www.arenasolutions.com).

6.4.2.1 SofTech ProductCenter® PLM

ProductCenter (sample screen capture shown in Figure 6.13) is a commercial software product that is an integrated suite of PLM software for managing product data. The software was engineered for the Microsoft Windows and UNIX operating systems. Along with core applications, it includes localized and web-based services. ProductCenter is suited for managing various types of CAD/CAE/CAM data, but it can be used for many forms of data management and product management. ProductCenter makes the use of spreadsheets for BOM management obsolete and provides organization for parts with various part types and attributes; in addition, all information managed can be accessed through the ProductCenter Hierarchy Explorer. This feature helps to facilitate small to mid-size manufacturers with a way to centralize product data, control the engineering change process, and share BOMs with suppliers. ProductCenter can be integrated with other CAD/CAE/CAM tools to help ease the management of product data from design to manufacturing.

6.4.2.2 Arena Cloud PLM

Arena pioneered cloud PLM applications. The company's products, including BOMControl, PartsList, and PDXViewer, enable engineering and manufacturing teams and their extended supply chains to speed up prototyping, reduce scrap, and streamline supply chain management. Arena cloud PLM applications simplify the BOMs and change management for companies of all sizes, and they offer the right balance of flexibility and control at every point in the product lifecycle—from prototype to full-scale production. These cloud-based applications enable manufacturers to manage BOMs, engineering change orders, product data exchange (PDX), and other key manufacturing files securely and efficiently.

PLM in the cloud is an internet-based system for managing a product and its associated information from concept to end of life. PLM in the cloud is growing in popularity with manufacturers around

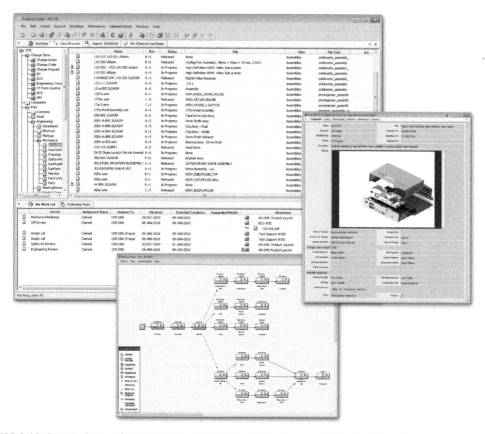

FIGURE 6.13 Sample Screen Shots of ProductCenter (en.wikipedia.org/wiki/ProductCenter).

the world as a way to manage the stages of product development in order to collaborate, track, and regulate changes to the product. BOMControl (see Figure 6.14) keeps BOM data centralized, controlled, and up-to-date, resulting in fewer errors, less scrap and rework, higher quality, and better cost control.

6.5 Product data exchange

The frequent needs from product data exchange (PDX) encountered in product development involve solid model translations between CAD software systems, for both part and assembly levels. In general, for part model translation, the established approach both in theory and in practice is geometric data exchange (GDE) and feature data exchange (FDE). In GDE, the boundary representation (B-rep) of the object is translated from a source to a target CAD system (Spitzy and Rappoport, 2004). The resulting part in the target CAD system is lumped into one single entity—that is, one single solid feature in the model tree of the target CAD system, often called dumb geometry. Individual solid features and parametric data of the solid model created in the source CAD system are lost in the

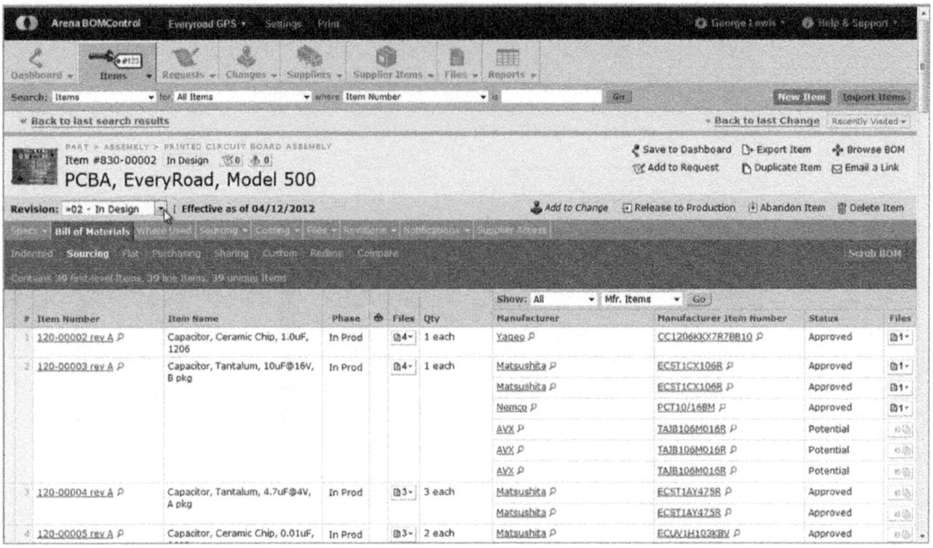

FIGURE 6.14 Sample Screen Shots of Arena Cloud Product Life Cycle Management (en.wikipedia.org/wiki/ProductCenter).

translation. On the other hand, in FDE, given a parametric history graph (or model tree) in a source system, the goal is to construct a graph in the target system that results in similar geometry while preserving as much parametric information as possible. FDE retains design intelligence and allows modifications at the receiving. However, it is not always technically possible to successfully exchange every feature. In the context of e-Design, FDE is much desirable than GDE when a design change is anticipated for the CAD models being translated.

In assembly model translation, which is generally more involved than part translation, mating constraints defined using geometric entities of solid features must be faithfully retained from the source to target CAD systems. For example, a concentric mating constraint is often defined by selecting an outer surface of a cylinder (e.g., on an extrude feature of the mating part) and an inner surface of a hole (e.g., an extrude cut feature of the base part). As a result, translating an assembly model, in which its constituent parts are translated using GDE approach, is fundamentally deficient because feature information is not retained in translation. Even if features and parametric information are retained for the constituent parts, assembly model translation may not be as successful as expected. This is because FDE only results in similar geometry while preserving as much parametric information as possible, implying that feature information may be altered or incomplete. Therefore, in this section, we mainly focus on part solid model translation and only briefly mention assembly model translations using simple examples. More details about part and assembly model translations can be found in tutorial lessons of Projects S1 and P1 for SolidWorks and Pro/ENGINEER, respectively.

In this section, we start by introducing the viable options in support of CAD model translations in Section 6.5.1. In Section 6.5.2, we discuss direct model translations; both part and assembly examples are included. In Section 6.5.3, we discuss data exchange using neutral formats, including two important standards for part translation—IGES and STEP. In Section 6.5.4, we briefly mention several third-party model translation software tools. In Section 6.5.5, we discuss a newly developed technology, solid feature recognition, which offers a better alternative for FDE.

6.5.1 Data exchange options

In general, there are three practical options of translating data from one CAD system to another: direct model translation, neutral file exchange, and third-party translators, as illustrated in Figure 6.15, in which System A is called the source and System B is called the target system.

Major CAD systems, such as SolidWorks, Pro/ENGINEER, NX, Unigraphics, and CATIA, directly read and/or write other CAD formats, simply by using File Open and File Save As options (Figure 6.15a). Because most CAD file formats and geometric modeling kernels are proprietary, this option is limited to selected CAD systems.

Another common method of translation is via an intermediate neutral format, as illustrated in Figure 6.15b. The source CAD system exports out to this format and the target CAD system reads in this format and converts data into its native form. Some formats are independent of the CAD vendors, being defined by standard organizations, such as IGES (IGES, 2001) and STEP (Nell, 2001). Others, such as VDA (en.wikipedia.org/wiki/VDA_6.1), although owned by a company, are widely used and are regarded as quasi-industry standards.

There are a number of companies that specialize in CAD data translations and provide software that can read one system and write the information in another CAD system format (Figure 6.15c).

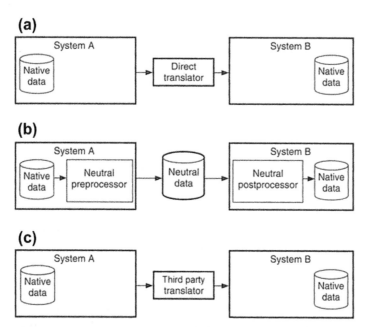

FIGURE 6.15 Three Methods of Data Exchange. (a) Direct Model Translation, (b) Neutral File Exchange, and (c) Third-Party Translators (Stokes, 1995).

These systems have their own proprietary intermediate format, some of which will allow reviewing the data during translation. Some of these translators work as stand-alone systems, whereas others require one or both of the CAD systems to be installed on the translation machine, as they use code such as application protocol interfaces (APIs) from these systems to read and write the data.

Because each CAD system has its own method of describing geometry, both mathematically and structurally, there is always some loss of information when translating data from one CAD system to another. The intermediate file formats are also limited in what they can describe, and they can be interpreted differently by both the source and target systems. It is therefore important in translating data between systems to identify what needs to be translated.

If the geometric model is required for the downstream process without anticipated design changes, then only the geometric description of the model needs to be translated. However, there are levels of detail. For example, is the data wireframe, surface, or solid sufficient? If a design change is anticipated, the feature information and model tree must be preserved between systems. In addition to geometric information, retaining the assembly structure may be required. In general, different data translation is required for different engineering activities.

From an e-Design perspective, GDE is generally sufficient, except for parts that anticipate changes, in which feature and parametric data must be available. When a fully integrated suite of CAD/CAE/CAM is not available or if engineering capabilities offered by certain software tools are not adequate, model translation is unavoidable. For example, to support motion analysis, geometry and coordinate systems of a solid part and subassembly are sufficient to support accurate calculation of mass properties and kinematic joint locations. To support finite element analysis, accurate solid or surface models of the respective CAD models are usually sufficient for finite element mesh generation. For CNC toolpath generation, solid part in CAD is directly useful. For some cases, even a surface model that represents the design surface (the part surface to which machining takes place) is sufficient, such as when using MasterCAM. In general, GDE is sufficient to support CAE and CAM activities.

6.5.2 Direct model translations

As the engineering capabilities offered by major CAD systems progresses, CAD models can be translated to and from more CAD systems. In order to support model translations, a target CAD system must be able to open the model file of the source system, parse data stored in the native format of the source system, interpret the data, and map data entities to convert them into the format of the target system.

In this subsection, we offer a more in-depth discussion on the subject by importing parts and assembly created in Pro/ENGINEER to SolidWorks, as well as presenting examples of importing SolidWorks parts and assembly to Pro/ENGINEER.

6.5.2.1 Importing Pro/ENGINEER parts to SolidWorks

SolidWorks offers two options for importing CAD models: importing solid features and importing geometry. We use the gear housing shown in Figure 6.16a as an example to illustrate both options. As shown in the Pro/ENGINEER model tree of Figure 6.16a, there are eight datum features and 14 solid features. SolidWorks will try to import these 14 solid features from Pro/ENGINEER.

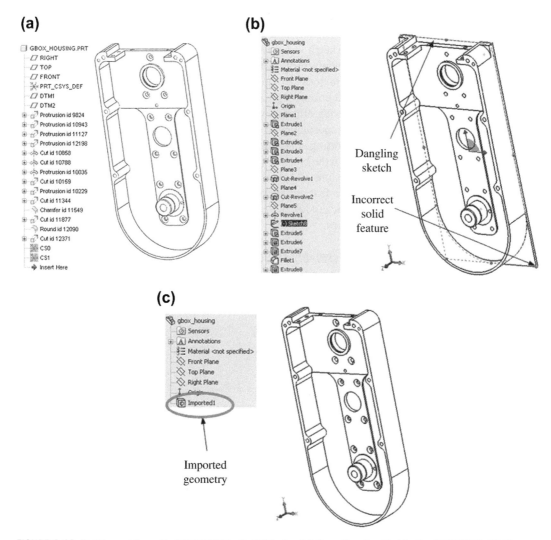

FIGURE 6.16 Part Import from Pro/ENGINEER to SolidWorks. (a) Gear Housing Part in Pro/ENGINEER, (b) The Translated Solid Model and Features in SolidWorks Using Option of Importing Solid Features, and (c) The Translated Solid Model in Solidworks Using Option of Importing Geometry.

Using the option of importing solid features, SolidWorks translates 12 out of 14 features. The converted model and features listed in the browser are shown in Figure 6.16b. As shown in Figure 6.16b, there is one dangling sketch, *Sketch8*, representing the unrecognizable solid feature in addition to the chamfer feature. In addition, the back plate (*Extrude1* in the browser) is recognized incorrectly. In general, SolidWorks is capable of importing some parts correctly and completely, especially when the solid features are relatively simple (but apparently not this gear housing part).

If we take a closer look at any of the solid features translated, such as *Extrude3*, the sketches (e.g., *Sketch3* of *Extrude3*) of the solid features do not have complete dimensions. A (−) symbol is placed in front of the sketch, indicating that the sketch is not fully defined.

Apparently, this translation is not satisfactory. Unfortunately, this translation represents a typical situation you will encounter for a large majority of the parts. In many cases, it may take a lesser effort to repair or recreate wrongly recognized or unrecognized solid features. However, when you translate an assembly with many parts, the repairing effort could be substantial.

Importing parts using the option of importing geometry is more straightforward, which has a higher successful probability than that of importing solid features. The model is imported as a single entity *Imported1*; a dumb geometry appeared in the browser (see Figure 6.16c). As mentioned earlier, there is no parametric solid feature with dimensions and sketch converted. However, the geometry converted seems to be accurate. All the geometric features in Pro/ENGINEER were included in this imported feature. This translation is considered successful. If we do not anticipate making any change to the gear housing, this imported part is satisfactory.

6.5.2.2 Importing Pro/ENGINEER assembly to SolidWorks

We import the input gear assembly shown in Figure 6.17a using both options. As shown in the left of Figure 6.17a (Pro/ENGINEER model tree), there are 11 parts in this assembly.

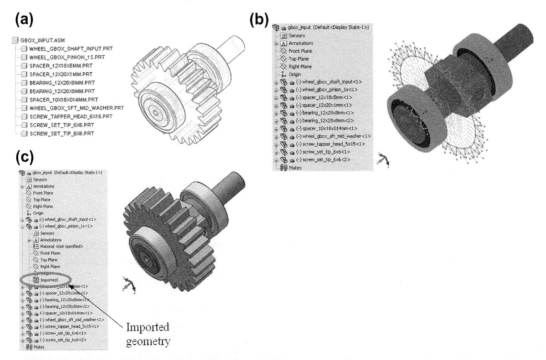

FIGURE 6.17 Assembly Import from Pro/ENGINEER to SolidWorks. (a) Input Gear Assembly in Pro/ENGINEER, (b) The Translated Assembly in SolidWorks Using Option of Importing Solid Features, and (c) The Translated Assembly in SolidWorks Using Option of Importing Geometry.

Using the option of importing solid features, parts are not completely converted, as shown in Figure 6.17b. Major solid features are missing, such as pinion 1 (*wheel_gbox_pinion_1s<1>*), where most solid features are not converted. In fact, there are only two extrude features successfully imported. The remaining entities are mostly sketches. Some parts seem to be imported fine. However, the *Mates* branch in the browser is completely empty, implying that no assembly mates have been imported.

Apparently, this translation is not satisfactory. A nontrivial effort will have to be devoted to reconstructing the solid features (therefore, solid models) as well as the final assembly.

The option of importing geometry is also more straightforward for assembly. In fact, the assembly and all 11 parts seem to be correctly imported, as shown in Figure 6.17c. By expanding any of the part listed in the browser, such as the gear (*wheel_gbox_pinion_1s<1>*), we see an imported feature listed, as depicted in Figure 6.17c. Again, there is no solid feature converted in any of the parts. In addition, the *Mates* branch is empty.

If we do not anticipate making any change to this input gear assembly, this imported assembly is satisfactory, except it does not have any assembly mates. Assembly of all 11 parts (maybe more, for other cases) will be a nontrivial effort. If you do not anticipate making changes in how these parts are assembled, you may merge all 11 parts into a single part, instead of assembling those using mating constraints.

A step-by-step detail of importing the Pro/ENGINEER part and assembly can be found in the tutorial lesson S1.3. You may go over the lesson to learn more about the model importing capabilities offered by SolidWorks.

6.5.2.3 Importing SolidWorks parts to Pro/ENGINEER

The capabilities of importing SolidWorks models offered by Pro/ENGINEER are primitive. Only geometry data is imported without parametric feature information. We use a simpler crankshaft example shown in Figure 6.18a to illustrate the translations. As shown in the SolidWorks browser of Figure 6.18a, there are three solid features, all boss-extrudes. After opening the Solid-Works model directly from within Pro/ENGINEER, the part is imported as one single feature (Imported Feature ID 4), as shown in Figure 6.18b. It is a model of dumb geometry. The part is not changeable.

(a)   **(b)**

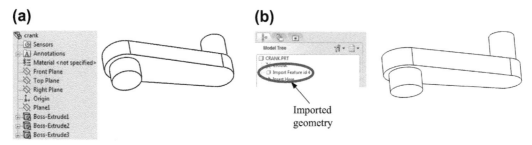

FIGURE 6.18 Part Import from SolidWorks to Pro/ENGINEER. (a) Crankshaft Part in SolidWorks, and (b) The Translated Solid Model in Pro/ENGINEER.

6.5.2.4 Importing SolidWorks assembly to Pro/ENGINEER

We import a simple slider-crank assembly shown in Figure 6.19a from SolidWorks to Pro/ENGI-NEER. As shown in the SolidWorks browser (left of Figure 6.19a), there are four parts in this assembly. The parts are parametric and assembly mating constraints are properly defined. After opening the SolidWorks assembly directly from within Pro/ENGINEER, the assembly is imported as four parts, as shown in Figure 6.19b. The assembly and parts seem to be fine. However, no mating constraints are imported properly. As a result, dragging a part (e.g., the piston) leads to a disassembled model shown in the lower half of Figure 6.19b. In addition, individual parts are imported as models of dumb geometry.

It is apparent that the capabilities offered by Pro/ENGINEER in importing SolidWorks models are not desirable. More details about the step-by-step process of importing the SolidWorks part and assembly discussed can be found in the tutorial lesson P1.3.

6.5.2.5 Data exchange between CAD and CAE/CAM

In addition to direct model translation between CAD systems, some CAE and CAM software reads CAD native files directly. For example, ANSYS reads CATIA and Pro/ENGINEER files, in addition to IGES, NX, SAT, and Parasolid. MasterCAM reads AutoCAD, Pro/ENGINEER, Rhino, SolidWorks, Unigraphics, and CATIA, in addition to IGES, STEP, SpaceClaim, ACIS, Parasolid, and VDA. The success rate of importing the native CAD models into CAE and CAM software is generally very good because the translation mostly involves parts only and requires only geometric data; in general no parametric features are involved.

(a)　　　　　　　　　　　　　　**(b)**

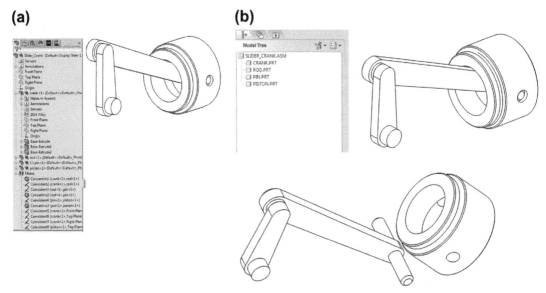

FIGURE 6.19 Assembly Import from Pro/ENGINEER to SolidWorks (a) Slider-crank assembly in SolidWorks, and (b) The translated assembly in Pro/ENGINEER.

6.5.3 **Neutral file exchange**

The most commonly employed neutral files for CAD model translations are IGES and STEP application protocols (APs). In addition, several geometric kernels, such as ACIS (www.spatial.com) and Parasolid (www.eds.com/products/plm/parasolid), are becoming popular in serving as neutral formats for CAD model translations. Other formats commonly supported, such as STL and VRML, simplify true geometric data into faceted boundary representation for different purposes. Moreover, a data eXchange file (DXF, en.wikipedia.org/wiki/AutoCAD_DXF) is the de facto format for drawing conversion, but less in support of solid model translation. In this subsection, we briefly introduce IGES and STEP APs with more detailed file formats and data structure included in appendices. Note that practically none of the neutral files are capable of supporting data exchange for parametric solid features.

6.5.3.1 IGES

The IGES (Initial Graphics Exchange Specification) project was started in 1979 by a group of CAD users and vendors, including Boeing, General Electric, Xerox, ComputerVision, and Applicon, with the support of the National Bureau of Standards (now known as NIST) and the U.S. Department of Defense (DoD). Soon after, it was adapted and recognized by American National Standard Institute as a standard tool format. Consequently, IGES has become an acceptable and widely used neutral format for translator development by many CAD/CAM software vendors. After the initial release of STEP (ISO 10303) in 1994, interest in further development of IGES declined, and Version 5.3 (1996) was the last published standard. IGES has been used in the automotive, aerospace, and shipbuilding industries. These part models may have to be used years after the vendor of the original design system has gone out of business. IGES files provide a way to access this data decades later. The structure of the IGES file, data format, and simple examples are provided in Appendix A for further reference.

6.5.3.2 STEP (ISO 10303)

The work with the ISO 10303 standard, informally called STandard for the Exchange of Product model data (STEP), was initiated in 1984 with the goal to standardize exchange of product data between product life cycle systems. After 10 years of work, the first parts of the STEP were published in 1994. The standard is a very comprehensive set of specifications, covering many different product types (electronic, electromechanical, mechanical, sheet metal, fiber composites, ships, architectural, furniture etc.) and many life cycle phases (design, analysis, planning, manufacturing, etc.). Using STEP-supporting tools, data can be exchanged by converting it from the native format of the source CAD system to the neutral ISO 10303-11 format, also known as an EXPRESS schema. Then, the target system imports the schema and converts it to its own native format. The EXPRESS schema defines not only the data types but also relations and rules applying to them. This makes it possible for the target system to validate the schema. EXPRESS is a textual and graphical data modeling language included in the STEP standard.

The STEP format is organized as a series of documents (referred to as "parts" in STEP terminology), with each part published separately. There are currently six series of STEP parts. The most important parts from an application perspective are the 200-series, also called APs, through which STEP meets the real world. The APs are the top parts, produced to meet specific data exchange requirements for a particular application. They cover a particular application and industry domain; hence, they are most relevant for users of STEP. AP202, AP203, and AP214 have reached the status of an International Standard (IS) version. Other APs include AP202DIS, AP209DIS and AP214DIS

(Draft International Standard), and AP214 CD II (Committee Draft). A complete list of APs can be found in Appendix B.

Among the APs, AP203 (configuration-controlled 3D designs of mechanical parts and assemblies) and AP214 (core data for automotive mechanical design processes) are the most popular and widely supported for CAD data exchange. AP203 defines the geometry, topology, and configuration management data of solid models for mechanical parts and assemblies. This file type does not manage colors and layers. AP214 has everything an AP203 file includes, but adds colors, layers, and geometric dimensioning and tolerance. AP214 is considered an extension of AP203 by many users.

It is worth noting that the initial release of ISO 10303 was aimed entirely at the exchange of explicit models, essentially a B-rep model, defined in terms of geometry, possibly with additional topological information providing connectivity relationships between geometric entities. This is because the state of the art was B-rep during the mid-1980s, when STEP development commenced. A B-rep model provides a complete representation of a solid shape, but retains no details of how that shape was created. As a result, no model tree and parametric feature information is retained using STEP for model translation. ISO 10303-108 on parameterization and constraints for explicit geometric product models (ISO, 2005) is a new STEP resource providing representations of parameters, explicit constraints, and explicit 2D sketches or profiles. More about STEP parts and APs can be found in Appendix B.

6.5.4 Third-party translators

Neutral formats such as IGES and STEP support part geometry translation well. Third-party translators focus on feature and parametric data translation as well as assembly. In this subsection, we briefly mention two software tools that offer better solutions to CAD model translation: Proficiency (www.transcendata.com/products/proficiency) and TransMagic (www.transmagic.com/products/features).

6.5.4.1 Proficiency

Proficiency is a feature-based translation solution developed by International TechneGroup Inc. (www.iti-oh.com), headquartered in Milford, OH. Proficiency enables the transfer of design intelligence between major CAD systems, such as geometry, features, sketches, manufacturing information, metadata, assembly information, and drawings in the conversion process. Accurate and usable models are achieved with up to 95% automation, as claimed by the software vendor.

6.5.4.2 TransMagic

TransMagic offers translators that convert CAD files from one native file format to another. During the translation, TransMagic performs "geometry mapping", mapping from one CAD kernel to another. TransMagic avoids what are known as "stitching errors" by repairing geometry via techniques such as correcting slightly overlapping or misaligned surfaces, removing duplicate control points, and duplicate vertices. To minimize translation errors, TransMagic typically—but not always—translates directly from one native CAD kernel to another. Still, "stitching errors" (gaps and overlaps) can occur while trying to import the file and reinterpret geometry. TransMagic is available as a stand-alone program. It is also available as a plug-in for many CAD programs so that the Open and Save dialog boxes are extended with TransMagic's functionality.

6.5.5 **Solid feature recognition**

Feature recognition (FR) has been an active research topic for decades. Earlier study focused on recognizing manufacturing features, such as pockets, holes, slots, and so forth, created in CAD solid models (Shah, 1995). This effort led to capabilities implemented in several commercial CAM software tools, such as CAMWorks that automatically recognizes manufacturing features in SolidWorks models.

In the context of product data exchange, FR refers to recognizing geometric features embedded in a solid model of dumb geometry. This is called solid (or geometric) FR. The geometric model can be an IGES, STEP, or STL models exported from another CAD system, with no parametric feature information.

Many methods have been proposed for solid FR from numerous source files, such as from STL (Sunil and Pande, 2008), STEP (Bhandarkar and Nagi, 2000), or a B-rep model (van der Velden et al., 2010). In this subsection, we include Venkataraman's FR algorithm (Venkataraman et al., 2001) to provide readers with a brief understanding of the underline technique that supports solid FR. This method is relevant because it was recently implemented in a number of CAD systems, including SolidWorks and CATIA, which are capable of recognizing basic features, such as extrude, revolve, and, more recently, sweep. This capability has been applied primarily for support of solid model translations between CAD systems with some success, in which not only geometric entities but also parametric features are translated.

Venkataraman's FR algorithm uses a simple four-step process: (1) simplify imported faces, (2) analyze faces for specific feature geometry, (3) remove recognized feature and update model, and (4) return to Step 2 until all features are recognized. The process is illustrated in Figure 6.20. The simplification step involves surface format conversion and face merging. For instance, several connected B-spline patches or triangular facets as in STL models with identical (or similar) surface normal vectors could be combined and represented as a single planar face. The next step is to match the simplified faces for geometry resembling a specific solid feature. That is, given a specific feature type, the algorithm searches for the surfaces that resemble geometry associated with that feature. For instance, a hole would be constructed from a base circle extruded a given length (Figure 6.20a), producing a cylindrical surface. In this case, the hole would be seen as a negative feature. If an enclosed cylindrical surface is not found, a failure error is returned. If the correct geometry is found,

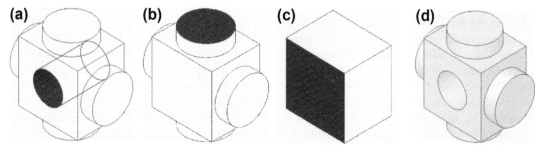

FIGURE 6.20 Illustration of Venkataraman's FR Algorithm. (a) Imported Surface Model with Hole Surface Selected, (b) Hole Recognized and Removed, with Extruded Face of Cylinder Selected, (c) Cylindrical Extrusions Recognized, with Base Block Extrusion Face Selected, and (d) All Features Recognized and Mapped to Solid Model (Chang, 2012).

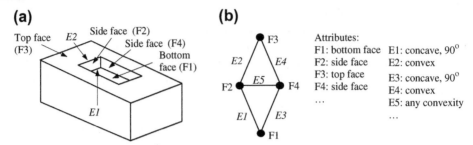

FIGURE 6.21 Feature Grammars for the Class of Simple Blind Pocket. (a) The Blind Pocket Feature, and (b) Face Adjacent Graph and Attributes.

the algorithm records the dimensions embedded in the composing surfaces (i.e., radius and depth in this case) and continues to the next step. Once a feature is recognized, it is temporally removed from the model. Continuing with the hole example, the hole feature would be filled, and the model would be updated to reflect the removal (Figure 6.20b). The algorithm would then analyze the faces of the updated part (Step 2) and continue the recognition process for the remaining features in the model (Figure 6.20c). Once all possible features are recognized, they are mapped to a new solid model of the part (Figure 6.20d), which is parametric with a feature tree that defines the feature regeneration sequence.

To recognize each feature, Venkataraman's FR algorithm abstracts a B-rep model as an attributed face adjacency graph. The faces of the feature are represented as nodes of the graph, while the edges in the feature are represented as lines that connect nodes of the graph. In addition, attributes are added to nodes and lines representing the topological and geometric characteristics of the corresponding faces and edges. For example, the blind pocket feature shown in Figure 6.21a consists of top, bottom, and side faces, as well as numerous edges. The partial face adjacency graph and attributes (e.g., edge convexity) are shown in Figure 6.21b. As a result, the problem of FR becomes a subgraph detection problem in which the feature graph is matched to similar instances in a predefined feature library. This is done by a graph matching algorithm following prescribed rules and grammars.

One of the potential issues revealed in commercial FR software is design intent recovery. For example, the flange of a tubing would be created as a single revolve feature, where a sketch is revolved about an axis (Figure 6.22a). However, current FR implementations are not flexible. As shown in Figure 6.22b, without adequate user interaction, the single sketch flange may be recognized as four or more separate features. Although the final solid parts are physically the same, their defining parameters are not. Such a batch mode implementation is not desired in recovering meaningful design intents.

In tutorial lesson S1.4, we use a housing example, an imported part shown in the left of Figure 6.23a, to illustrate the steps of FR using FeatureWorks of SolidWorks. The FeatureWorks module of SolidWorks recognizes solid features on an imported object in a SolidWorks part document. Recognized features are (almost) the same as features that are created using parametric feature-based CAD software. Designers may edit the definition of recognized features and change their attributes and dimension parameters. For example, the fillet radius is changed from 0.0625 to 0.15 in, as shown on the right of Figure 6.23a. For features that are based on sketches, designers can edit the sketches to change the geometry of the features. We introduce both the automatic and interactive options. Overall,

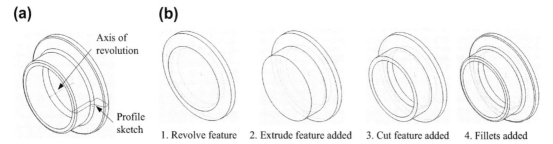

FIGURE 6.22 Feature Recognition for a Tubing Flange. (a) A Single Revolved Feature, and (b) Four Features: Revolve, Extrude, Cut, and Fillet (Chang, 2012).

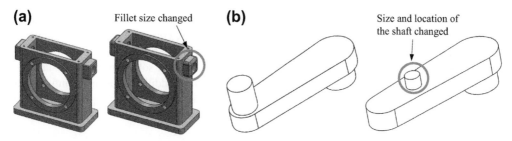

FIGURE 6.23 Examples for Solid Feature Recognition. (a) The Housing Example Employed for Tutorial Lesson S1.4, and (b) The Crankshaft Example Employed for Tutorial Lesson P1.3.

one possible strategy for using FeatureWorks in solid FR is to use automatic FR to recognize as many features as possible, and then recognize the remaining features interactively.

Also in tutorial lesson P1.3, we use a very simple example—a crankshaft, as shown in Figure 6.23b—to illustrate the steps of FR using Pro/ENGINEER. Note that the FR capability implemented in Pro/ENGINEER is primitive and far less useful than that of FeatureWorks. Two shafts (extrusion features) are successfully recognized. For example, the size and location of one of the shafts are changed, as shown on the right of Figure 6.23b. However, the crank body (the first extrusion feature) cannot be recognized because it is considered as the base feature, which is not recognizable by Pro/ENGINEER.

6.6 Case studies

Three case studies are presented in this section: SolidWorks Workgroup PDM, integrated testbed using Windchill, and infrastructure of tool integration for e-Design. The case study of SolidWorks PDM offers engineering students a quick overview about a viable solution for their needs in using PDM for support of design projects. The second study presents a case in which Windchill was integrated to support an engineering team working on reverse engineering projects. This case study is intended to show readers one possible scenario of using a commercial PDM system to streamline engineering activities. The third case study illustrates the concept and implementation of a software infrastructure for tool integration that supports e-Design.

6.6.1 SolidWorks Workgroup PDM

There are two PDM systems offered by SolidWorks: SolidWorks Workgroup PDM and SolidWorks Enterprise PDM. Workgroup PDM is part of the SolidWorks Premium or Professional version and is considered a mid-range PDM system that is intended for small engineering workgroups, usually <10 team members. Enterprise PDM offers more robust capabilities and is better suited for larger teams. Because Workgroup PDM is part of SolidWorks, which many students may have access to, we offer a brief discussion of the Workgroup PDM as a case study.

The Workgroup PDM application is PDM software that runs inside the SolidWorks environment or as a stand-alone application inside SolidWorks Explorer. Workgroup PDM controls projects with procedures for check-in, check-out, revision control, and other administration tasks. A Workgroup PDM structure is illustrated in Figure 6.24, in which a "vault" sits at the center.

Workgroup PDM is very simple to install; installation can be completed in a few minutes by a nonexpert. During installation, a vault location is defined. The vault can be on a local drive or on a network server. In addition to the vault, user and administrator client software is also installed. The PDM license is managed by the SolidWorks license server, so no additional license configuration is required provided that SolidWorks is already installed. To support a project team, a global vault may be located on a network server, with carefully restricted administrative rights. Only one vault can be installed on a given computer.

Workgroup PDM associates metadata with CAD documents, Metadata is a series of text files that contain server options (user information, revision schemes, etc.) and file information (revision history, owner, etc.). Workgroup PDM allows management and viewing of metadata associated with both local documents and documents in the vault.

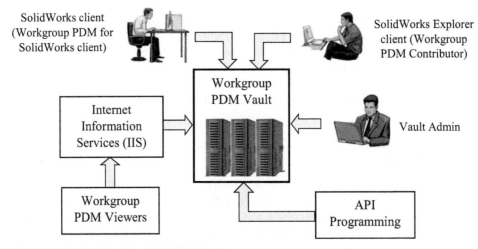

FIGURE 6.24 SolidWorks Workgroup PDM Structure.

When a file in the vault is opened and checked out, the file is downloaded from the vault to a local folder. If an assembly or drawing is opened, all of the referenced files are downloaded as well. Workgroup PDM compares the documents on the local drive to the documents in the vault, giving the user the option to overwrite previous versions with the latest versions from the vault. A file can be opened only by its owner. If the owner wishes to release a file for someone else to work on, he or she releases ownership. This file can then be checked out by others. Because Workgroup PDM is a mid-range PDM system best suited to small workgroups, it is easy to administer and use for a small team.

6.6.2 Integrated testbed using Windchill

In this case study, we present an integrated testbed that supports defense logistics centers to conduct reverse engineering of aging systems and components (Chang et al., 2006). This testbed, which was constructed using commercial off-the-shelf software and equipment, supports three major engineering tasks: the reverse engineering that supports recovering of technical data from worn sample parts, reengineering that alters design for better performance or lesser cost, and fast prototyping that incorporates advanced manufacturing technologies to produce functional or physical prototype of the part in small quantity in a short turnaround time.

Most reverse engineering solutions involve multidisciplinary design activities. Consequently, design collaboration is essential for a typical reverse engineering project to allow designers in different disciplines to perform their roles. In the integrated system, the design collaboration is based on two kinds of designers' interactions: asynchronous and synchronous. Asynchronous interactions involve email, notification, forums, and sharing documents where the designer is not required to respond in real-time. During synchronous interactions, the designer is required to respond in real-time. These synchronous interactions include whiteboard, chat room, model viewer, and video and audio communication. To meet these requirements, the integrated testbed supports the following:

- Appropriate distribution of activities to members of the team;
- Tools that can support real-time collaboration among team members with engineering information;
- An environment that organizes and provides easy access to engineering and other information related to the project for the team;
- A knowledge base that includes information related to different reverse engineering processes, tools, and techniques;
- A reverse engineering template that can be modified to support different reverse engineering processes and reduce the initial effort to setup products.

The testbed is intended to provide a software environment that supports multiple geographically dispersed designers. This principle extends to all reverse engineering activities, data, and collaborative activities, as well as to the infrastructure design. The testbed is set up using simple client-server architecture. The Windchill and communication module is housed in the server and is connected to the Internet. Multiple clients (users) access product and reverse engineering information from the servers using a web browser. Some product management functions supported by the servers are: (1) managing the product data and model in a structure through which a designer can easily locate the product data; (2) keeping the data secure and restrict illegal operations through basic file access controls; (3) providing functions to manage the file operation privileges based on designers' roles in the team; and

(4) supporting file status control to prevent the file inconsistency which may occur when two users modify the same file simultaneously.

To support real-time collaboration, a web-based tool has been developed (Figure 6.25). This collaborative tool supports text messaging, audio, video, sketching, and viewing of 3D models in real-time to facilitate activities required for meetings. To enhance collaboration among different members of the team, the collaborative 3D model viewer allows users to have real-time synchronous view of the model, add notes to the 3D model, and exchange text and audio information in real-time. Collaborative

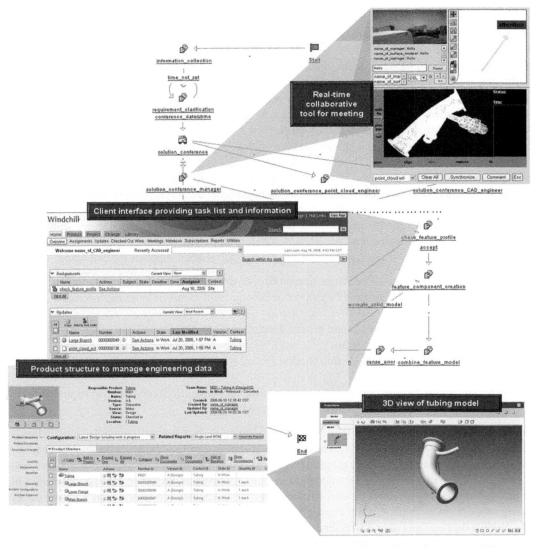

FIGURE 6.25 Reverse Engineering Template and Activities in the Integration Testbed (Chang et al., 2006).

meetings, if needed, can be scheduled in an ad-hoc manner. When a meeting is scheduled, appropriate group members are sent an email that has the web link to the collaborative tool and the scheduled meeting time. During the scheduled time, all group members can log into the collaborative tool to discuss issues related to the project using the testbed.

In order to demonstrate the testbed, a case scenario was created. The reverse and reengineering scenario highlights (1) a systematic reverse engineering approach, (2) an enhanced ability of team member collaboration, and (3) a customized Windchill product management system. The reverse engineering of an airplane anti-icing tubing scenario involves an engineering team consisting of four members who are geographically distributed: manager, CAD engineer, and two point-cloud engineers. A template with a flow of activities (see Figure 6.25), along with appropriate instructions, has been setup in the Windchill environment. This template is the starting point for the manager to initiate a reverse engineering project. The initial steps for the manager involve gathering information, design constraints, and point-cloud information for the product. Once the information has been gathered, the manager creates the team and calls a meeting in the integration framework using the real-time collaborative tools (Figure 6.25) to discuss details of the project. After the meeting, the appropriate reverse engineering process can be selected and modified according to the requirements and needs of the project. The integration framework then supports accomplishing these tasks by appropriate users. Information and instruction on how to complete the different tasks are also available to the users from the testbed. Information created from each activity is uploaded in the testbed for other members of the team to view, access, evaluate, and use. These data are organized in a set of defined folders that follow the product structure to reduce the effort of finding the files. The progress of the project can be monitored by any member of the team at any given time. After each task is completed, the testbed sends appropriate notification to relevant team members to proceed to the next steps.

6.6.3 Tool integration for e-Design

In this case study, we discuss an integration infrastructure that supports tool integration for e-Design (Tsai et al., 1995). The infrastructure supports engineers in creating CAD and simulation models of the mechanical system, accessing CAE tools to perform multidisciplinary engineering analyses, using planning tools to create and manage design processes, communicating and exchanging engineering data, conducting design trade-off analyses, and making informed decisions to yield a robust optimum design.

The infrastructure was designed to correlate various simulation models with a common product representation derived from a CAD model (see Figure 6.26). A base definition is created from the CAD model to serve as the common ground for design data sharing and collaboration. Engineering views are then derived from the base definition to support additional analysis requirements in each engineering discipline. Engineering view models are correlated, or mapped, with the base definition to support design collaboration and can be shared among the design team. Engineering tool wrappers provide service to their respective tools, including accessing the analysis model from the global database, converting model data into tool specific data formats, transmitting data to a specified location, and retrieving and displaying analysis results. Finally, to create an e-Design tool environment, design process management has to be employed to define, disseminate, coordinate, and track design activities.

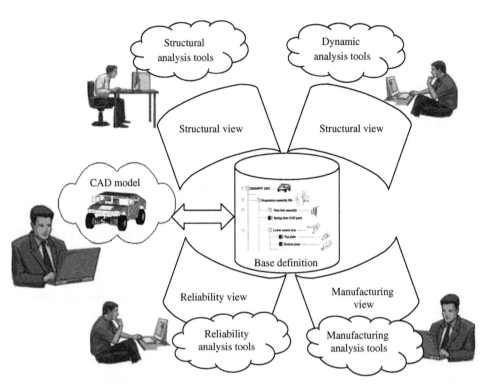

FIGURE 6.26 Concept of the Software Infrastructure for Support of Tool Integration for e-Design (Tsai et al., 1995).

6.6.3.1 CAD and base definition

To support multidisciplinary CAE analyses, a base definition has to be built as the common ground among the CAE team members. The base definition contains two major types of information: an entity hierarchy and entity attributes. The entity hierarchy describes how the components of the system are grouped together. The hierarchy of an engine example is depicted in Figure 6.27. If an entity in the hierarchy is an assembly, it can be expanded to display its components or collapsed without showing its components. The entity attributes for a part include mass, center of gravity (CG), moments of inertia, material properties, and geometry information. The default coordinate system defined in the CAD model is used as the local coordinate system for the part; the CG is reported relative to it. Geometry information of the mechanical system is kept in the original CAD format and later transferred to different formats to support various simulation model creations. In addition, parameters used to build the CAD geometry need to be extracted and later used to support design trade-offs. Attribute information for an assembly differs from that for a part, with the addition of assembly information describing the position and orientation of individual components relative to a local reference frame. Once all the hierarchy information and assembly information are available, the global position and orientation of the individual part or assembly can be automatically calculated.

In the e-Design environment, design parameters are associated with the dimensions of features in the parameterized CAD models. The design parameters are considered to be attributes of entities in the

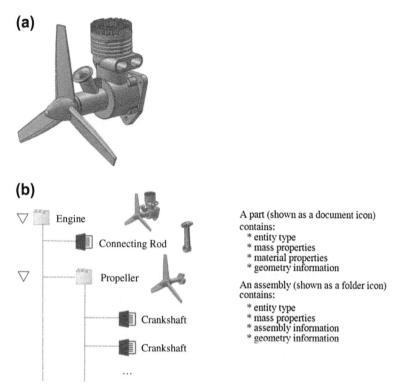

(a)

(b)

Engine

Connecting Rod

Propeller

Crankshaft

Crankshaft

...

A part (shown as a document icon)
contains:
  * entity type
  * mass properties
  * material properties
  * geometry information

An assembly (shown as a folder icon)
contains:
  * entity type
  * mass properties
  * assembly information
  * geometry information

FIGURE 6.27 Engine Example. (a) Computer-Aided Design Model, and (b) Product Base Definition (Tsai et al., 1995).

base definition; they stay with the entities when they are regrouped in engineering views to create assemblies. The feature-based design parameters serve as a common language to support design trade-offs across various engineering disciplines where relevant performance of the mechanical systems is measured.

6.6.3.2 Disciplines and views

In addition to establishment of a common base definition, the integration infrastructure has to support engineers from different disciplines (e.g. dynamics and structure) to create their own simulation models and to perform engineering analyses. Due to the fact that data requirements vary from discipline to discipline, the infrastructure has to allow engineers to augment the model data in the base definition with discipline specific data. While allowing diverse data to be added, the infrastructure has also to maintain the consistency among these data so that the common ground is not broken and design trade-offs across different disciplines can still be performed.

To address these issues, a key concept—engineering views—is introduced in the infrastructure. The engineering views are associated with their corresponding disciplines to support the data

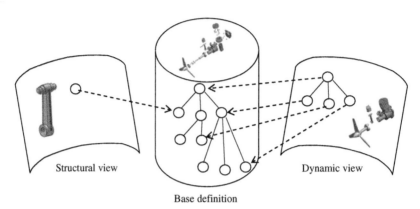

Structural view

Dynamic view

Base definition

FIGURE 6.28 Mappings between Base Definition and Engineering Views (Tsai et al., 1995).

augmentation in a natural way. Furthermore, the data created in the engineering views can be shared among the engineers that perform engineering analyses in the same discipline. Therefore, effort is saved in creating engineering models.

Another important function the engineering views need to support is maintaining a consistent product data set for the mechanical system being evaluated. The mappings between each view model and the base definition has to be established (see Figure 6.28). All the engineering models, along with their simulation results and the CAD model (brought in as the base definition), are correlated through these mappings, allowing meaningful communication among the CAE team members and design trade-offs across disciplines.

Another benefit of establishing these mappings is that they can be used as the foundation of automating the engineering model (re)creation that is required during design iterations. Once a design change is proposed, each engineering discipline has to re-evaluate the performance of the new design. However, if the engineering model has to be regenerated from scratch, with engineers heavily involved, then the effort previously spent in model creation is wasted. Therefore, a mechanism for retaining the mapping relationships between the engineering model and the base definition would greatly speed up the design cycle. As the CAD model is modified, the engineering model could be automatically modified with very little effort. The goal of concurrence design then can be achieved.

As an example, in the dynamics view, the assembly hierarchy defined in the CAD model might not be suitable for multibody mechanical system definition. For that, parts or assemblies need to be regrouped into bodies and then connection joints, allowing relative motion between bodies, need to be defined. Once the regrouping is performed, the composite mass, CG, moments of inertia, and assembly information of the body can be calculated automatically based on individual component mass properties and assembly information. The dynamic view of the engine example shown in Figure 6.27 is shown in Figure 6.29, in which parts under the engine assembly in CAD (and base definition) are grouped and mapped to dynamic view, consisting of three subassemblies and one part that correspond to the motion model of the example. Once the data required in the dynamics view are created, the tool wrapper can be invoked to export the mechanical system definition into the local working environment and create a motion simulation model (e.g., using DADS). The results of the

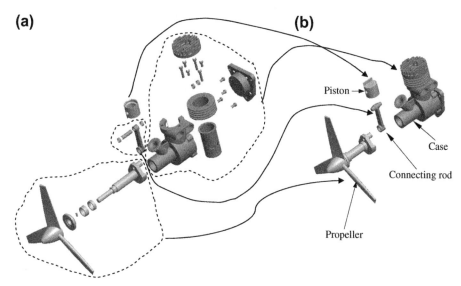

(a)   **(b)**

Piston

Case

Connecting rod

Propeller

FIGURE 6.29 Dynamic View of the Engine Example. (a) Computed-Aided Design, (CAD) Assembly Hierarchy in Base Definition, and (b) CAD Model in Exploded View Showing Four Bodies for Motion Analysis.

dynamics analyses can also be retrieved via the tool wrapper to the global database to support other simulations.

6.6.3.3 Engineering tool wrappers

After the engineering view models are created in the e-Design environment, engineers are ready to perform various analyses using engineering tools. To interface with each engineering tool, the infrastructure needs to provide services to prepare the analysis model from global database, translate it into the tool specific data format, and transmit data to the dedicated location specified by the tool. Each wrapper provides services to a specific tool; in other words, it is customized to properly interface with that tool. For instance, the engineer can use the PATRAN wrapper to transfer the PATRAN hyperpatch model from the structural view and visualize the dynamics simulation results to determine a peak load, which is required by the structural analysis tools. Another service that wrappers provide is retrieving analysis results from each tool. The analysis results are interpreted by the wrappers and then stored in the global database for later use.

6.6.3.4 Design process management

Effective utilization of the integrated e-Design environment and collaboration among a CAE team is contingent on a number of data generation and communication factors intrinsic to the operational requirements of the engineering disciplines and the product design process in general. As described above, the e-Design environment consists of a number of individual engineering views integrated via a common base definition and a suite of tool wrappers. Enhanced collaboration among engineers of this environment, vis-à-vis the design process management, will occur when engineers can identify data sources that meet their analysis requirements and communicate their input data needs and simulation results.

The design process management envisioned for the integrated e-Design environment employs process definition and analysis, task identification and dissemination, and progress tracking to provide enhanced collaboration among the CAE team. Process definition enables CAE teams to specify and capture data generation, design, analysis, and design trade-off activities in the e-Design process and represent data flow between process activities and perspectives. Process analysis allows the chief engineer to identify potential bottlenecks in a process and aids in the definition of an optimal design process. Process activities can be characterized by user data requirements, by operational parameters such as time and resource requirements, and by activity dependencies, thus providing information supporting the determination of a project plan in compliance with the time frame and resources specified for completion of the design project. The design project plan can then be displayed to all team members to provide them with an awareness of where data comes from and where it goes, thus defining responsibilities and obligations in the design project. Finally, progress tracking allows the CAE team to review and update the design project plan and to correlate the design project plan with the design process. By this approach, CAE team members will be provided with a frame of reference, with respect to project planning and environment operations, supporting communication and collaboration to achieve project objectives and adhere to project schedules and milestones.

6.6.3.5 Design collaboration

Lastly, design collaboration includes the communication board, design process management, design parameterization, and design trade-off. The communication board provides a means for CAE team members to communicate about design tasks. The design parameterization module assists engineers in identifying design parameters to facilitate effective design evaluation. The design trade-off module collects performance evaluation information from the engineering tools and assists engineers in obtaining optimal design.

6.7 Summary

In this chapter, we addressed two issues that are very relevant to product design: PDM and product data exchange. We provided an overview on the practical means for file management and an introduction to PDM technology and systems. We presented fundamentals of PDM, including the data that PDM manages, the capabilities of PDM, and the benefits and successful stories of PDM technology. We also discuss commercially available PDM systems. For product data exchanges, we discussed the numerous approaches and translators available, as well as their strengths and limitations. We include examples to demonstrate the part and assembly model translations between Pro/ENGINEER and SolidWorks, as well as FR in both Pro/ENGINEER and SolidWorks. Finally, we offered case studies that illustrate the practical use of SolidWorks Workgroup PDM, a practical case of using PDM for support of reverse engineering project, and infrastructure for support of tool integration for e-Design.

We discussed lots of topics and offered more diverse materials in this chapter. We hope this chapter is not too difficult for you to read and digest. After reading this chapter, you should have acquired basic knowledge and good understanding of these two important issues in product design.

Together with the topics discussed in the previous chapters of this book, you should have a very good understanding of the various subjects involved in product data modeling, such as geometric modeling, CAD theories, mechanical assembly, design parameterization, and PDM. More importantly, we hope this book provided you with a fundamental understanding of product modeling principles and modern engineering tools for solid and assembly modeling, so that you can apply the principles and software tools to support practical design applications.

With a good understanding of product modeling, you should feel competent in using CAD tools in support of your design projects and move into other important topics for e-Design, such as product performance evaluation, product manufacturing and cost estimating, and design theory and methods, which are discussed in other books of the Computer-Aided Engineering Design series.

Appendix A **IGES file structure and data format**

Similar to most CAD systems, IGES is based on the concept of entities. Entities could range from simple geometric objects, such as points, lines, plane, and arcs, to more sophisticated entities, such as dimensions. Entities in IGES are divided into three categories:

1. Geometric entities such as arcs, lines, and points that define the object,
2. Annotation entities, such as dimensions and notes that aid in the documentation and visualization of the object,
3. Structure entities that define the associations between other entities in IGES file.

An IGES file is a sequential file consisting of a sequence of records. The file formats treat the product definition to be exchanged as a file of entities, with each entity being represented in a standard format, to and from which the native representation of a specific CAD/CAM system can be mapped.

An IGES file consists of five sections, which must appear in the following order: Start section, Global section, Directory Entry (DE) section, Parameter Data (PD) section, and Terminate section, as shown in Figure A.1 (Stokes, 1995). In fact, an IGES file is composed of 80-character ASCII records. Each of the sections can be identified by the letters S, G, D, P, and T, respectively appearing in the 73rd column of each record or line in the IGES file. The role of these sections is summarized in the following.

Start section

The start section is for free-form text generated by the user or the CAD/CAM system to tell the receiver basic information about the IGES file, commonly described as a "prologue" to the IGES file. It is essentially a human-readable introduction to the file. This section contains information such as the names of the sending (source) and receiving (target) CAD/CAM systems and a brief description of the product being converted.

(a)   Start

Global

Directory
entry

Parameter
data

Terminate

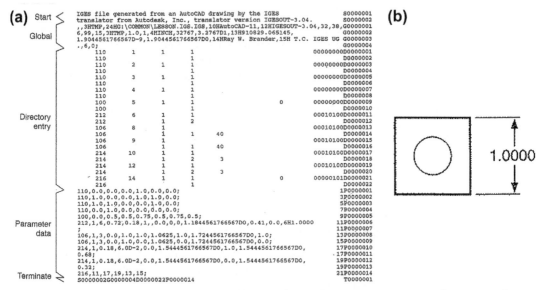

(b)

FIGURE A.1 Anatomy of a Sample IGES File. (a) IGES File with Sections Labeled, and (b) Equivalent Graphics (Stokes, 1995).

Global section

The global section provides information that pertains to the entire file. It is a fairly short section, typically in three or four lines. This information describes the preprocessor and information needed by the postprocessor to interpret the file. Some of the parameters that are specified in this section are:

1. Characters used as delimiters between individual entries and between records (usually commas and semicolons respectively),
2. The name of the IGES file itself,
3. Vendor and software version of sending (source) system,
4. Number of significant digits in the representation of integers and single and double precision floating point numbers on the sending systems,
5. Date and time of file generation,
6. Model space scale,
7. Model units,
8. Minimum resolution and maximum coordinate values,
9. Name of the author of IGES file.

Directory entry section

The directory entry (DE) section is an index listing each entity in the file, together with certain attributes associated with them. The entry for each entity occupies two 80-character

Table A.1 IGES Entities (Stokes, 1995)

IGES Entity Number	Form Numbers	IGES Entity Name
0		Null entry
100		Circular arc
102		Composite curve
104	0–3	Conic arc
106	1–63	Copious data
108	(–1)–1	Plane
110		Line
112		Parametric spline curve
114		Parametric spline surface
116		Point
118	0–1	Ruled surface
120		Surface of revolution
122		Tabulated cylinder
123		Direction (G)
124	0–12	Transformation matrix
125	0–4	Flash
126	0–5	Rational B-spline curve
128	0–9	Rational B-spline surface
130		Offset curve
132		Connect point
134		Node
136		Finite element (G)
138		Nodal displacement and rotation
140		Offset surface
141		Boundary (G)
142		Curve on a parametric surface
143		Bounded surface (G)
144		Trimmed surface
146		Nodal results (G)
148		Elements results (G)
150		Block
152		Right angular wedge
154		Right circular cylinder
156		Right circular cone frustum
158		Sphere
160		Torus
162	0–1	Solid of revolution
164		Solid of linear extrusion

Continued

Table A.1 IGES Entities (Stokes, 1995)—cont'd

IGES Entity Number	Form Numbers	IGES Entity Name
168		Ellipsoid
180		Boolean tree
182		Selected component (G)
184		Solid assembly
186		Manifold solid B-rep object (G)
190		Plane surface (G)
192		Right circular cylindrical surface (G)
194		Right circular conical surface (G)
196		Spherical surface (G)
198		Toroidal surface (G)
202		Angular dimension
204		Curve dimension (G)
206		Diameter dimension
208		Flag note
210		General label
212	0–105	General note (F, G)
213		New general note (G)
214	1–12	Leader (arrow)
216	0–2	Linear dimension (G)
218	0–1	Ordinate dimension (G)
220		Point dimension
222	0–1	Radius dimension (G)
228	0–3	General symbol (G)
230	0–1	Sectioned area (G)
302		Associativity definition
304	1–2	Line font definition
306		Macro (G)
308		Subfigure definition
310		Text font definition
312	0–1	Text display template
314		Color definition
316		Units data (G)
320		Network subfigure definition
322	0–2	Attribute table definition
402	1–21	Associativity instance (F, G)
404	0–1	Drawing (G)
406	1–31	Property (F, G)

Table A.1 IGES Entities (Stokes, 1995)—cont'd

IGES Entity Number	Form Numbers	IGES Entity Name
408		Singular subfigure instance
410	0–1	View (G)
412		Rectangular array subfigure instance
414		Circular array subfigure instance
416	0–4	External reference (G)
418		Nodal load/constraint
420		Network subfigure instance
422	0–1	Attribute table instance
430		Solid instance
502		Vertex (G)
504		Edge (G)
508		Loop (G)
510		Face (G)
514		Shell (G)

Notes: 1. All information is based upon IGES version 5.1, September 1991.
2. F = Some or all forms of this entity have been obsoleted by newer entities.
3. G = Some or all forms of this entity have not been fully tested.

records that are divided into a total of 20 8-character fields as shown in Figure A.1. The first and the eleventh (beginning of the second record of any given entity) fields contain the entity type number such as 100 for circle, 110 for lines, etc. The second field contains a pointer to the parameter data entry for the entity in the PD section. The pointer of an entity is simply its sequence number in the DE section. Some of the entity attributes specified in this section are line font, layer number, transformation matrix, line weight, and color. A list of IGES entities is provided in Table A.1 for reference.

Parameter data section

The parameter data (PD) section contains the actual data defining each entity listed in the DE section. For example, a straight line entity is defined by the six coordinates of its two endpoints. Although each entity has always two records in the DE section, the number of records required for each entity in the PD section varies from one entity to another (the minimum is one record) and depends on the amount of data. Parameter data are placed in free format in columns 1–64. The parameter delimiter (usually a comma) is used to separate parameters and the record delimiter (usually a semicolon) is used to terminate the list of parameters. Both delimiters are specified in the Global section of the IGES file. Column 65 is left blank. Columns 66–72 on all PD records contain the entity pointer specified in the first record of the entity in the DE section.

(a)   **(b)**

```
                                                            S    1
1H,,1H;,4HSLOT,37H$1$DUA2:[IGESLIB.BDRAFT.B2I]SLOT.IGS;,     G    1
17HBravo3 BravoDRAFT,31HBravo3->IGES V3.002 (02-Oct-87),32,38,6,38,15,  G    2
4HSLOT,1.,1,4HINCH,8,0.08,13H871006.192927,1.E-06,6.,       G    3
31HD. A. Harrod, Tel. 313/995-6333,24HAPPLICON - Ann Arbor, MI,4,0;   G    4
    116     1     0     1     0     0     0     0    1D    1
    116     1     5     1     0                      0D    2
    116     2     0     1     0     0     0     0    1D    3
    116     1     5     1     0                      0D    4
    100     3     0     1     0     0     0     0    1D    5
    100     1     2     1     0                      0D    6
    100     4     0     1     0     0     0     0    1D    7
    100     1     2     1     0                      0D    8
    110     5     0     1     0     0     0     0    1D    9
    110     1     3     1     0                      0D   10
    110     6     0     1     0     0     0     0    1D   11
    110     1     3     1     0                      0D   12
116,0.,0.,0.,0,0,0;                                 1P    1
116,5.,0.,0.,0,0,0;                                 3P    2
100,0.,0.,0.,1.,0.,-1.,0,0;                         5P    3
100,0.,5.,5.,-1.,5.,1.,0,0;                         7P    4
110,0.,-1.,0.,5.,-1.,0.,0,0;                        9P    5
110,0.,1.,0.,5.,1.,0.,0,0;                         11P    6
S    1G    4D    12P    6                            T    1
```

FIGURE A.2 Sample IGES File. (a) IGES File Contents, and (b) Equivalent Graphics (en.wikipedia.org).

Terminate section

The Terminate section contains a single record that specifies the number of records in each of the four preceding sections for checking purposes.

Figure A.2 shows another IGES sample file containing only two POINT (Type 116), two CIRCULAR ARC (Type 100), and two LINE (Type 110) entities. It represents a slot, with the points at the centers of the two half-circles that form the ends of the slot, and the two lines that form the sides.

As stated earlier, the file is divided into five sections: Start, Global, Directory Entry, Parameter Data, and Terminate, indicated by the characters S, G, D, P, or T in column 73. The characteristics and geometric information for an entity are split between two sections—one is in a two-record, fixed-length format (the DE section), whereas the other is in a multiple-record, comma-delimited format (the PD section), as can be seen in a more human-readable representation of the file. When displayed, the user should see two yellow points, one located at the origin of model space [0,0,0], two red circular arcs, and two green lines.

For a more in-depth discussion on IGES, readers are referred to books such as Kennicott (1996).

Appendix B Step data structure and applications protocols

STEP consists of several hundred documents called parts, as illustrated in Figure B.1. Every year new parts are added or new revisions of older parts are released. This makes STEP the biggest standard within ISO. Each part has its own scope and introduction. These parts are assigned a name and number and grouped together with common functions within a specific range.

The 10 series parts comprise the computer-interpretable area of STEP. This area allows all users to operate by the same guidelines and rules necessary to maintain consistent, accurate data

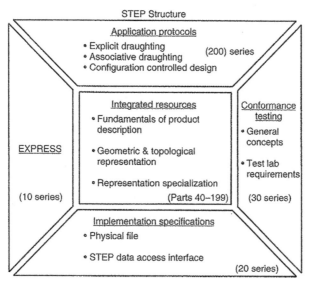

FIGURE B.1 The Structure of STEP (Dincau, 1995).

exchange. EXPRESS is the data modeling language used to make STEP computer-interpretable. The language can be compiled to produce "C" structures, SQL statements, or other similar types of information. This language is an important advantage of STEP over IGES, which offers nothing comparable.

The 20-series parts define the physical file and database-sharing exchange area and are the enabling tools for STEP data translation.

The 30-series parts define conformance testing requirements and are used for data and application verification.

The 40-series parts are considered to be the bread and butter of STEP. These parts contain such generic resource information as raw geometry and display attributes, among other things. These and the 100-series parts are the tools used to create application protocols (APs).

The 100-series parts are similar in concept to the 40-series parts in their use to create application protocols. The difference between the two is that the 100-series is specific to an application area.

The 200-series are where STEP meets the real world. Each AP includes a scope describing its purpose, an activity diagram describing the functions that an engineer needs to perform within that scope, and an application requirement model describing the information requirements of those activities. These information requirements are then mapped into the common set of integrated resources. The result is a data exchange standard for the activities within the scope.

The ultimate goal is for STEP to support the product life cycle, from conceptual design to final disposal, for all kinds of products. However, it will be a number of years before this goal is reached. The most tangible advantage of STEP to users today is the ability to exchange design data as solid models and assemblies of solid models. Other data exchange standards, such as the newer versions of IGES, also support the exchange of solid models, but not as well.

Table B.1 A List of STEP Application Protocols

Part	Description
201	Explicit drafting
202	Associative drafting
203	Configuration-controlled design
204	Mechanical design using boundary representation
205	Mechanical design using surface representation
206	Mechanical design using wireframe representation
207	Sheet metal dies and blocks
208	Life cycle product change process
209	Design through analysis of composite and metallic structures
210	Electronic printed circuit assembly, design, and manufacturing
211	Electronics test diagnostics and remanufacture
212	Electrotechnical plants
213	Numerical control process plans for machined parts
214	Core data for automotive mechanical design processes
215	Ship arrangement
216	Ship molded forms
217	Ship piping
218	Ship structures
219	Dimensional inspection process planning for CMMS
220	Printed circuit assembly manufacturing planning
221	Functional data and schematic representation for process plans
222	Design engineering to manufacturing for composite structures
223	Exchange of design and manufacturing DPD for composites
224	Mechanical product definition for process planning
225	Structural building elements using explicit shape rep
226	Shipbuilding mechanical systems
227	Plant spatial configuration
228	Building services
229	Design and manufacturing information for forged parts
230	Building structure frame steelwork
231	Process engineering data
232	Technical data packaging
233	Systems engineering data representation
234	Ship operational logs, records and messages
235	Materials information for products
236	Furniture product and project
237	Computational fluid dynamics
238	Integrated CNC machining
239	Product life cycle support
240	Process planning

A list of the STEP APs is given in Table B.1. The ability to support many protocols within one framework is one of the key strengths of STEP. All the protocols are built on the same set of integrated resources, so they all use the same definitions for the same information. For example, AP203 and AP214 use the same definitions for three-dimensional geometry, assembly data, and basic product information. Therefore, CAD vendors can support both with one piece of code.

Questions and exercises

1. When you open or save a CAD model using Pro/ENGINEER or SolidWorks (or other CAD systems), you may notice that the CAD software is able to open and save the model into many different formats. It is important for a CAD user and a designer to know these file formats, where they come from, and their use in solid modeling and product design. In this exercise, you are asked to create a list of file formats that are supported by either Pro/ENGINEER or SolidWorks (or the CAD system you have access to) and report the following:
   a. File suffix and the name of the file format.
   b. Source of the file format (for example, the CAD system that creates the file or a neutral format). For neutral format, please report the nature of the format, its use in data exchange, and pros and cons.
   c. Provide the sources of your information, web links, technical report, etc.
2. Pick a commercial PDM (or PLM) system, carry out a case study on the system, and report the following:
   a. Brief information about the software company who develops and commercializes the PDM system.
   b. Major functions of the system, its strengths, and its weaknesses.
   c. Major companies who are using the system.
   d. Provide sources of your information, web links, technical report, etc.
3. In Section 6.3.4 we discussed two stories regarding the use of PDM systems in industry. Please conduct a similar study to report a similar successful story and report the following:
   a. Name of the company or organization, and the nature of its product or project.
   b. The PDM system the company uses.
   c. How did the company or organization use the PDM system? What was the driving factor that propelled the company to adopt PDM?
   d. What is the benefit that company is able to obtain (for example, reduced time-to-market, increased engineering efficiency, reduced prototype costs, reduced late changes to parts, etc.)? Provide quantitative data if possible.
   e. Provide the sources of your information, web links, technical report, etc.

References

Abramovici, M., Sieg, O.C., 2002. Status and development trends of product lifecycle management systems. In: Proceedings of the IPPD 2002 Wroclaw. Wroclaw, Poland.

ACIS, Spatial Technology, Inc. www.spatial.com.

AutoCAD Mechanical, 2009. Autodesk, Inc., Sausalito, CA. 203.53.66.237/Data/Attachments/AutoCAD% 20Mechanical%202009%20Detail_HR.pdf.

Bhandarkar, M.P., Nagi, R., 2000. STEP-based feature extraction from STEP geometry for Agile manufacturing. Computers in Industry 41, 3–24.

Buchal, R.O. The use of product data management (PDM) software to support student design projects. In: The 3rd CDEN/RCCI International Design Conference, Toronto, Ontario, July 24–26, 2006.

Caldwell, B., Mocko, G.M. Product data management in undergraduate education, DETC2008–50015. In: Proceedings of the ASME 2008 International Design Engineering Technical Conferences & Computers and Information in Engineering Conference, August 3–6, 2008, Brooklyn, New York, USA.

Chandrasegaran, S.K., Ramania, K., Sriram, R.D., Horváth, I., Bernard, A., Harik, R.F., Gao, W., 2013. The evolution, challenges, and future of knowledge representation in product design systems. Computer-Aided Design 45, 204–228.

Chang, K.H., March 2012. A review on shape engineering and design parameterization in reverse engineering. Reverse Engineering, InTech 268-4, 162–186. ISBN 979-953-307.

Chang, K.H., Choi, K.K., Wang, J., Tsai, C.S., Hardee, E., June 1998. A Multi-level product model for simulation-based design of mechanical systems. Concurrent Engineering Research and Applications (CERA) Journal 6 (2), 131–144.

Chang, K.H., Siddique, Z., Edke, M., Chen, Z., 2006. An integrated testbed for reverse engineering of aging systems and components. Computer-Aided Design and Applications 3 (1–4), 21–30.

CIMdata (1998), Product Data management: the Definition. An Introduction to Concepts, Benefits, and Terminology.

Crnkovic, I., Asklund, U., Persson Dahlqvist, A., 2003. Implementing and Integrating Product Data Management and Software Configuration Management. Artech House, London.

Dassault Systèmes, 2010. SolidWorks Corp., White Paper: PDM vs PLM: It All Starts with PDM.

Dassault Systèmes, 2013. Enovia V6R2013x Enovia Factsheet. www2.3ds.com/fileadmin/PRODUCTS/Enovia/ PDF/Datasheets/Enovia-V6R2013x-factsheet.pdf.

Dincau, M.A., 1995. Solid modeling and product data exchange using step (PDES/STEP). In: LaCourse, D.E. (Ed.), Handbook of Solid Modeling. McGraw-Hill, Inc.

Hepplemann, J., 1998. PDM for the enterprise. Mechanical Engineering 120 (10), 74–79.

Horváth, I., 2004. A treatise on order in engineering design research. Research in Engineering Design 15, 155–181.

IGES, January 2001. In: Lide, D.L. (Ed.), A Century of Excellence in Measurements, Standards, and Technology—a Chronicle of Selected NBS/NIST Publications, 1901–2000. NIST Special Publication 958.

ISO, 2005. Industrial Automation Systems and Integration—Product Data Representation and Exchange—Part 108: Integrated Application Resource: Parameterization and Constraints for Explicit Geometric Product Models, ISO 10303–108:2005. International Organization for Standardization (ISO), Geneva, Switzerland.

Kennicott, P.R., 1996. Initial Graphics Exchange Specification, IGES 5.3. U.S. Product Data Association.

Lee, S.G., Ma, Y.-S., Thimm, G.L., Verstraeten, J., 2008. Product lifecycle management in aviation maintenance, repair and overhaul. Computers in Industry 59, 296–303.

Lee, D., Shin, H., Choi, B.K., 2010. Mediator approach to direct workflow simulation. Simulation Modeling Practice and Theory 18 (5), 650–662.

Liu, D.T., Xu, X.W., 2001. A review of web-based product data management systems. Computers in Industry 44, 251–262.

Miller, E., 1998. PDM moves to the mainstream. Mechanical Engineering 120 (10), 74–79.

Nell, J., 2001. STEP on a Page, NIST. www.nist.gov/sc5/soap.

Owen, R., Horváth, I., 2002. Towards product-related knowledge asset warehousing in enterprises. In: Proceedings of the 4th International Symposium on Tools and Methods of Competitive Engineering. TMCE, pp. 155–170.

Parasolid, EDS, Corporate Headquarters, Plano, Texas: www.eds.com/products/plm/parasolid.

Qiu, Z.M., Wong, Y.S., June 2007. Dynamic workflow change in PDM systems 58 (5), 453–463.

Shah, J.J., 1995. Parametric and Feature-based CAD/CAM: Concepts, Techniques, and Applications. John Wiley & Sons.

Spitzy, S., Rappoport, A., 2004. Integrated feature-based and geometric CAD data exchange. In: Elber, G., Patrikalakis, N., Brunet, P. (Eds.), ACM Symposium on Solid Modeling and Applications.

Stark, J., 2011. Product Lifecycle Management: 21st Century Paradigm for Product Realization (Decision Engineering), second ed. Springer-Verlag. 10: 0857295454.

Stokes, H., 1995. Solid modeling and the initial graphics exchange specification (IGES). In: LaCourse, D.E. (Ed.), Handbook of Solid Modeling. McGraw-Hill, Inc.

Sunil, V.B., Pande, S.S., 2008. Automatic recognition of features from freeform surface CAD models. Computer-Aided Design 40, 502–517.

Tang, D., Qian, X., 2008. Product lifecycle management for automotive development focusing on supplier integration. Computers in Industry 59, 288–295.

Tsai, C.S., Chang, K.H., Wang, J. Integration infrastructure for a simulation-based design environment. In: Proceedings of the Computers in Engineering Conference and the Engineering Data Symposium, ASME Design Theory and Methodology Conference, pp. 9–20, Boston, MA, August 1995.

van der Velden, C., Zhang, H.-L., Yu, X., Jones, T., Fieldhouse, I., Bil, C. Extracting engineering features from b-rep geometric models. In: 27th International Congress of the Aeronautical Sciences, ICAS2010, 19–24 September 2010, Nice France.

Venkataraman, S., Sohoni, M.A., Kulkarni, V. A graph-based framework for feature recognition. In: Sixth ACM Symposium on Solid Modeling and Applications, pp. 194–205, Ann Arbor, Michigan, June 4–8, 2001.

Project S1 Solid Modeling with SolidWorks

CHAPTER OUTLINE

Computer-aided design (CAD) assembly was discussed in Chapter 4, in which both theoretical and practical aspects of the subject were discussed. There is no need to emphasize the importance of understanding the concept and theory employed for creating assembly. It is equivalently important for engineers to learn to use CAD software for creating assembly. In addition, in Chapter 6, we discussed an important and practical issue in product data management—that is, CAD model translation between heterogeneous CAD systems. We discussed the numerous means available for addressing this issue. It is critical for engineers to learn how to handle CAD model translations using existing tools.

In Project S1, we introduce SolidWorks for creating assemblies and supporting CAD model translations. We include three examples to help you get started on learning and using the software: a single-piston engine, in which you will learn the basics of creating an assembly using mating constraints; a gear train assembly imported from Pro/ENGINEER to SolidWorks; and a simple housing part, illustrating the details of solid feature recognition. Example models are available for download at the book's companion website (http://booksite.elsevier.com/9780123985132).

Overall, the objective of this project is to enable readers to use SolidWorks for constructing assembly models and carrying out model translations. For those who are interested in learning more about part and assembly modeling to elevate yourself to an intermediate level (and beyond), you may want to go over more examples offered by tutorial books on the subject, such as the *SolidWorks 2013*

Bible by Matt Lombard (www.wiley.com/WileyCDA/WileyTitle/productCd-1118508408.html) or the *Beginner's Guide to SolidWorks 2013—Level I* by Alejandro Reyes (www.sdcpublications.com/Textbooks/Beginners-Guide-SolidWorks-2013-Level/ISBN/978-1-58503-774-2).

Note that the lessons included in this project were developed using SolidWorks 2012 SP4.0. If you are using a different version of SolidWorks, you may see slightly different menu options or dialog boxes. Because SolidWorks is fairly intuitive to use, these differences should not be too difficult to figure out.

S1.1 Introduction to SolidWorks

SolidWorks® is a computer software tool that supports users in creating solid models and beyond. The SolidWorks part mode allows users to create parts, whereas its assembly mode supports the assembling of parts to create an assembly. You can build complex assemblies consisting of many components, which can be parts or subassemblies. For most operations, the behavior of components is the same for both types. Adding a component to an assembly creates a link between the assembly and the component. When SolidWorks opens the assembly, it finds the component file and brings the component to the assembly. Changes in the component are automatically reflected in the assembly, and vice versa. The document name extension for assemblies is *.SLDASM*.

The main objective of this tutorial project is to help you, as a new user, become familiar with SolidWorks assembly capabilities and capabilities offered by SolidWorks for support of solid model translations. The discussion on SolidWorks software in this section is brief. For more information about capabilities and use of menus and buttons, refer to SolidWorks Help by selecting from the pull-down menu: Help → SolidWorks Help.

S1.1.1 User interface

A typical user interface window in SolidWorks (Figure S1.1) consists of a graphics area, a FeatureManager design tree (in the Browser), CommandManager (with toolbar), and filters. The graphics area displays the solid model with which you are working on. The CommandManager above the graphics area is a context-sensitive toolbar that dynamically updates based on the toolbar you want to access. By default, it has toolbars embedded in it based on the model type. When you click a tab below the CommandManager, it updates to show that toolbar. For example, if you click the Sketches tab, the Sketch toolbar appears. The buttons in the CommandManager of the assembly are listed in Table S1.1. Click some of the buttons and try to become more familiar with their functions.

The FeatureManager design tree to the left of the graphics area displays these items for assemblies:

- Top-level assembly (the first item)
- Various folders, such as Annotations A and Mates 🔘🔘
- Assembly planes and origin
- Components (subassemblies and individual parts)
- Assembly features (cuts or holes) and component patterns

You can expand or collapse each component to view its details by clicking ⊞ beside the component name. To collapse all the items in the tree, right-click anywhere in the tree and select Collapse Items.

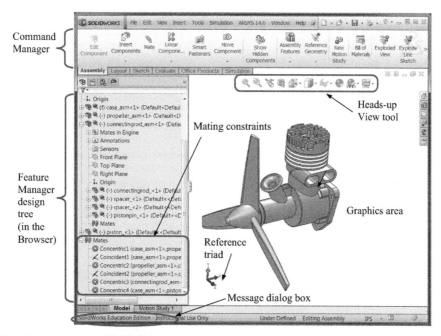

FIGURE S1.1 User Interface of SolidWorks.

Table S1.1 The Shortcut Buttons in the Assembly Toolbar of CommandManager		
Symbol	**Name**	**Function**
	Edit component	To edit a part while in an assembly
	Insert component	To insert a component to an assembly
	New part	To create a new part in an assembly
	New assembly	To create a new assembly
	Copy with mates	To include mates when you create additional instances of a component in an assembly
	Mate	To position two components relative to each other
	Linear component pattern	To pattern components in one or two linear directions
	Circular component pattern	To create a circular pattern of components in an assembly

Continued

Table S1.1 The Shortcut Buttons in the Assembly Toolbar of CommandManager—cont'd

Symbol	Name	Function
	Feature-driven component pattern	To create a pattern of components based on an existing pattern
	Mirror components	To add components by mirroring existing part or subassembly components
	Smart fasteners	To add fasteners to the assembly using the SolidWorks toolbox library of standard hardware
	Move component	To move a component within the degrees of freedom defined by its mates
	Rotate component	To rotate a component within the degrees of freedom defined by its mates
	Show hidden components	To temporarily show all hidden components and make the selected component visible
	Assembly features	To create various assembly features
	Reference geometry	To create various assembly reference geometry
	New motion study	To insert a new motion study
	Bill of materials	To add a bill of materials
	Exploded view	To separate the components into an exploded view
	Explode line sketch	To add or edit a three-dimensional (3D) sketch showing the relationships between exploded components
	Instant 3D	To quickly create and modify model geometry using drag handles and rulers

You can use the same component multiple times within an assembly. For each occurrence of the component in the assembly, the suffix $<n>$ is incremented.

Also, in the FeatureManager design tree, a component name can have a prefix that provides information about the state of its relationships to other components. The prefixes are:

($-$) underdefined
($+$) overdefined
(f) fixed
(?) not solved

The absence of a prefix indicates that the component's position is fully defined. Right-clicking on a node in the FeatureManager design tree will bring up command options that you can choose to modify or adjust the entity.

The Message dialog box at the bottom of the SolidWorks window provides information related to the function you are performing.

S1.1.2 **Examples**

Three simple examples are included in this tutorial project, which illustrates the step-by-step details of creating an assembly, importing Pro/ENGINEER parts and assemblies to SolidWorks, and recognizing solid features in an imported part. We start with a very simple single-piston engine example, in which the propeller is allowed to rotate. We then use a gear train assembly to show how to import a Pro/ENGINEER part and assembly to SolidWorks. As for importing assemblies, you will learn the reality in terms of translating the assembly mating constraints from one CAD software to another. We use a simple housing part to illustrate the details in solid feature recognition. All examples and topics to be discussed in each lesson are summarized in Table S1.2.

Table S1.2 Examples Employed in This Project

Section	Example	Solid Models	Things to Learn
S1.2	Single-piston engine		**1.** This is an introductory tutorial lesson, showing detailed steps for creating an assembly using four components. **2.** We review the mating constraints that are defined between parts and subassemblies. **3.** You will learn how to create an assembly using basic mating constraints that allows the propeller to rotate.
S1.3	Gear train assembly		**1.** In this tutorial lesson, we focus on importing Pro/ENGINEER parts and assemblies to SolidWorks. **2.** We discuss numerous options for CAD model translations offered by SolidWorks. **3.** You will learn two options in bringing Pro/ENGINEER parts and assemblies into SolidWorks—Option 1: importing solid features and Option 2: importing just geometry—as well as their pros and cons.
S1.4	Hosing		**1.** In this tutorial lesson, we focus on the solid feature recognition capabilities offered by FeatureWorks **2.** We discuss both automatic and interactive options and a best possible strategy for feature recognition using FeatureWorks. **3.** You will learn both automatic and interactive options, as well as their pros and cons.

S1.2 Single-piston engine

In this lesson, you will learn how to create an assembly model for a single-piston engine, as shown in Figure S1.2. You will learn how to select mating constraints (called assembly mates in SolidWorks) to assemble parts and subassemblies. After the assembly is created, you may drag the propeller to check the kinematics of the assembly. We start this lesson with a brief overview about the engine assembly to be created in SolidWorks, then show you the detailed steps for creating the assembly.

S1.2.1 The single-piston engine example

The engine example consists of four major components: case (case_asm), propeller (propeller_asm), connecting rod (connectingrod_asm), and piston, as shown in Figure S1.3a. For this lesson, the parts and subassemblies have been created in SolidWorks. There are 23 model files provided, including three assemblies: case_asm, propeller_asm, and connectingrod_asm. In addition, the final assembled engine example (Engine) is included for your reference. You may open this Engine assembly file to preview it.

The connecting rod is assembled to the propeller (at the crankshaft) using concentric and coincident mates, as shown in Figure S1.3b. The connecting rod is free to rotate relative to the propeller (at the crankshaft) along the x-direction. The piston is assembled to the connecting rod (at the piston pin) using a concentric mate. The piston is also assembled to the engine case using another concentric mate. This mate restricts the piston movement along the y-direction, which in turn restricts the top end of the connecting rod to move vertically. Finally, the propeller is assembled to the case using concentric and coincident mates. The components are assembled so that the propeller is free to rotate along the x-direction.

FIGURE S1.2 The Single-Piston Engine Example in a CAD Assembly.

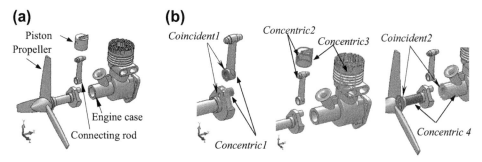

FIGURE S1.3 The Single-Piston Engine Example. (a) Exploded View, and (b) Mating Constraints Defined between components (parts or subassemblies).

S1.2.2 **Using SolidWorks**

Start SolidWorks and choose *File > New*. In the *New SolidWorks Document* dialog box (Figure S1.4), choose *Assembly* and click *OK*. The *Begin Assembly* dialog box will appear to the left of the graphics area (overlapping with the browser), as shown in Figure S1.5.

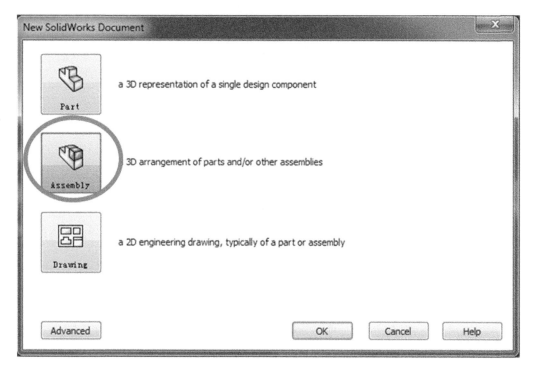

FIGURE S1.4 The *New SolidWorks Document* Dialog Box.

From the *Begin Assembly* dialog box, click the *Browse* button. In the *File Open* dialog box, navigate to the folder *Single-Piston Engine Tutorial* (or the folder where you put the parts and assemblies), pull down the *File Type* list, and then choose *Assembly (*.asm; *.sldasm)* (see Figure S1.6). You should see a list of SolidWorks assembly files in the dialog box. Pick *case_asm.SLDASM*, and click *Open*. The engine case subassembly will appear in the graphics area and will move with your

FIGURE S1.5 The *Begin Assembly* Dialog Box.

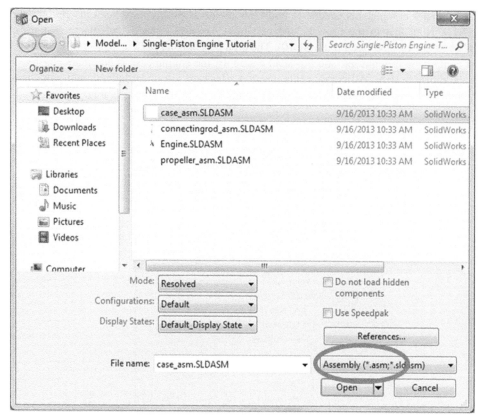

FIGURE S1.6 The *File Open* Dialog Box.

mouse pointer. Left-click at any location within the graphics area to insert the subassembly. Note that the case subassembly is now listed in the browser, and you can expand it to view the details and individual parts by clicking ⊞ beside *(f) case_asm<1>*, as shown in Figure S1.7.

Next, insert other parts and subassemblies. Choose from the pull-down menu:

Insert > Component > Existing Part/Assembly

to bring up the *Insert Component* dialog box, which is similar to the *Begin Assembly* dialog box shown in Figure S1.5. Click on the *Browse* button in the *File Open* dialog box, choose *propeller_asm.SLDASM*, and click *Open* to insert. Left-click in the graphics area to temporarily place the propeller subassembly. Repeat the same steps to bring in the subassembly *connectingrod_asm.SLDASM* and part *piston_.sldprt*. Note that when inserting the piston, you need to choose *Part (*.prt; *.sldprt)* in the *File Type* pull-down list, as shown in Figure S1.8.

Now all major components needed to build the engine assembly have been inserted, as shown in Figure S1.9. Note that in the browser, the first item *Assem1* is the default name of the top-level assembly you just created (see Figure S1.10). The prefix (f) in front of *case_asm<1>* indicates that, as the first component brought into the top-level assembly, the engine case is automatically fixed in the global coordinate system. All components inserted afterward are underdefined, and you can use the mouse pointer to drag and move those floating components in the graphics area.

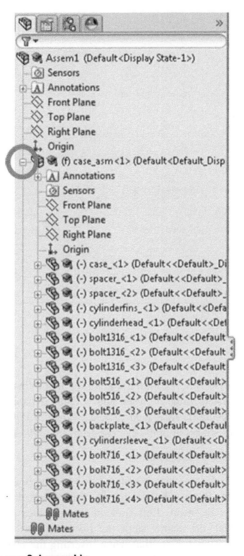

FIGURE S1.7 Expand the *Case_asm* Subassembly.

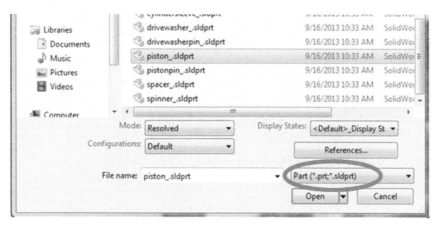

FIGURE S1.8 **Change the File Type from Assembly to Part** *(*.prt; *.sldprt).*

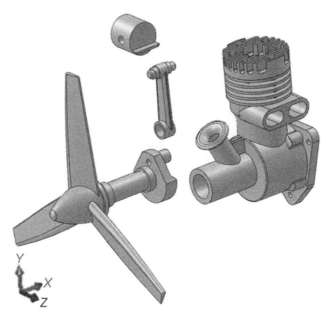

FIGURE S1.9 **Components Inserted.**

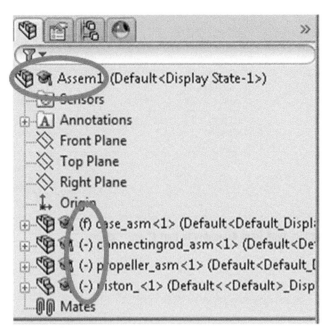

FIGURE S1.10 The FeatureManager Design Tree.

Choose *File* > *Save* from the pull-down menu and save the current assembly as *Engine_Assembly.SLDASM* in the same folder with the components. Once the assembly is saved, the name of the top-level assembly in the browser will change accordingly.

Next, define mating constraints to assemble the parts and subassemblies. From the pull-down menu, choose *Insert* > *Mate*. The *Mate* dialog box will appear, in which the *Mate Selections* box is automatically activated, as shown in Figure S1.11. In the graphics area, pick the back fan-shaped face on the propeller, and then pick the front ring-shape face on the connecting rod, as illustrated in Figure S1.12. You may need to rotate the view by holding down and dragging in the graphics area using the middle mouse button so that you can see and select both faces (see Figure S1.12b).

As shown in Figure S1.13, the two chosen surfaces will be listed in the *Mate Selections* box, and the *Coincident* mate option under *Standard Mates* will be automatically selected based on the types of the two entities you chose to define the mate. At the same time, in the graphics area, the position and orientation of corresponding components (in this case, the propeller and the connecting rod) will be adjusted to comply with the mate. Click the *OK* button ✔ at the top left corner of the *Mate* dialog box to accept the coincident mate. The *Mate Selections* box will be cleared for you to pick the entities for the next mate.

FIGURE S1.11 The *Mate* Dialog Box.

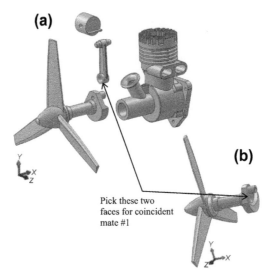

(a)

(b)

Pick these two
faces for coincident
mate #1

FIGURE S1.12 Coincident Mate #1. (a) Face to Pick in the Connecting Rod, and (b) Face to Pick in the crankshaft.

FIGURE S1.13 The Two Chosen Surfaces Listed.

FIGURE S1.14 Options to Undo the Selection.

If you accidently selected a wrong surface when defining a mate, right-click on the *Mate Selections* box and you will be able to delete the highlighted item or clear all selections, as shown in Figure S1.14. If you accepted a wrong mate, you can click the undo button 🔄 (Figure S1.14) to cancel the last mate.

Now, define the second mate by selecting the cylindrical surface of the crankshaft on the propeller and the inner surface of the hole on the connecting rod (Figure S1.15), and then choosing *Concentric* under *Standard Mate* (Figure S1.15). Click the *OK* button ✓ to accept.

Next, pick the cylindrical surface of the piston pin on the connecting rod and the inner surface of the hole on the piston, as shown in Figure S1.16. Choose *Concentric* as the mate type and click ✓ to accept.

Following the same steps, define a concentric mate between the outer cylindrical surface of the piston and the inner cylindrical surface of the engine case. As shown in Figure S1.17, to select the inner face of the engine case, you may need to rotate the view and click on the surface through the oval-shaped hole on the engine case.

Next, define a coincident mate between the front ring shaped face on the engine case and the back ring shaped face on the propeller, as shown in Figure S1.18.

Finally, as shown in Figure S1.19, pick the surface of the propeller shaft and the inner surface of the shaft tube on the engine case. Select *Concentric* as the mate type in the *Mate* dialog box and click *OK* ✓. Note that all the mates you have defined so far are listed under *Mates* in the *Mate* dialog box (Figure S1.20). Click the *OK* button ✓ again to accept all mates and close the dialog box.

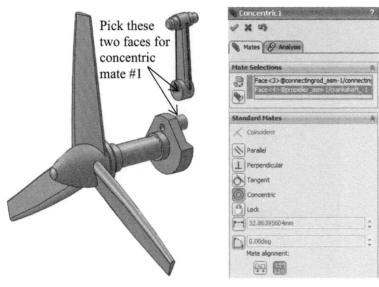

FIGURE S1.15 Concentric Mate #1.

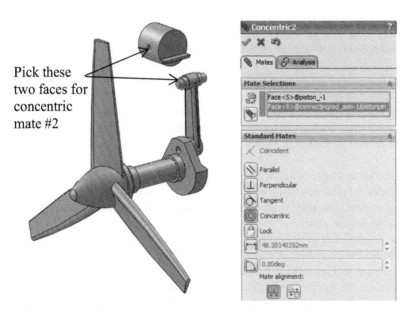

FIGURE S1.16 Concentric Mate #2.

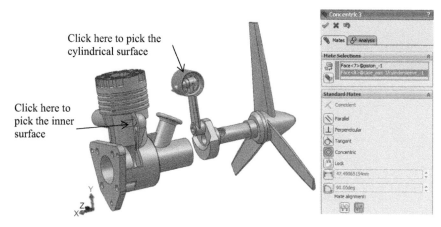

Click here to pick the cylindrical surface

Click here to pick the inner surface

FIGURE S1.17 Concentric Mate #3.

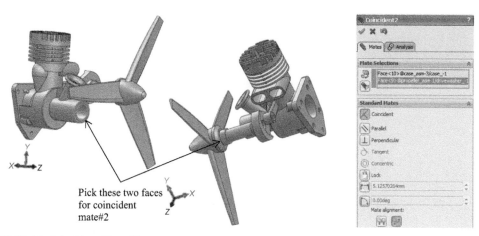

Pick these two faces for coincident mate#2

FIGURE S1.18 Coincident Mate #2.

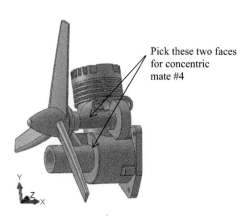

Pick these two faces for concentric mate #4

FIGURE S1.19 Concentric Mate #4.

FIGURE S1.20 List of Mating Constraints Defined.

Now the single-piston engine assembly has been created, as shown in Figure S1.21. We used two coincident and four concentric mates to assemble the components. You can expand *Mates* in the browser to view all the assembly mates (see Figure S1.22). When you click on any assembly mates, the corresponding mating entities will be highlighted in the graphics area.

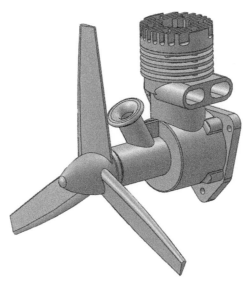

FIGURE S1.21 The Single-Piston Engine Assembly.

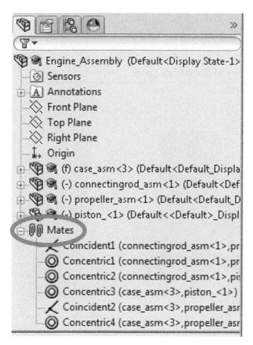

FIGURE S1.22 Mates in the Assembly.

In this assembly, the propeller is free to rotate along the *x*-direction. You can drag the propeller in the graphics area to check the kinematics of the engine. If you right-click on *case_asm<1>* in the browser and choose *Change Transparency* (see Figure S1.23), the engine case will turn partially transparent, and you can see the motion of the piston when the propeller is rotating, as shown in Figure S1.24. Finally, save your model.

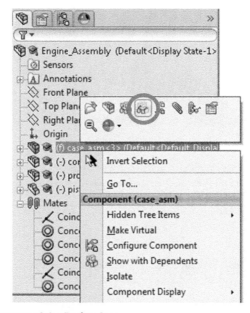

FIGURE S1.23 Adjust Transparency of the Engine Case.

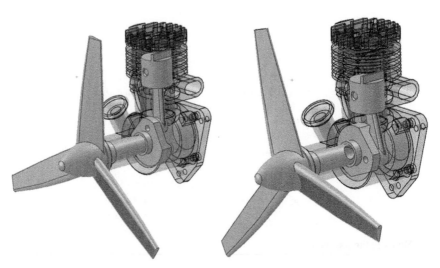

FIGURE S1.24 Dragging the Propeller to Create Motion for the Assembly.

S1.3 **Importing Pro/ENGINEER parts and assemblies to SolidWorks**

From time to time when conducting engineering design using SolidWorks, you may encounter the need to import solid models from other CAD software, such as Pro/ENGINEER. SolidWorks provides an excellent capability that supports importing solid models from a broad range of software and formats, including Parasolid, ACIS, Initial Graphics Exchange Standards (IGES), Standard for Exchange of Product Data (STEP), SolidEdge, and Pro/ENGINEER. For a complete list of supported software and formats in SolidWorks, please refer to Figure S1.25. You may access this list by choosing *File > Open* from the pull-down menu, and pull down the *Files of type* in the *File Open* dialog box.

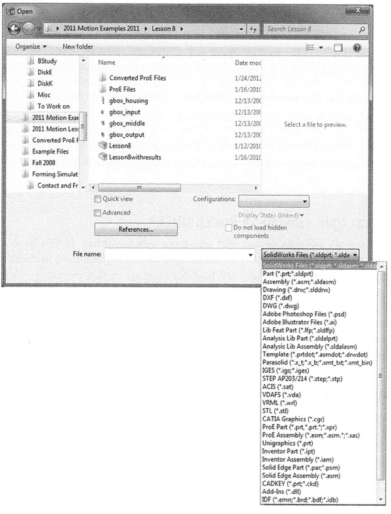

FIGURE S1.25 The *File Open* Dialog Box.

In this tutorial lesson, we focus on importing Pro/ENGINEER parts and assemblies. Hopefully, the methods and principles you learn from this lesson will offer viable ideas for importing solid models from other software and formats.

SolidWorks provides capabilities for importing both parts and assemblies. Users can choose two options when importing solid models: Option 1 to import solid features and Option 2 to import just geometry. Importing solid features may bring you a parametric solid model that you will be able to modify just like a SolidWorks part. On the other hand, if you choose to import geometry only, you will end up with an imported feature that you cannot change because all solid features are lumped into a single imported geometry without any solid features or dimensions.

Importing geometry is relatively straightforward. In general, SolidWorks does a good job of bringing in a Pro/ENGINEER part as a single imported geometry. In fact, several other translators, such as IGES and STEP, support such geometric translations as well. IGES and STEP are especially useful when there is no direct translation from one CAD program to another.

Importing solid models with solid features is a lot more challenging, in which solid features embedded in the part geometry, such as holes and chamfers, must be identified first. In addition, sketches that were employed for generating the solid features must be recovered and the feature types (e.g., revolve, extrude, sweep) must be identified. With a virtually infinite number of possibilities in creating solid features, it is almost certain that you will encounter problems when importing solid models with feature conversion. Therefore, if you do not anticipate making design changes in the parts imported to SolidWorks, it is highly recommend that you import them as a single geometric feature.

We will discuss the approaches for importing parts and assemblies. In each case, we will try both options—that is, importing solid features vs importing geometry. We will use the gear train example as the test case and as an example for illustrations.

S1.3.1 The gear train example in Pro/ENGINEER

The gear train assembly consists of one part and three subassemblies. If you have access to Pro/ENGINEER, you may want to open the final assembly, *gear_train_final.asm*, to check the assembled gear train shown in Figure S1.26. There are four components in this assembly: *gbox_housing.prt*, *gbox_input.asm*, *gbox_middle.asm*, and *gbox_output.asm*. The input and output gear assemblies consist of one gear each, *Pinion 1* and *Gear 2*, respectively. The middle gear assembly has two gears, *Gear 1* and *Pinion 2*. The four spur gears form two gear pairs: *Pinion 1* and *Gear 1*, and *Pinion 2* and *Gear 2*, as illustrated in Figure S1.26. *Gear 1* and *Pinion 2* are mounted on the same shaft. There are 22 distinct parts in this assembly, as listed in Table S1.3.

S1.3.2 Importing Pro/ENGINEER parts

We will import the gear housing (*gbox_housing.prt*) shown in Figure S1.27 using both options. We will try the first option of importing solid features. If you have access to Pro/ENGINEER, you may choose from the pull-down menu *Tools > Model Player* to see the sequence of feature creation. As shown in Figure S1.27 (Pro/ENGINEER model tree), there are eight datum features and 14 solid features. *SolidWorks* will try to import these 14 solid features from Pro/ENGINEER.

Option 1: Importing Solid Features

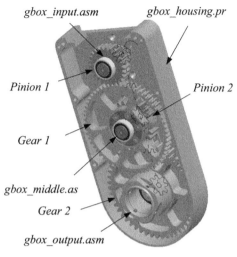

gbox_input.asm   *gbox_housing.pr*

Pinion 1   *Pinion 2*

Gear 1

gbox_middle.as

Gear 2

gbox_output.asm

FIGURE S1.26 The Gear Train Assembly.

Table S1.3 List of Part and Assembly Files		
Part/Subassemblies	**Part Names**	**Remarks**
gbox_housing.prt		
gbox_input.asm	wheel_gbox_shaft_input.prt	
	wheel_gbox_pinion_1s.prt	Pinion 1
	spacer_12 × 18 × 5 mm.prt	
	spacer_12 × 20 × 1 mm.prt	
	bearing_12 × 18 × 8 mm.prt (2)	
	spacer_10 × 18 × 014 mm.prt	
	wheel_gbox_sft_mid_washer.prt	
	screw_tapper_head_5 × 15.prt	
	screw_set_tip_6 × 6.prt (2)	
gbox_middle.asm	wheel_gbox_pinion_2s.prt	Pinion 2
	wheel_gbox_gear_1s.prt	Gear 1
	wheel_gbox_shaft_mid_pinion.prt	
	wheel_gbox_shaft_mid_gear.prt	
	bearing_12 × 18 × 8 mm.prt (2)	
	screw_tapper_head_5 × 28.prt (6)	
	wheel_gbox_sft_mid_washer.prt (2)	
	screw_tapper_head_5 × 15.prt (2)	
	align_pin_4 × 27 mm.prt (2)	
gbox_output.asm	wheel_gbox_gear_2s.prt	Gear 2
	wheel_gbox_connect_wheel.prt	
	bear_tap_roller25 × 47 × 15 mm.prt	
	screw_straight_head_4 × 15.prt (10)	
	align_pin_4 × 20 mm.prt (2)	
	wheel_gbox_connect_wh_setscrew.prt (4)	

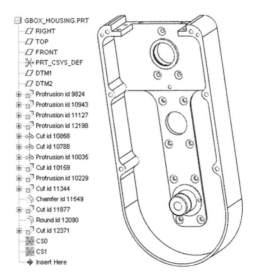

FIGURE S1.27 Gear Housing Part in *Pro/ENGINEER*.

Start *SolidWorks* and choose *File > Open*. Change to the folder where these gear train parts and assemblies reside. In the *File Open* dialog box (Figure S1.25), pull down the *Files of type* list and choose *ProE Part (*.prt, *.prt.*, *.xpr)*. You should see a list of Pro/ENGINEER parts in the dialog box. Click *gbox_housing.prt*, and click *Open*. In the *Pro/ENGINEER to SolidWorks Converter* dialog box (Figure S1.28), choose *Analyze the model completely* (default), and click *Import material properties*, *Import sketch/curve entities*, and *Import geometry from hidden sections*. Click *OK*.

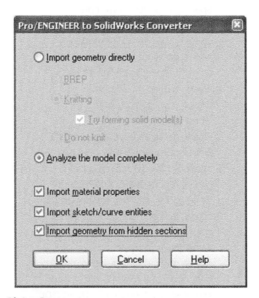

FIGURE S1.28 The Converter Dialog Box.

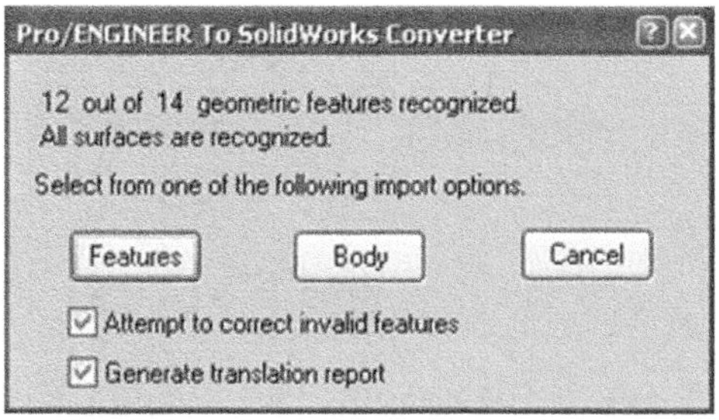

FIGURE S1.29 The Converter Dialog Box.

In the next Converter dialog box (Figure S1.29), click the *Features* button. The conversion process will start. You will see sketches and solid features appear in the graphics window. After about a minute, the *Translation Report* dialog box appears, as shown in Figure S1.30, summarizing the results of the translation. The report indicates that 12 out of 14 features were recognized and translated. The converted model and features listed in the browser are shown in Figure S1.31.

As shown in Figure S1.31, there is one dangling sketch, *Sketch8*, representing the unrecognizable solid feature in addition to the chamfer feature. You may identify the sketch in the graphics area by clicking its name listed in the browser. In addition, the back plate (*Extrude1* in the browser) is recognized incorrectly. Certainly, *SolidWorks* is capable of importing some parts correctly and completely, especially when the solid features are relatively simple (but not this gear housing part).

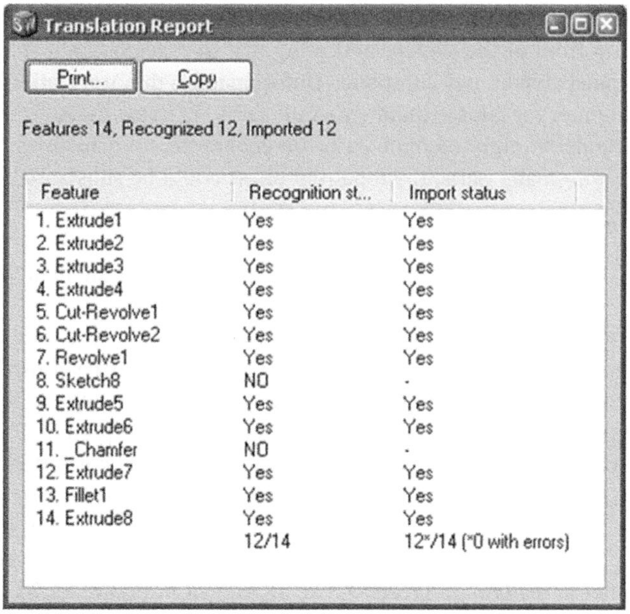

FIGURE S1.30 The Translation Report.

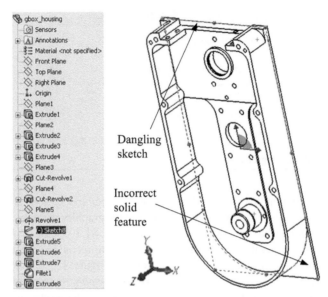

FIGURE S1.31 The Converted Solid Model and Features.

If you take a closer look at any of the successful solid features, for example *Extrude3*, you will see that the sketches (e.g., *Sketch3* of *Extrude3*) of the solid features do not have complete dimensions. A (−) symbol is placed in front of the sketch, indicating that the sketch is not fully defined.

Apparently, this translation is not complete. Unfortunately, this translation represents a typical situation you will encounter for a large majority of the parts. In many cases, it may take only a small effort to repair or recreate wrongly recognized or unrecognized solid features. However, when you translate an assembly with many parts, the repairing effort could be substantial.

Option 2: Importing Geometry

Importing geometry is more straightforward and has a higher success rate than that of importing solid features.

Repeat the same steps to open the gear housing part, *gbox_housing.prt*. In the *Pro/ENGINEER to SolidWorks Converter* dialog box (Figure S1.32), choose *Import geometry directly* (default), and then *Knitting* (default) in order to import solid models instead of just surface models. Note that if you choose *BREP* (Boundary Representation), only boundary surfaces will be imported. Click *OK*.

The conversion process will start. After about a minute or two, the converted model will appear in the graphics window, as shown in Figure S1.33. In addition, an entity *Imported1* will appear in the browser (Figure S1.33). As mentioned earlier, there will be no parametric solid feature with dimensions and sketch converted if you choose *Option 2*. However, the geometry converted seems to be accurate. All the geometric features in Pro/ENGINEER shown in Figure S1.27 were included in this imported feature. This translation is successful. If you do not anticipate making any changes to the gear housing, this imported part is satisfactory.

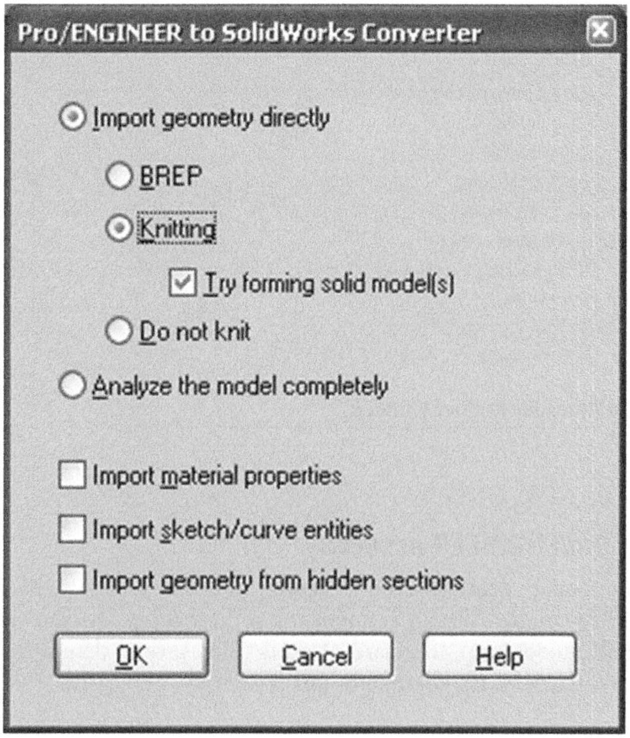

Pro/ENGINEER to SolidWorks Converter

- ⦿ Import geometry directly
  - ○ BREP
  - ⦿ Knitting
    - ☑ Try forming solid model(s)
  - ○ Do not knit
- ○ Analyze the model completely

- ☐ Import material properties
- ☐ Import sketch/curve entities
- ☐ Import geometry from hidden sections

OK   Cancel   Help

FIGURE S1.32 The *Converter* Dialog Box.

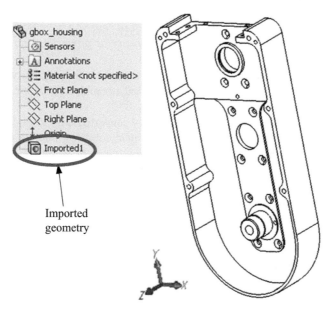

- gbox_housing
  - Sensors
  - Annotations
  - Material <not specified>
  - Front Plane
  - Top Plane
  - Right Plane
  - Origin
  - Imported1

Imported
geometry

FIGURE S1.33 The Gear Housing Model Converted as a Single Solid Feature, *Imported1*.

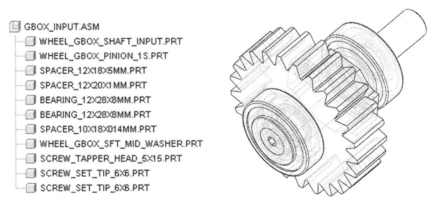

FIGURE S1.34 Input Gear Assembly in Pro/ENGINEER.

S1.3.3 Importing Pro/ENGINEER assembly

We will import the input gear assembly (*gbox_input.asm*) shown in Figure S1.34 using both options. We will try *Option 1* first; i.e. importing solid features. As shown in the left of Figure S1.34 (Pro/ENGINEER model tree), there are 11 parts (plus several datum features, not shown) in this assembly. SolidWorks will try to import this assembly as well as the 11 parts from Pro/ENGINEER.

Option 1: Importing Solid Features

Repeat the same steps to open the input gear assembly, *gbox_input.asm*. In the *Pro/ENGINEER to SolidWorks Converter* dialog box (Figure S1.35), choose *Use feature import for all parts*, and choose *Overwrite* for *If same name SolidWorks file is found* (just in case you have SolidWorks files with the same file names in the same folder). Choose *Import material properties* and *Import sketch/curve entities*. Click *Import*. The conversion process will start.

You will see sketches, solid features, and solid models appear in the graphics window. After about a minute, the translation process is completed. The converted assembly and the browser with parts listed are shown in Figure S1.36.

As shown in Figure S1.36, the parts are not completely converted. Major solid features are missing, such as pinion 1 (*wheel_gbox_pinion_1s<1>*), where most solid features are not converted. If you expand the part, you will see that only two extrude features were converted; the remaining entities are mostly sketches. The remaining parts were converted. However, the *Mates* branch in the browser is completely empty, implying that no assembly mates have been imported.

Apparently, this translation is not satisfactory. A nontrivial effort will be have to be devoted to reconstructing the solid features (therefore, solid models) as well as the final assembly.

Option 2: Importing Geometry

Importing geometry is also more straightforward for assembly and has a higher rate of success.

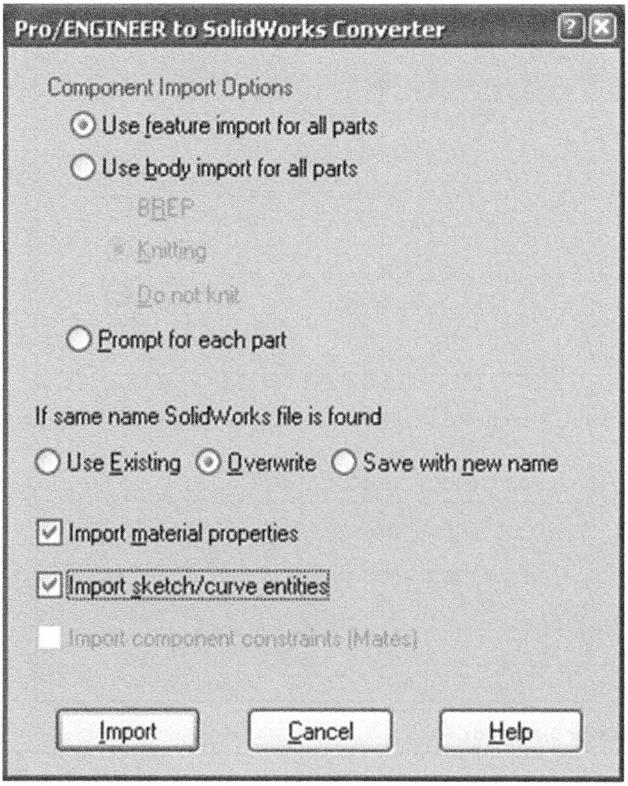

Pro/ENGINEER to SolidWorks Converter [?][X]

Component Import Options

⊙ Use feature import for all parts

◯ Use body import for all parts

~~BREP~~

~~Knitting~~

~~Do not knit~~

◯ Prompt for each part

If same name SolidWorks file is found

◯ Use Existing   ⊙ Overwrite   ◯ Save with new name

☑ Import material properties

☑ Import sketch/curve entities

☐ Import component constraints (Mates)

[Import]   [Cancel]   [Help]

FIGURE S1.35 The *Converter* Dialog Box.

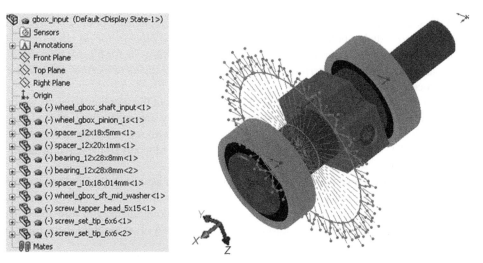

gbox_input (Default<Display State-1>)
- Sensors
- Annotations
- Front Plane
- Top Plane
- Right Plane
- Origin
- (-) wheel_gbox_shaft_input<1>
- (-) wheel_gbox_pinion_1s<1>
- (-) spacer_12x18x5mm<1>
- (-) spacer_12x20x1mm<1>
- (-) bearing_12x28x8mm<1>
- (-) bearing_12x28x8mm<2>
- (-) spacer_10x18x014mm<1>
- (-) wheel_gbox_sft_mid_washer<1>
- (-) screw_tapper_head_5x15<1>
- (-) screw_set_tip_6x6<1>
- (-) screw_set_tip_6x6<2>
- Mates

FIGURE S1.36 The Converted Assembly with Constituent Parts.

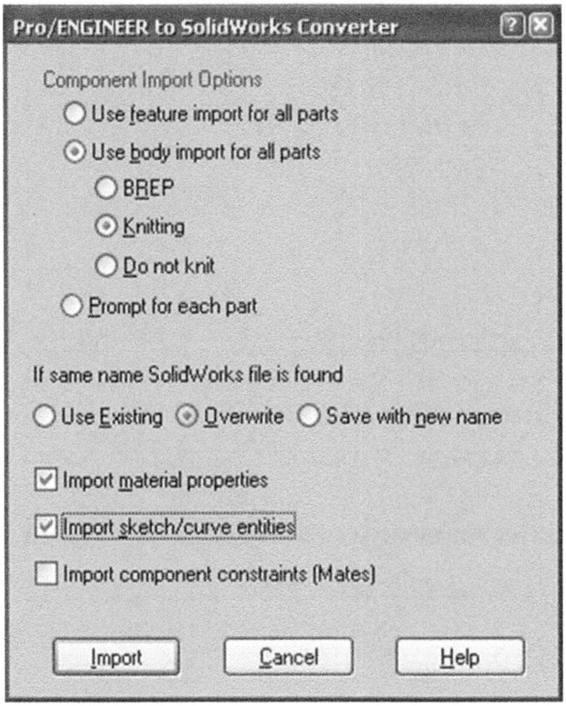

FIGURE S1.37 The *Converter* Dialog Box.

Repeat the same steps to open the input gear assembly, *gbox_input.asm*. In the *Pro/ENGINEER to SolidWorks Converter* dialog box (Figure S1.37), choose *Use body import for all parts* (default), and then *Knitting* (default) in order to import solid models. Choose *Overwrite* for *If same name SolidWorks file is found*, and choose *Import material properties* and *Import sketch/curve entities*. Click *Import*. The conversion process will begin.

After about a minute or two, the converted assembly appears in the graphics window, as shown in Figure S1.38. The assembly and all 11 parts seem to be correctly imported. If you expand any of the part branch, such as the gear (*wheel_gbox_pinion_1s<1>*), you will see an imported feature listed, as depicted in Figure S1.38. Again, there is no solid feature converted in any of the parts. In addition, the *Mates* branch is empty.

If you do not anticipate making any change to this input gear assembly, this imported assembly is satisfactory, except it does not have any assembly mates. Assembling all 11 parts (or more, in some cases) will take a nontrivial effort. If you do not anticipate making changes in how these parts are assembled, you may merge all 11 parts into a single part, instead of assembling them using mating constraints.

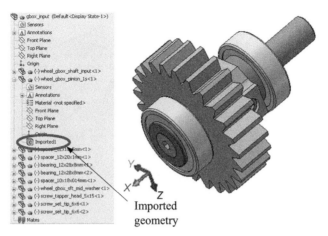

Imported
geometry

FIGURE S1.38 The Converted Assembly with Constituent Parts.

In SolidWorks, you can join two or more parts to create a new part in an assembly. The merge operation removes surfaces that intrude into each other's space and merges the parts into a single solid volume. We will insert a new part into the assembly and merge all 11 parts into this new part.

Choose from the pull-down menu *Insert > Component > New Part*. A new part with an assigned name is listed in the browser (Figure S1.39). SolidWorks is expecting you to select a plane or a flat face to place a sketch for the new part.

New part

FIGURE S1.39 A New Part Added.

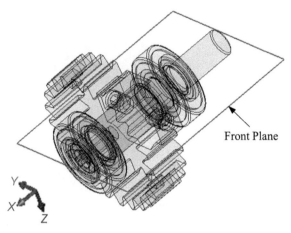

FIGURE S1.40 The Assembly *Front* Plane.

Click a plane or planar face on a component. For example, pick the assembly *Front* plane from the browser; the *Front* plane will appear in the graphics window (Figure S1.40). In the new part, a sketch opens on the selected plane. Close the sketch. Because you are creating a joined part, you do not need a sketch.

Next, select all 11 parts and merge them into the new part.

From the browser, click the first part *wheel_box_shaft_input<1>*, press the *Shift* key, and then click the last part, *screw_set_tip_6×6<2>*. All 11 parts will be selected.

From the pull-down menu, choose *Insert > Features > Join*. The *Join* dialog box will appear (overlapping with the browser) as shown in Figure S1.41. In the *Join* dialog box, all 11 parts are listed.

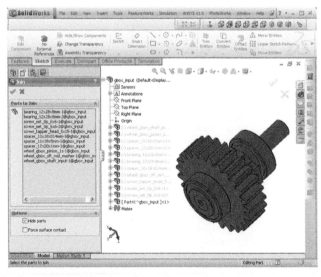

FIGURE S1.41 The *Join* Dialog Box (Overlapping with the Browser).

All you have to do is to click the checkmark on top to accept the parts. Save the part (choose the part in the *Resolve Ambiguity* dialog box shown in Figure S1.42, then click *OK*). In the warning dialog box (Figure S1.43), click *OK*. Enter the name *gbox_input.sldprt* in the *Save As* dialog box, and click *Save*. Save one more time for the assembly. Close the entire assembly model.

Now open the part *gbox_input*. Make sure you open *gbox_input.sldprt* instead of *gbox_input.sldasm*. The part *gbox_input* will appear in the graphics window. In addition, all entities belong to this part will be listed in the browser, as shown in Figure S1.44. Note that there is an arrow

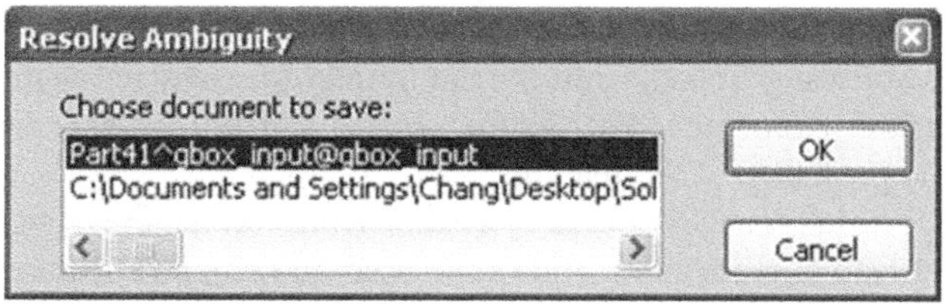

FIGURE S1.42 The Resolve Ambiguity Dialog Box.

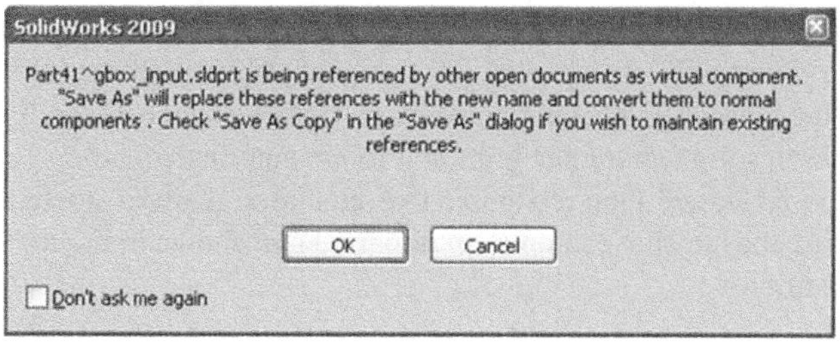

FIGURE S1.43 The Warning Dialog Box.

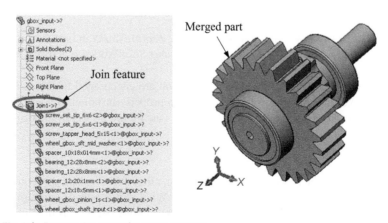

Merged part

Join feature

FIGURE S1.44 The Join Feature Created with Parts Merged.

symbol -> to the right of the root entity, *gbox_input*. This symbol indicates that these entities enclosed in this part refer to other parts or assembly. Note that the *Join1* branch has the same symbol. Expand the *Join1* branch. You will see 11 parts listed, all with arrows, pointing to the actual parts currently in the same folder.

S1.4 Solid feature recognition using *FeatureWorks*

The FeatureWorks module of SolidWorks recognizes solid features on an imported object in a SolidWorks part document. Recognized features are (almost) the same as features that you create using the SolidWorks software. You can edit the definition of recognized features and change their attributes and dimension parameters. For features that are based on sketches, you can edit the sketches to change the geometry of the features.

There are basically two approaches offered by FeatureWorks for feature recognition: automatic and interactive. Automatic feature recognition allows the users to recognize features in a batch mode in accordance with the preset options. Features that were created last (also called outer features) are usually recognized first, essentially in a reverse order of feature creation sequence recorded in the FeatureManager design tree. Whenever a problematic feature is encountered, the process aborts, and whichever features recognized prior to this point are organized in a partial FeatureManager design tree. On the other hand, interactive feature recognition allows users to recognize one or one type of features at a time. In this case, users will have to pick the geometric entities in the imported object from the graphics window for the features to be recognized in the right sequence. Overall, one possible strategy for using FeatureWorks in solid feature recognition is to use automatic feature recognition to recognize as many features as possible, and then recognize the remaining features interactively.

In this lesson, we use the housing example, an imported part shown in Figure S1.45, to illustrate the steps in feature recognition.

S1.4.1 Setting key options

First, we need to activate the FeatureWorks add-in module by choosing from the pull-down menu.

Tools > Add-Ins

In the Add-Ins dialog box shown in Figure S1.46, click *FeatureWorks* in both boxes (Active Add-ins and Start Up), and then click *OK*. You may need to restart SolidWorks to activate the FeatureWorks module.

Certainly, before going over this tutorial project, you are encouraged to check with your system administrator to make sure that FeatureWorks has been properly installed to your computer.

Next, we clear the *Use fully defined sketches* option (see Figure S1.47) by choosing from the pull-down menu.

Tools > Options

and clicking *Sketch* under the System Options tab. Then, click *OK*. Clearing this option allows you to create features from sketches that are not fully defined. What will happen if you turn on the *Use fully defined sketches* option? A fully defined sketch must be available before it is used to create a feature.

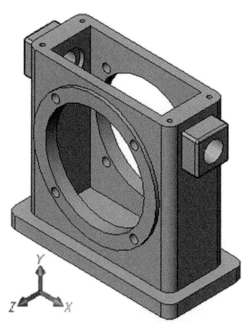

FIGURE S1.45 The Housing Example for Solid Feature Recognition.

FIGURE S1.46 The Add-Ins Dialog Box.

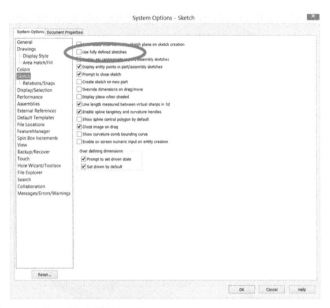

FIGURE S1.47 The *System Options-Sketch* Dialog Box.

This will cause problems in feature recognition because, in this case, a feature must have a fully defined sketch before it is recognized.

S1.4.2 Using FeatureWorks

Open the housing example (filename *housing.SLDPRT*). In the FeatureWorks message dialog box (Figure S1.48), choose *No* for the time being, to not start the recognition process. Instead, you will open the model and then start feature recognition. As shown under the FeatureManager design tree, only one solid feature, *Imported1*, is listed (Figure S1.49).

Next we set up FeatureWorks options by choosing from the pull-down menu:

Insert > FeatureWorks > Options

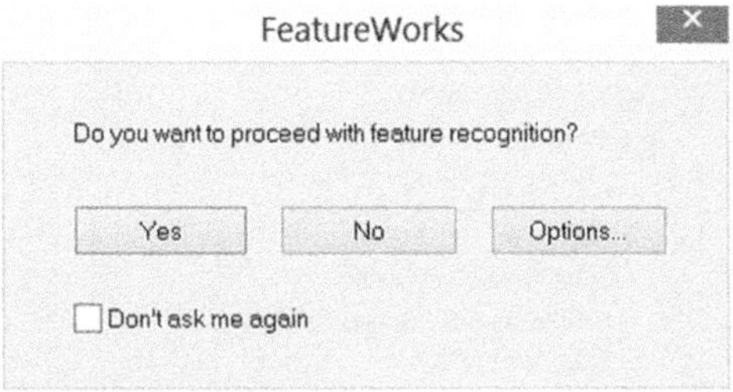

FIGURE S1.48 The FeatureWorks Message Dialog Box.

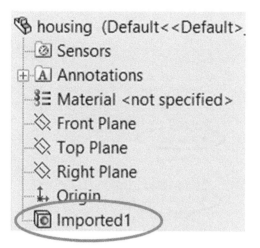

FIGURE S1.49 *Imported1* **Feature Listed in the FeatureManager Design Tree.**

In the FeatureWorks Options dialog box (Figure S1.50), select *Dimensions/Relations*, then choose *Enable Auto Dimensioning of Sketches* to enable auto dimensioning of the recognized sketches, choose *Baseline under Scheme*, and click *Add constraints* to sketch under relations. Click *OK* to accept the selections.

S1.4.2.1 Automatic feature recognition

Next, start feature recognition by choosing from the pull-down menu:

Insert > FeatureWorks > Recognize Features

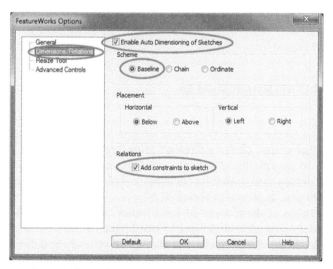

FIGURE S1.50 The FeatureWorks Options Dialog Box.

FIGURE S1.51 The FeatureWorks Message Dialog Box.

In the FeatureWorks dialog box, choose *Automatic, Standard features*, select all feature types, and click the checkmark to accept the selections and start the automatic feature recognition (see Figure S1.51). After a minute or two, 16 solid features and 3 planes will be recognized and listed in the FeatureManager design tree, as shown in Figure S1.52.

Note that most sketches of the recognized solid features are fully defined. For example, if you expand the *Cut-Extrude1* and double-click *Sketch5* (see Figure S1.53a), the dimensions of the sketch appear (Figure S1.53b). Right-click *Sketch5* and choose *Edit Sketch*, and choose *Normal view*, and sketch relations and dimensions are shown (Figure S1.53c). All entities are in black color, implying that the sketch is fully defined. Save your model under the default name *Part1*.

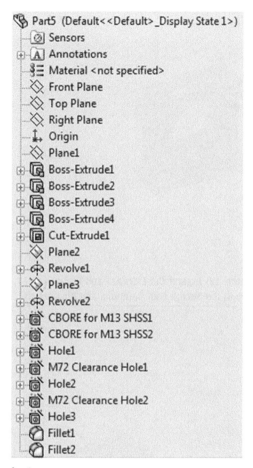

FIGURE S1.52 Features Recognized.

S1.4.2.2 Interactive feature recognition

Close the model *Part1* to go back to the imported model, in which *Imported1* is the only solid feature listed in the FeatureManager design tree. This time, use the interactive option for feature recognition.

Like before, start feature recognition by choosing from the pull-down menu:

Insert > FeatureWorks > Recognize Features

In the FeatureWorks dialog box, choose *Interactive, Standard features,* select all feature types, and choose *Chain* fillet faces (see Figure S1.54). In the graphics window, pick any of the round faces on the two ears. The faces picked and chain faces of the selected round faces are listed in the *Selected* entities box, as shown in Figure S1.54. Click *Recognize* to start feature recognition.

(a)   **(b)**   **(c)**

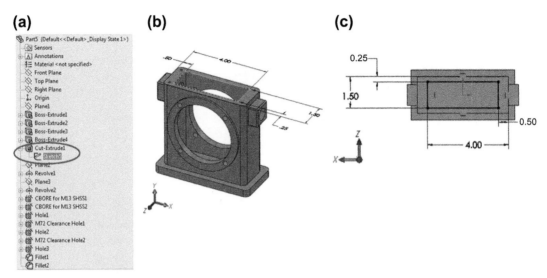

FIGURE S1.53 Fully Defined Sketch. (a) Expand *Cut-Extrude1* and Click *Sketch1*, (b) Sketch Highlighted with dimensions, and (c) Normal View of the Sketch with Relations and Dimensions.

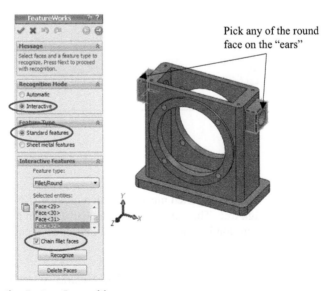

Pick any of the round face on the "ears"

FIGURE S1.54 Interactive Feature Recognition.

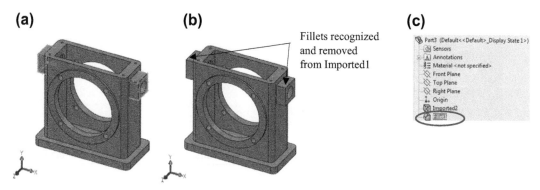

FIGURE S1.55 Interactive Feature Recognition. (a) Before Recognition, (b) After Recognition, and (c) Recognized Features, *Fillet1*, Added to the FeatureManager Design Tree.

Almost immediately, the selected faces are recognized as a fillet feature, which is listed below the imported feature in the FeatureManager design tree, as shown in Figure S1.55c.

You may right click *Fillet1* in the FeatureManager design tree and select *Edit* (see Figure S1.56a). In the *Fillet1* dialog box (Figure S1.56b), enter a different fillet radius, such as 0.625. Click the checkmark to accept the change. The fillet feature will be changed to a radius of 0.625 because the fillet is recognized as a parametric solid feature.

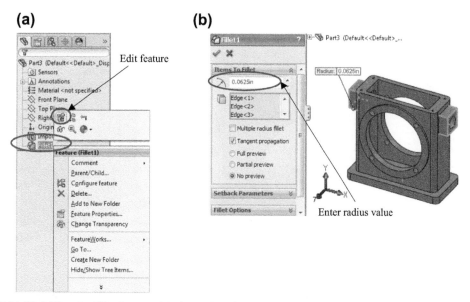

FIGURE S1.56 Editing the Fillet Feature. (a) Right-click *Fillet1* and Choose the *Edit* feature, and (b) Enter a Different Value for the Radius.

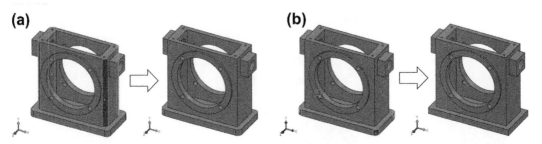

FIGURE S1.57 Recognizing More Fillet Features. (a) Vertical Edge Walls, and (b) Corners at the Bottom Plate.

Following the same steps, you may select more round faces to recognize them as fillet features, as shown in Figure S1.57.

Next, choose *Hole* from the FeatureWorks dialog box (Figure S1.58a). Pick the inner face of the hole of the right ear. Click *Recognize*. The hole is recognized and temporarily removed from the solid model, as shown in Figure S1.58b. Repeat the same for the hole on the other side. Note that multiple hole selection is not supported in SolidWorks 2012 or earlier.

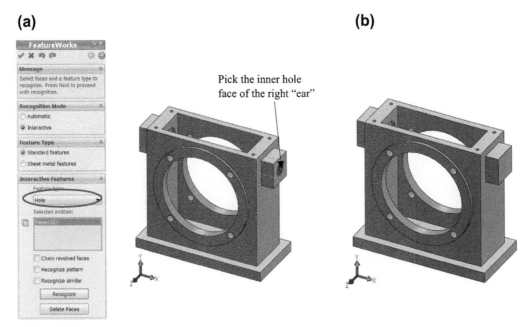

FIGURE S1.58 Recognizing the Hole Feature in the Right "Ear". (a) Selections, and (b) Hole Recognized and Temporarily Removed.

(a)   **(b)**

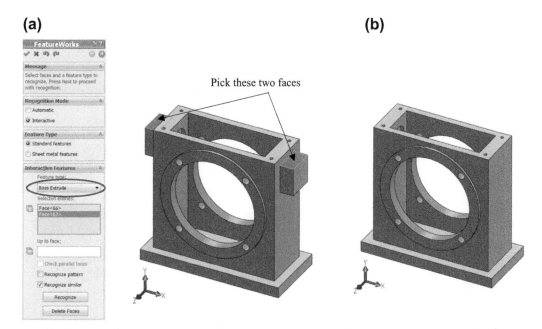

Pick these two faces

FIGURE S1.59 Recognizing the Boss Feature of the Two "Ears". (a) Selections, and (b) Boss Features Recognized and Temporarily Removed.

Next, choose *Boss Extrude* from the FeatureWorks dialog box (Figure S1.59a). Pick the two top faces of the ears from the model. Click *Recognize*. The two rectangular blocks are recognized and temporarily removed from the solid model (Figure S1.59b).

Next, we recognize the four small vertical holes on the top face as a pattern feature. Choose *Hole* from the FeatureWorks dialog box (Figure S1.60a). Pick a vertical hole (Figure S1.60a), choose *Recognize*, select *Rectangular*, under Recognize pattern and then click *Recognize*. The holes are recognized as a pattern feature and temporarily removed from the solid model, as shown in Figure S1.60b.

Click the checkmark on the top left of the FeatureWorks dialog box. The recognized features are listed in the FeatureManager design tree, as shown in Figure S1.61. Click these features to see more information, such as sketches. Are these sketches fully defined?

Save your model, and continue recognizing the remaining features in Exercise S1.2.

(a)

(b)

Pick a small vertical
hole on the top face

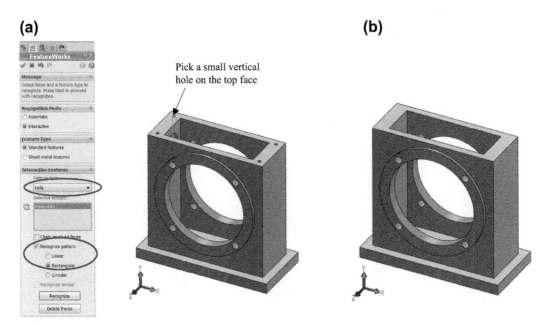

FIGURE S1.60 Recognizing the Boss Feature of the Two "Ears". (a) Selections, and (b) Boss Features Recognized and Temporarily Removed.

FIGURE S1.61 Recognized Features listed in the FeatureManager Design Tree.

Exercises

S1.1. Open the *gbox_middle.asm* (see Figure ES1.1) in the same folder as Lesson S1.3. Report the results of the assembly importing, using both options (*Use body import for all parts* and *Use feature import for all parts*). Discuss the pros and cons of both options.

S1.2. Continue to recognize the remaining features from the tutorial lesson S1.4 (see Figure ES1.2). Report the following:

   a. A write-up for the feature recognition type and recognition order
   b. Screen captures for feature model tree, and solid models before and after each feature recognition
   c. Discuss failed feature recognition, if any
   d. Describe at least two lessons learned from this exercise that may benefit your fellow classmates or co-workers.

FIGURE ES1.1 The *gbox_middle.asm*.

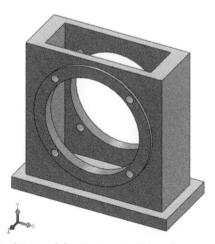

FIGURE ES1.2 The Housing Model with Remaining Features to Recognize.

Project P1 Solid Modeling with Pro/ENGINEER

CHAPTER OUTLINE

Several important topics relevant to product design, such as direct modeling, assembly, design parameterization, and product data exchange, were discussed in Chapters 3–6, respectively. There is no need to emphasize the importance of understanding the concept and theory behind these topics. It is equivalently important for engineers to learn to use computer-aided design (CAD) software for creating assemblies, capturing design intents, addressing CAD model translations, and directly modifying part designs. It is essential for engineers to learn how to handle these issues using existing tools.

In Project P1, we introduce Pro/ENGINEER (Wildfire 5.0) for creating assemblies, capturing design intents, and supporting CAD model translations. In addition, we include a short introduction to the direct modeling capability offered by Creo Direct. We include four tutorial lessons to help you get started on learning and using the software. These lessons use the example of a single-piston engine, with which you will learn the basics of creating an assembly using mating constraints, a slider-crank mechanism for capturing design intents in assembly, the same slider-crank example for learning CAD model translation and feature recognition, and a crankshaft example that shows the capabilities of direct modeling offered in Creo Direct. Example models are available for download at the book's companion website (http://booksite.elsevier.com/9780123985132).

Overall, the objective of this project is to enable readers to use Pro/ENGINEER and Creo Direct for assembly modeling, design intent capturing, model translations, and direct modeling. If you are interested in learning more about assembly modeling and elevating yourself to an intermediate level (and beyond), you may want to review more assembly examples offered by tutorial books on the subject, such as Creo Parametric 1.0 Tutorial and MultiMedia DVD by Roger Toogood and Jack Zecher (http://www.sdcpublications.com/Textbooks/Creo-Parametric-1-0-Tutorial/ISBN/978-1-58503-692-9) and Parametric Modeling with Pro/ENGINEER Wildfire 5.0 by Randy H. Shih (http://www.sdcpublications.com/Textbooks/Parametric-Modeling-ProENGINEER-Wildfire-5/ISBN/978-1-58503-539-7).

Note that the lessons included in this project were developed using Pro/ENGINEER Wildfire 5.0 M040 and Creo Direct 2.0. If you are using a different version of Pro/ENGINEER and Creo Direct, you may see slightly different menu options or dialog boxes. Because the software tools are fairly intuitive to use, these differences should not be too difficult to figure out.

P1.1 Introduction to Pro/ENGINEER

Pro/ENGINEER® is an integrated three-dimensional (3D) CAD/CAM/CAE solution. Using Pro/ENGINEER, you can create parametric, feature-based, associative solid models and assemble individual components to build complex assemblies using mating constraints (called placement constraints in Pro/ENGINEER). Each component in an assembly can be opened in its own Pro/ENGINEER window for users to perform many operations. Changes in components will be automatically reflected in the assembly. The document name extension for Pro/ENGINEER assemblies is .asm.

The main objective of this tutorial project is to help you, as a new user, to become familiar with Pro/ENGINEER assembly capabilities, capabilities offered by Pro/ENGINEER for support of solid model translations, as well as direct model editing supported by Creo 2.0 (in Section P1.5).

P1.1.1 User interface

A typical user interface window of Pro/ENGINEER, shown in Figure P1.1, consists of the menu bar, graphics area, model tree, toolchests, and message area. The graphics area displays the solid model on which you are working. The message area lists guiding information related to the function you are performing. The menu bar contains commands for creating, saving, and modifying models, and for setting your Pro/ENGINEER environment and configuration options. You can customize the menu bar by adding, removing, copying, or moving commands. Top toolchests provides buttons for general file and model operations, such as open, close, zoom in/out, and display options. In addition, the side bar to the right of the graphics area (side toolchests) shows a list of feature buttons that support solid and assembly modeling.

The Model Tree to the left of the graphics area displays a list of assembly file names and the corresponding constituent components or subassemblies. You can expand or collapse each subassembly to view its components by clicking ⊞ in front of the name of the subassembly. A component name with a little square prefix indicates that the component is not fully constrained. Right-clicking on any component in the model will bring up command options that you can choose to modify or adjust the entity. When a part file is opened, the Model Tree shows the part file name, datum entities, coordinate system, and part features. If multiple Pro/ENGINEER windows are opened, the Model Tree contents reflect the file in the current window.

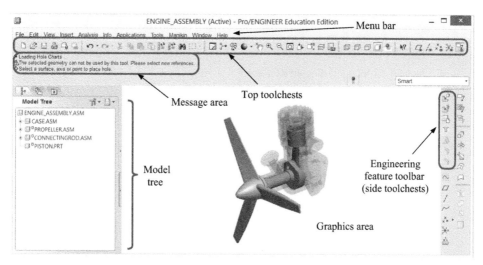

FIGURE P1.1 User Interface of Pro/ENGINEER.

There are three buttons associated with assembling on the Engineering Features toolbar in the right toolchests (Figure P1.1):

- Add component to the assembly
- Add a manikin to the assembly
- Create a component in assembly mode

When you insert a component to the current assembly, the Component Placement dashboard appears. As shown in Figure P1.2, the dashboard consists of the dialog bar, slide-up panels, and buttons for tool options.

Some of the buttons on the Component Placement dashboard are listed in Table P1.1. In addition to the buttons, from the dialog bar, you can choose the mating constraints you want to assemble the component being inserted. In Pro/ENGINEER, you can choose from two types of constraints: predefined constraint sets and user-defined constraint sets, as shown in Figure P1.3. Detailed constraint options within each category are listed in Tables P1.2 and P1.3, respectively. You can use the button on the dialog bar to convert a user-defined set to a predefined set or vice versa, because the two types of constraints are equivalent in terms of constraining the relative motion between components. For example, you can replace a *Pin* constraint in the predefined set with an *Insert* constraint and a *Mate* (or

FIGURE P1.2 Component Placement Dashboard.

Table P1.1 Shortcut Buttons on the *Component Placement* Dashboard

Icon	Function
	Places a component using an interface
	Places a component manually
	Converts a user-defined set to a predefined set or vice versa
	Displays the component in its own window as you define constraints
	Displays the component in the graphics window and updates component placement as you define constraints (default)
	Pauses component placement so you can use a tool
	Resumes component placement following a pause
	Applies component placement and quit the dashboard
	Cancels component placement

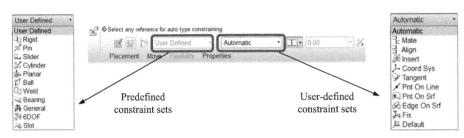

FIGURE P1.3 Predefined and User-Defined Constraint Sets.

Align) constraint that belong to the user-defined set. In the following tutorial lessons, we focus only on user-defined constraint sets when building the assemblies. However, predefined constraint sets are required if the assembled model is to be used for motion analysis in *Mechanism Design*.

The placement status of the component is also listed in the dialog bar to the right of the pull-down lists (Figure P1.2). Component status can switch between *No Constraints*, *Partially Constrained*, *Fully Constrained*, and *Constraints Invalid* depending on the constraints defined.

Table P1.2 Predefined Constraint Sets

Icon	Name	Function
	Rigid	Allows no movement in the assembly
	Pin	Contains a rotational movement axis and translation constraints
	Slider	Contains a translational movement axis and rotation constraints
	Cylinder	Contains a 360° rotational movement axis and translational movement
	Planar	Contains a planar constraint to allow rotation and translation along the reference planes
	Ball	Contains a point alignment constraint for 360° movement
	Weld	Contains a coordinate system and an offset value to "weld" the component in a fixed position to the assembly
	Bearing	Contains a point alignment constraint to allow rotation along a straight trajectory
	General	Creates a user-defined set of two constraints
	6DOF	Contains a coordinate system and an offset value, to allow movement in all directions
	Slot	Contains a point alignment to allow rotation along a nonstraight trajectory

Table P1.3 User-Defined Constraint Sets

Icon	Name	Function
	Mate	Positions two references of the same type so that they face each other
	Align	Positions two planes on the same plane (coincident and facing the same direction), two axes coaxial, or two points coincident
	Insert	Inserts a revolved component surface into an assembly revolved surface
	Coordinate system	Aligns the components coordinate system with an assembly coordinate system
	Tangent	Positions two references of different types so that they face each other. The point of contact is a tangent
	Pnt on line	Positions a point on a line
	Pnt on Srf	Positions a point on a surface
	Edge on Srf	Positions an edge on a surface
	Fix	Fixes the current location of a component that was moved or packaged
	Default	Aligns the component coordinate system with the default assembly coordinate system

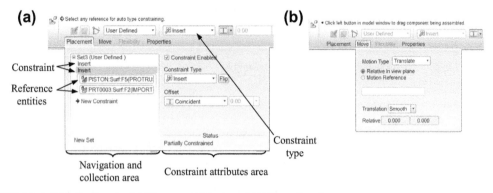

FIGURE P1.4 Slide-Up Panels. (a) Placement Panel, and (b) Move Panel.

Four slide-up panels are hidden beneath the dialog bar (Figure P1.2). You can open each panel by clicking its name. The two most used panels are *Placement* and *Move*. As shown in Figure P1.4a, the *Placement* panel enables and displays component placement and connection definitions. The *Navigation and Collection* area (left-hand side) displays sets and constraints. Reference entities selected for each constraint are listed under the constraint name. A new constraint will be activated after a valid pair of references is selected for the current constraint, until the component is fully constrained. The *Constraint Attributes* area (right-hand side) contains context-sensitive options for constraints selected in the *Navigation and Collection* area. The *Move* panel (Figure P1.4b) is used to move the component being assembled for easier access. The component to be moved must be packaged or configured with a predefined constraint set.

P1.1.2 Examples

Four lessons are included in this tutorial project, which illustrate step-by-step details for creating assemblies, capturing design intents, importing SolidWorks models to Pro/ENGINEER, and direct modeling using Creo Direct. We start with a very simple single-piston engine example, in which the propeller is allowed to rotate. We then use a slider-crank example to show how to capture design intents in assembly after making changes to its components. The same slider-crank model will be used to illustrate how SolidWorks parts and assemblies can be imported to Pro/ENGINEER, as well as the feature recognition capability in Pro/ENGINEER. Finally, we use a crankshaft model to demonstrate basic direct modeling capabilities offered in Creo Direct. All examples and topics to be discussed in each lesson are summarized in Table P1.4.

P1.2 Single-piston engine

In this lesson, you will learn how to create an assembly model for a single-piston engine shown in Figure P1.5. You will learn how to select placement constraints to assemble parts and subassemblies. After the assembly is created, you may drag the propeller to check the kinematics of the assembly. We will start this lesson with a brief overview about the engine assembly to be created in Pro/ENGINEER, then show you the detailed steps in creating the assembly.

Table P1.4 Examples Employed in this Project

Section	Example	Solid Models	Things to Learn
P1.2	Single-piston engine		**1.** This is an introductory lesson, showing detailed steps in creating an assembly using four components. **2.** We will review the mating constraints that are defined between parts and subassemblies. **3.** You will learn how to create an assembly that allows the propeller to rotate.
P1.3	Slider-crank mechanism		**1.** This tutorial lesson shows detailed steps in capturing design intents for the slider-crank mechanism.
P1.4	Importing SolidWorks parts and assemblies to Pro/ENGINEER		**1.** In this tutorial lesson, we will focus on importing SolidWorks parts and assemblies to Pro/ENGINEER. **2.** You will learn how to recognize features on an imported model using the feature recognition tool in Pro/ENGINEER
P1.5	Direct modeling using Creo direct		**1.** This tutorial lesson shows steps in modifying part geometry imported as an Initial Graphics Exchange Standards (IGES) model using direct modeling capabilities in Creo direct.

P1.2.1 The single-piston engine example

The engine example consists of four major components: case (*case.asm*), propeller (*propeller.asm*), connecting rod (*connectingrod.asm*), and piston (*piston.prt*), as shown in Figure P1.6a. For this lesson, the parts and subassemblies have been created in Pro/ENGINEER. There are 22 model files created, including three assemblies: *case*, *propeller*, and *connectingrod*. In addition, the final assembled engine example (*single_piston_engine.asm*) is included for your reference. You may open this *single_-piston_engine* assembly file to preview it.

The propeller is assembled to case using mate and insert constraints, as shown in Figure P1.6b. We will assemble the components so that the propeller is free to rotate along the *x*-direction.

FIGURE P1.5 The Single-Piston Engine Example in Computer-Aided Design Assembly.

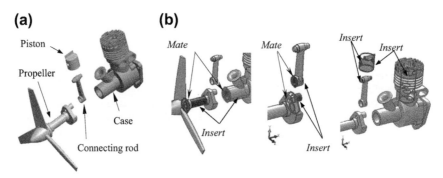

FIGURE P1.6 The Single-Piston Engine Example. (a) Exploded View, and (b) Placement Constraints Defined between Components (parts or subassemblies).

The connecting rod is assembled to the propeller (at the crankshaft) using mate and insert constraints. The connecting rod is free to rotate relative to the propeller (at the crankshaft) along the *x*-direction. Finally, the piston is assembled to the connecting rod (at the piston pin) using an insert constraint. The piston is also assembled to the engine case using another insert constraint. This constraint restricts the piston movement along the *y*-direction, which in turn restricts the top end of the connecting rod to move vertically.

P1.2.2 Using Pro/ENGINEER

Start Pro/ENGINEER and choose *File > New*. In the *New* dialog box, choose *Assembly*, enter assembly name *Engine_Assembly*, and then click *OK*, as shown in Figure P1.7.

Now we start to insert components into the assembly. Choose from the pull-down menu:

Insert > Component > Assemble

FIGURE P1.7 New Dialog Box.

FIGURE P1.8 Open Dialog Box.

to bring up the *Open* dialog box. Navigate to the folder *Single-Piston Engine Tutorial* (or the folder where you put the parts and assemblies), select subassembly *case.asm*, and click *Open*, as shown in Figure P1.8. The engine case will appear in the graphics area, and you should see the *Component Placement* dialog bar above the graphics area (Figure P1.9).

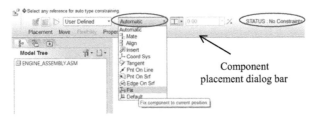

FIGURE P1.9 Component Placement Dialog Bar.

Pull down the *Constraint* list, and then choose *Fix* to fix the location of the engine case in the global coordinate system (Figure P1.9). Note that the component status on the dialog bar switches from *No Constraints* (Figure P1.9) to *Fully Constrained* when the *Fix* constraint is defined. Click the *OK* ✓ bottom on the right of the dialog bar to accept the inserted component and constraint. Once a part or subassembly is brought in, it will be listed in the *Model Tree* to the left of the graphics area. You can expand the subassembly to view individual parts by clicking ⊞ in front of the name of the inserted component, as shown in Figure P1.10.

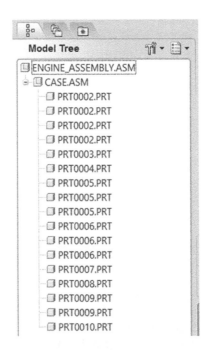

FIGURE P1.10 Model Tree.

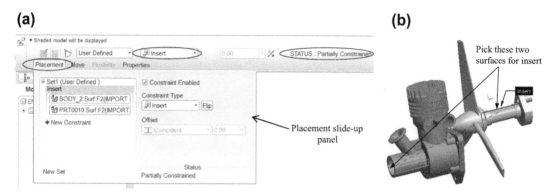

FIGURE P1.11 Assembling the Propeller to the Engine Case Using Insert Constraint. (a) Placement Slide-up Panel, and (b) Surfaces to Pick in the Graphics Area.

Next, we insert the second component. From the pull-down menu, choose *Insert > Component > Assemble*, and open subassembly *propeller.asm*. Then we define placement constraints for the propeller. In the *Component Placement* dialog bar, click *Placement* to bring up the *Placement Slide-up Panel* (Figure P1.11a). The default constraint type is *Automatic*. From the *Constraint* pull-down list, choose *Insert*, then pick the surface of the propeller shaft and the inner surface of the shaft tube on the engine case as references for the constraint, as shown in Figure P1.11b. Once the constraint is defined, in the graphics area, the position and orientation of corresponding components (in this case, the propeller) will be adjusted automatically to comply with the constraint. In the meantime, the component status in the *Component Placement* dialog bar turns from *No Constraints* to *Partially Constrained* (Figure P1.11a).

Next, we define the second constraint by clicking *New Constraint* in the *Placement* slide-up panel. Choose *Mate* from the *Constraint* pull-down list, then pick the front ring-shaped face on the propeller and the back ring-shaped face on the engine case, as shown in Figure P1.12. Note that the component status becomes *Fully Constrained* (Figure P1.12), which means the propeller will not be able to rotate. This is because when the *Allow Assumptions* option is turned on (by default) in the slide-up panel (Figure P1.12), Pro/ENGINEER automatically makes assumptions in terms of the orientation of the constraint. Clear the *Allow Assumptions* check box, and you should see that the component status returns to *Partially Constrained*. Click *OK* ✔ to accept the inserted component and close the dialog bar.

Before inserting the next component, we need to temporarily hide the engine case because part of the propeller is now hidden inside the case, which may cause problems when assembling other components to the propeller. Right-click *CASE.ASM* in the model tree and choose *Hide* to hide the engine case. Next, repeat the same steps to bring in the component *connectingtod.asm*. In the *Component Placement* dialog bar, create an *Insert* constraint between the cylindrical surface of the crankshaft on the propeller and the inner surface of the hole on the connecting rod. And then define a *Mate* constraint by selecting the back fan-shaped face on the propeller and the front ring-shape face on the connecting rod (Figure P1.13). Click *OK* ✔ to accept the component and constraints.

(a)

(b)

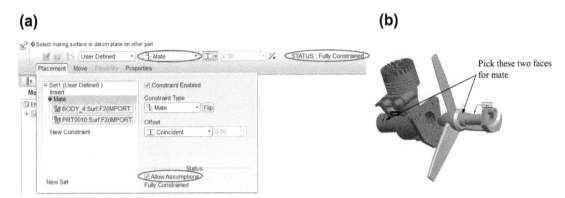

FIGURE P1.12 Assembling the Propeller to the Engine Case Using Mate Constraint. (a) Placement Slide-up Panel, and (b) Faces to Pick in the Graphics Area.

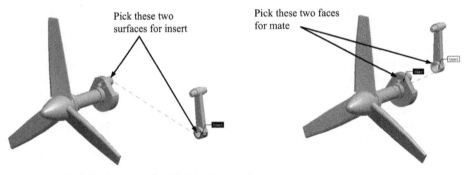

FIGURE P1.13 Assembling the Connecting Rod to the Propeller.

Now we bring in the last component *piston.prt*. From the *Constraint* pull-down list, choose *Insert*, and then pick the cylindrical surface of the piston pin on the connecting rod and the inner surface of the hole on the piston, as shown in Figure P1.14. The last constraint to be defined is between the outer cylindrical surface of the piston and the inner cylindrical surface of the chamber of the engine case. Therefore, we first click *OK* ✔ in the *Component Placement* dialog bar to accept the inserted piston and close the dialog bar, and then right click *CASE.ASM* in the model tree and choose *Unhide* to show the engine case in the graphics area. In addition, before defining the next constraint, we need to adjust the orientation of the piston to avoid confusion of the software when solving the constraint. As shown in Figure P1.15, from the *View* toolbar, choose *Drag Components*, and drag the piston at the tip of the boss feature so that it is roughly facing upwards. Click *Close* to exit the *Drag* dialog box.

To define the last constraint, right click *PISTON.PRT* in the model tree and choose *Edit Definition* to bring up the *Component Placement* dialog bar. Pull down the *Placement* slide-up panel and

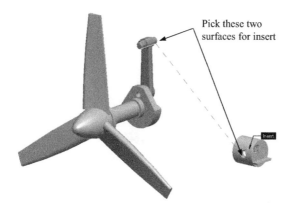

Pick these two
surfaces for insert

FIGURE P1.14 Assembling the Piston to the Connecting Rod.

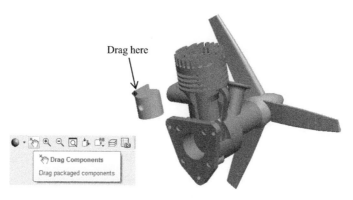

Drag here

FIGURE P1.15 Drag the Piston to Adjust Orientation.

click *New Constraint*. Select *Insert* as the constraint type and pick the reference surfaces as shown in Figure P1.16. Note that to select the inner face of the chamber, you may need to rotate the view and click the surface through the oval-shaped hole on the chamber (Figure P1.16). Click the *OK* ✔ button to accept the constraint.

Now the single-piston engine has been completely assembled, as shown in Figure P1.17a. Note that in the model tree, there is a little square in front of the name of each component, except for *CASE.ASM* (Figure P1.17b). This is because only the engine case is fully constrained in this assembly. To review the placement constraints defined for each component, right-click on the component name in the model tree, choose *Edit Definition*, and then expand the *Placement* slide-up panel in the dialog bar. When you click on each constraint, the corresponding reference entities will be highlighted in the graphics area.

In this assembly, the propeller is free to rotate along the *x*-direction. You can drag the propeller in the graphics area using the *Drag Components* tool 🖐 to check the kinematics of the engine. To check the motion of the piston hidden inside the case, pull down the *Appearance Gallery* button from the *View* toolbar, and choose *ptc-glass* under *My Appearances*, as shown in Figure P1.18a. The mouse

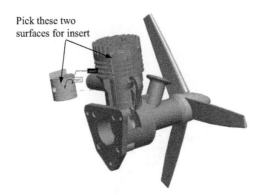

FIGURE P1.16 Assembling the Piston to the Engine Case.

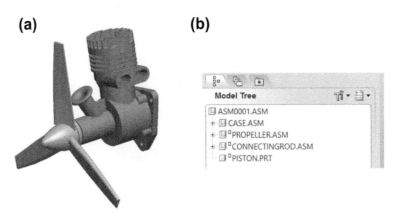

FIGURE P1.17 The Single-Piston Engine Assembly. (a) Assembled Model, and (b) Model Tree.

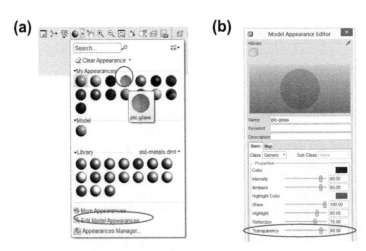

FIGURE P1.18 Changing Appearance of the Engine Case. (a) Appearance Gallery Pull-down Panel, and (b) Model Appearance Editor Dialog Box.

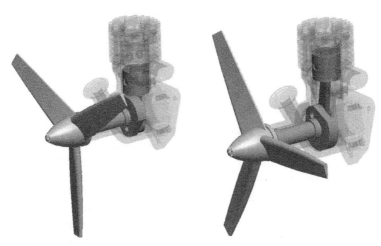

FIGURE P1.19 Single-Piston Engine Assembly with Partially Transparent Engine Case.

pointer will then turn into a brush. Click the engine case in the graphics area and click *OK* in the *Select* dialog box. You will see that the engine case becomes partially transparent. To further adjust transparency, choose *Edit Model Appearance* in the *Appearance Gallery* pull-down panel (Figure P1.18a), increase *Transparency* to 90 (Figure P1.18b), and close the *Model Appearance Editor* dialog box. In the graphics area, you should see clearly the motion of the piston and the connecting rod while the propeller is rotating, as shown in Figure P1.19. Save your model for future reference.

P1.3 Slider-crank mechanism

In this lesson, we will learn how to capture design intents for a slider-crank mechanism shown in Figure P1.20a, as mentioned in Chapter 5. Because the detailed steps in using placement constraints in

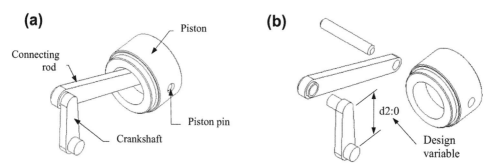

FIGURE P1.20 The Slider-Crank Mechanism. (a) CAD Solid Model, and (b) Exploded View with a Design Variable d2:0.

Pro/ENGINEER to create assembly have been presented in Section P1.2, we will not repeat the steps in this lesson. Instead, we focus on capturing the design intents for the assembly. We will start this lesson with a brief overview about the mechanism, then discuss steps in creating two design variables for the assembly.

P1.3.1 **The slider-crank example**

This mechanism consists of four parts: crankshaft, connecting rod, piston pin, and piston, as shown in Figure P1.20b. Our goal is to assemble the crankshaft with the datum planes for a particular orientation, like the one shown in Figure P1.21a, then capture two design variables: crankshaft length d2: 0 and rod length d3:2, as shown in Figure P1.21b. Currently, the dimension values are: d2:0 = 3 and d3: 2 = 8. Note that when we change the design variables, we expect to see all four parts are intact (i.e., properly assembled) and the piston pin axis stays on the *ASM_TOP* datum plane (see Figure P1.21b).

In order to capture the intent, we create one additional datum plane in the assembly. The datum plane *ASM_DTM1* will be created by rotating *ASM_TOP* along the axis *A_2* (in crankshaft with an angle $\sin^{-1}(3/8)$. Note that this angle dimension can be defined by entering the dimension value of 22 (approximately) and later adding the following relation:

$$\text{d1: 1} = \sin^{-1}(3/8) \tag{P1.1}$$

Note that *ASM_DTM1* will be used to properly orient the connecting rod. In addition, the rotation angle of *ASM_DTM1*, which determines the configuration of the assembly, will be related to the crankshaft length through the following equation:

$$\text{d1: 1} = \sin^{-1}(\text{d2: 0}/\text{d3: 2}) \tag{P1.2}$$

Equation P1.2 defines the relation of angle (d1:1) to the design variables d2:0 (crankshaft length) and d3:2 (rod length). Dimension d1:1 actually rotates the datum plane *ASM_DTM1* according to the change of d2:0 and d3:2. This is how the design intent is captured for this example.

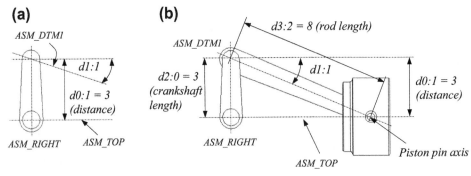

FIGURE P1.21 Design Variables to Define in the Assembly. (a) Crankshaft Orientation in Assembly, and (b) Side View of the Assembly with Design Variables d2:0 and d3:2.

P1.3.2 **Using Pro/ENGINEER**

Create a new assembly in Pro/ENGINEER with the name *Slider_Crank* and first bring in the *crank-shaft* component. The assembly datum planes *ASM_RIGHT*, *TOP*, and *FRONT* are employed to assemble the *crankshaft* using *Mate* and *Align* placement constraints:

- Mate: datum plane *ASM_RIGHT* and *TOP* in *crankshaft*
- Align: datum plane *ASM_TOP* and *RIGHT* in *crankshaft*
- Align: datum plane *ASM_FRONT* and *FRONT* in *crankshaft*

Note that during assembling you can use the datum plane ⟁ and datum axis ⟋ buttons in the *Datum Display* toolbar to turn on and off the display of the datum planes and axes in the graphics area. The *crankshaft* assembled to the assembly datum planes is shown in Figure P1.22.

Next, create a new assembly datum plane *ASM_DTM1*. Choose from the pull-down menu:

Insert > Model Datum > Plane

to bring up the *Datum Plane* dialog box. In the graphics area, pick the datum plane *ASM_TOP* and axis *A_2* in *crankshaft* while holding the *Ctrl* key. In the dialog box, enter 22 as the rotation angle and click *OK*, as shown in Figure P1.23. As discussed earlier, 22 is just an approximate value; therefore, you need to assign a precise angle for the datum plane *ASM_DTM1* using the *Relations* tool.

From the pull-down menu, choose *Tools > Relations* and the *Relations* dialog box will appear. If you click the datum plane *ASM_DTM1* in the graphics area, you can see all the related dimensions. In this case, the only dimension is d1:1, which is the rotation angle of *ASM_DTM1* with respect to the datum plane *ASM_TOP*, as highlighted in Figure P1.24. Click on the dimension d1:1 to insert it into the text box in the *Relations* dialog box as d1 (Figure P1.24). Continue to enter the relation as $d1 = \text{asin}(3/8)$ in the text box and click *OK* to accept. Click the *Regeneration* button ⚲ to the top right of the graphics area, and then click *Regenerate* in the pop-up *Regeneration Manager* dialog box to rebuild the model. Note that the angle of the datum plane *ASM_DTM1* will be updated in the graphics area, and the *Regeneration* button icon will switch from yellow light ⚲ to green light ⚲.

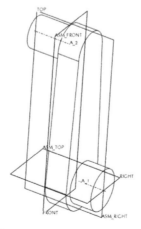

FIGURE P1.22 Assembling the Crankshaft.

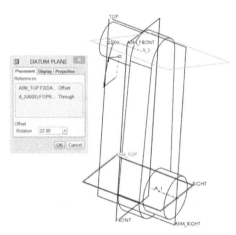

FIGURE P1.23 Creating Datum Plane *ASM_DTM1*.

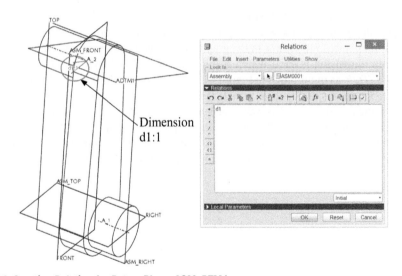

FIGURE P1.24 Creating Relation for Datum Plane *ASM_DTM1*.

The next component to be inserted to the assembly is the *rod*. The *rod* is assembled to *crankshaft* and the datum planes using the following (see Figure P1.25):

- Align: axis *A_1* in *rod* and *A_2* in *crankshaft*
- Mate: front face of the boss in *rod* and back face in *crankshaft*
- Align: datum plane *ASM_DTM1* and *FRONT* in *rod*

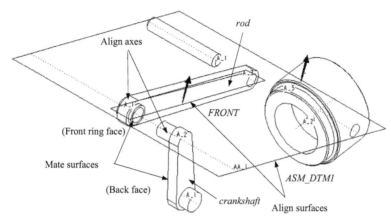

FIGURE P1.25 Assembling the Rod.

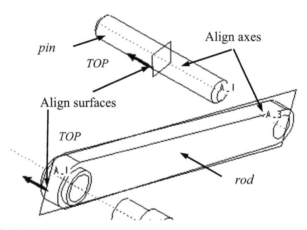

FIGURE P1.26 Assembling the Pin.

Next, we bring in the *pin* and assemble it to the *rod* using two constraints (see Figure P1.26):

- Align: axis *A_1* in *pin* and *A_3* in *rod*
- Align: *TOP* in *pin* and *TOP* in *rod*.

Finally, the *piston* is inserted and assembled using (see Figure P1.27):

- Align: axis *A_5* in *piston* and *A_1* in *pin*
- Mate: *TOP* in *piston* and *TOP* in *pin*

In addition to the two constraints above, we will define the *FRONT* plane in *piston* to be parallel with the datum plane *ASM_TOP* using an *Align* constraint, as shown in Figure P1.28. Note that the *Offset* option is set to be *Oriented* instead of *Coincident* in the slide-up panel.

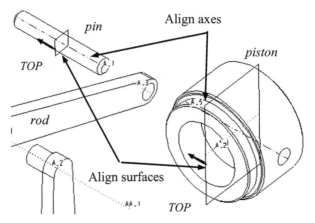

FIGURE P1.27 Assembling the Piston.

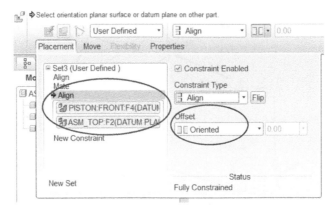

FIGURE P1.28 Align Constraint between *FRONT* in Piston and *ASM_TOP*.

So far, the slider-crank mechanism has been assembled, in which all components are fully constrained. The assembly placement constraints are summarized in Table P1.5.

Now, adjust the first design variable d2:0 to see its impact to the assembly. Right-click *CRANK.PRT* in the model tree and choose *Open* to open the crankshaft in a new window. In the *crank* model tree, expand feature *Protrusion id 9*, right-click sketch *S2D0002*, and then choose *Edit Definition*. In the graphics area, double-click the length dimension 3.00 and change it to 5 as shown in Figure P1.29. Press the *Enter* key to apply the change and click *Done* ✔ on the sketch toolbar to exit the sketch mode. Note that the length of the crankshaft has been updated in the graphics area. Choose from the pull-down menu *Window > SLIDER_CRANK* to return to the assembly window, and then click 🔄 to regenerate the model. As shown in Figure P1.30, the piston pin axis is no longer in the ASM_TOP plane. Why? Because the orientation of the datum plane *ASM_DTM1* is fixed as $\sin^{-1}(3/8)$, and is not changing with the length of the crankshaft.

Table P1.5 Placement Constraints Defined for the Slider-Crank Mechanism

	ASM_RIGHT, TOP, FRONT	Crankshaft	ASM_DTM1	Rod	Pin
Crankshaft	Mate: ASM_RIGHT/Top Align: ASM_TOP/Right Align: ASM_FRONT/Front				
Rod		Align: A_1/A_2 Mate: faces	Align: ASM_DTM1/ Front		
Pin				Align: A_1/A_3 Align: Top/Top	
Piston	Align: ASM_TOP/Front				Align: A_5/A_1 Mate: Top/Top

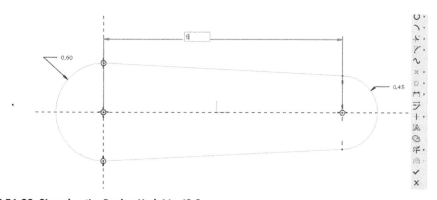

FIGURE P1.29 Changing the Design Variable d2:0.

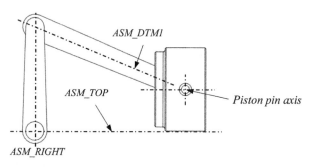

FIGURE P1.30 Design Intent not Captured After Changing Crankshaft Length.

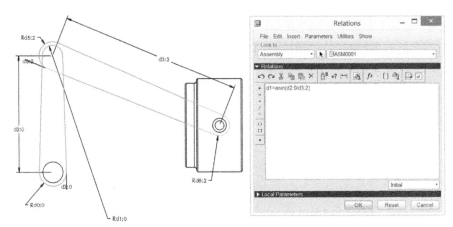

FIGURE P1.31 Create a Relation for Datum Plane _ASM_DTM1_.

To capture the design intent, we need to relate the orientation of _ASM_DTM1_ to both the length of

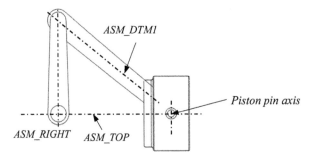

FIGURE P1.32 Design Variable d2:0 Properly Defined.

the _crank_ and the _rod_. To do that, choose _Tools_ > _Relations_ and modify the relation defined earlier to $d1 = \mathrm{asin}(d2{:}0/d3{:}2)$, as shown in Figure P1.31. Click _OK_ to exit the dialog box and then regenerated the assembly. Now you can see that the angle of the datum plane _ASM_DTM1_ has changed, and the piston pin returns to the _ASM_TOP_ datum plane (Figure P1.32).

Certainly, the procedure above is not the only way of capturing the design intent in this case. For example, if we make the _FRONT_ plane in _piston_ coincide (instead of parallel) with the datum plane _ASM_TOP_, then the datum plane _ASM_DTM1_, as well as the second _Align_ constraint for _rod_, are no longer needed.

Change the length of the crankshaft back to 3 and regenerate the assembly.

Next, we adjust the second design variable d3:2. Following the same steps, change the _rod_ length (sketch _S2D0002_ under feature _Protrusion id 10_) from 8 to 12 at the part level and regenerate the

(a)

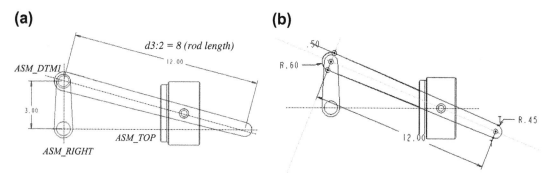

(b)

FIGURE P1.33 **Problems Encountered in Design Variable d3:2.**

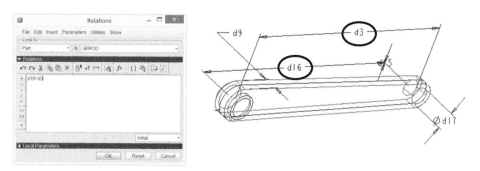

FIGURE P1.34 **Create Relation to Relate Dimensions d3 and d16.**

assembly. How will the design be regenerated? Like the one in Figure P1.33a and b? Or something else? Is the design intent adequately captured?

There are two problems. The smaller hole in the rod does not move together with the dimension d3 (d3:2 in assembly). The piston and piston pin do not move together with the dimension d3:2.

The reason that you see problems is that, at part level, the location of the smaller hole in the rod is related not to the length of the rod but to the location of the larger hole. When the rod is elongated after the design change, the distance between the two holes in the rod remains constant. Therefore, you need to define a relation for the *rod* part (see Figure P1.34), as:

d16 = d3 (why not d3 = d16?)

Then, regenerate the part to make sure that the design intent is captured at the part. After that, go back to the assembly window and regenerate the model. As can be seen in Figure P1.35, due to the relation defined in the assembly, all components are properly assembled, and the piston pin axis stays within the *ASM_TOP* datum plane.

Save your models for future reference.

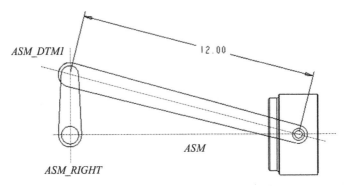

FIGURE P1.35 Design Variable d3:2 Properly Defined.

P1.4 Importing SolidWorks parts and assemblies to Pro/ENGINEER

In Project S1, we demonstrated how to import Pro/ENGINEER parts and assemblies to SolidWorks. In this lesson, we discuss the approaches of importing solid models from other CAD software, such as SolidWorks, into Pro/ENGINEER. Pro/ENGINEER supports importing solid models from a broad range of software and formats, including Parasolid, ACIS, Initial Graphics Exchange Standards (IGES), Standard for Exchange of Product data (STEP) and SolidWorks. For a complete list of supported software and formats in Pro/ENGINEER, please refer to Figure P1.36. You may access this list by choosing *File > Open* from the pull-down menu, and pull down the *Files of type* in the *File Open* dialog box.

In this tutorial lesson, we focus on importing SolidWorks parts and assemblies. We hope that the methods and principles you learn from this lesson are applicable to importing solid models from other software and formats.

Pro/ENGINEER provides capabilities for importing both part and assembly. However, unlike SolidWorks, it does not offer the functionalities for importing feature history or dimensions. The parts and assemblies of any format will be imported as a dumb geometry only, which means all solid features are lumped into a single imported body without dimensions.

Certainly, if you do not anticipate making design changes in Pro/ENGINEER, there is no problem with importing parts as a single geometric feature. However, when it is required to make design changes to the model in Pro/ENGINEER, you will need to recognize solid features embedded in the imported object. Prior to Pro/ENGINEER Wildfire 4.0, the feature recognition tool (FRT) can only be installed in Pro/ENGINEER as a third-party plug-in application. For Pro/ENGINEER versions later than Wildfire 4.0, this plug-in application is integrated in the software packaged as the FRT. This tool identifies and replaces geometry representing features in imported models with true parametric features, and enables you to selectively edit geometry of features in imported solid models without completely rebuilding the model manually.

Compared to the feature recognition capability in SolidWorks (discussed in Project S1.4), the FRT in Pro/ENGINEER is much simpler and less desired. Only six types of major features are supported: extrusion, slot, hole, chamfer, fillet, and pattern. In addition, the feature recognition process in

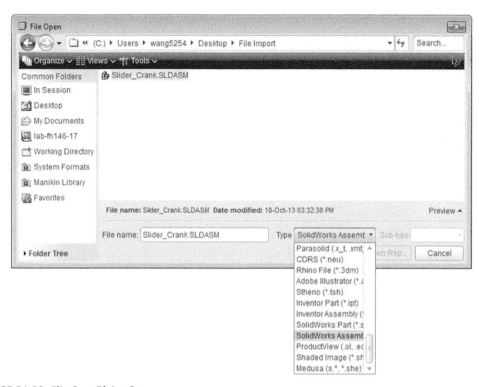

FIGURE P1.36 *File Open* Dialog Box.

Pro/ENGINEER is not automated, which means users need to manually recognize features one at a time, while determining the right sequence of picking the geometric entities on the imported object.

Moreover, during SolidWorks feature recognition, each feature will be temporarily removed from the solid model once the feature is recognized to facilitate identifying the next feature. However, in Pro/ENGINEER, all recognized features remain in display, including those embedded in the part geometry, such as holes and chamfers, thus making it difficult in most cases to continue the recognition process. Furthermore, in Pro/ENGINEER, the innermost feature (the feature that was created first) will not be recognized due to the lack of reference. Because of that, the best a user can achieve from the feature recognition process in Pro/ENGINEER will be a partially parametric model.

Overall, the feature recognition capability in Pro/ENGINEER offers design engineers an easy and useful tool to make some design changes to the imported geometry. However, the capability is not powerful enough to convert the imported object to a fully parametric model.

In this lesson, we will first import a slider-crank assembly created in SolidWorks (Figure P1.37) to Pro/ENGINEER. Then, we will use one of the components in this assembly to illustrate the steps of using the FRT in Pro/ENGINEER.

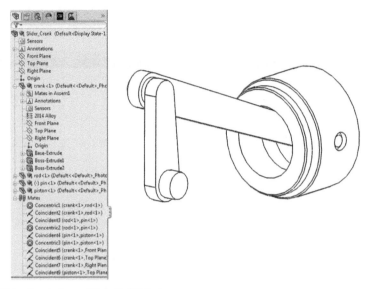

FIGURE P1.37 Slider-Crank Assembly in SolidWorks.

P1.4.1 The slider-crank example in SolidWorks

The slider-crank mechanism consists of four parts. If you have access to SolidWorks, you may want to open the final assembly, *slider_crank.SLDASM*, to review the assembled mechanism, as shown in Figure P1.37. The four parts in this assembly are: *crank.SLDPRT, pin.SLDPRT, piston.SLDPRT,* and *rod.SLDPRT*. The mating constraints used to assemble the components are listed in the feature tree shown in Figure P1.37. Pro/ENGINEER will try to import this assembly as well as the four parts from SolidWorks.

P1.4.2 Importing SolidWorks assembly

Start Pro/ENGINEER and choose *File > Open*. In the *File Open* dialog box, change to folder *File_import* (or the folder where you have these parts and assemblies). Pull down the *Files Type* list, and choose *SolidWorks Assembly* (**.sldasm*). Select the only SolidWorks assembly file in the dialog box, *Silder_Crank.SLDASM*, and click *Open*. In the *Import New Model* dialog box, you should see that the *Type* option is automatically set to be *Assembly*, as shown in Figure P1.38. Note that if you are opening a SolidWorks part file, you can choose either *Part* or *Assembly* as the import type. If the *Assembly* option is chosen for a SolidWorks part, you will end up with a Pro/ENGINEER assembly that contains one single component.

Accept the default import options by clicking *OK* in the *Import New Model* dialog box. You should see the assembly model in the graphics area, as shown in Figure P1.39. Note that in the model tree, four parts in the assembly are imported successfully, and all of them are fixed in the global

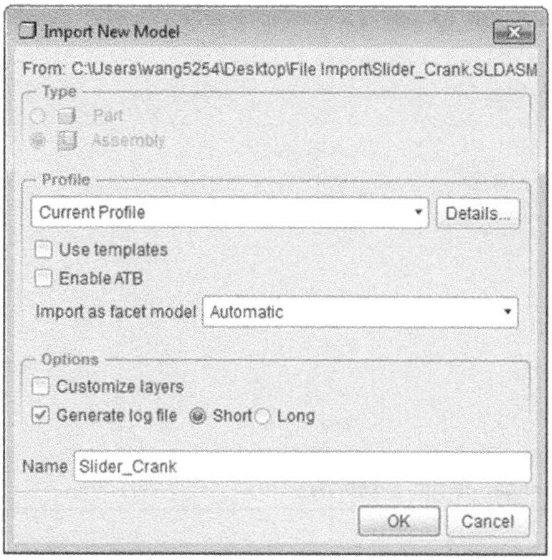

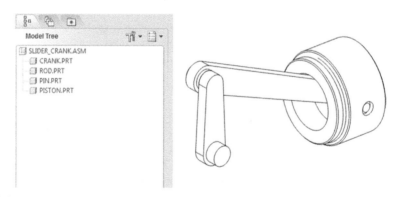

FIGURE P1.38 *Import New Model* **Dialog Box.**

FIGURE P1.39 Imported Assembly.

coordinate system (you will not be able to drag them to move in the graphics area using the *Drag Component* tool).

Are the individual parts constrained? Right-click on any part (in this case, we use the piston) in the model tree and choose *Edit Definition* to bring up the *Component Placement* dialog bar. As shown in Figure P1.40, no constraint is defined for the part, and the component status is *No Constraints*. Now, close the toolbar without changing anything, and you will see that the part *piston* becomes non-fully-constrained (little square in front of the part name in the model tree, as shown in Figure P1.41). Also, you can now move the piston in the graphics area using the *Drag Components* tool (Figure P1.41).

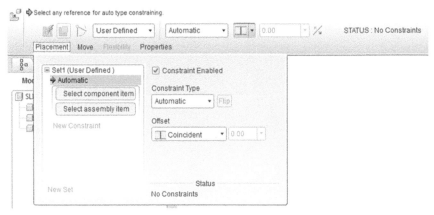

FIGURE P1.40 No Constraint Defined in the Assembly.

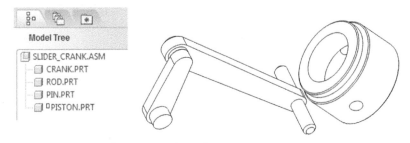

FIGURE P1.41 Imported Assembly with Unconstrained Piston.

Apparently, a SolidWorks assembly will be imported to Pro/ENGINEER as individual parts without any mating constraint. If the assembly consists of a large number of parts, reassembling all parts in Pro/ENGINEER will take a nontrivial effort.

Now, open individual parts in the imported assembly. Right-click *crank* in the model tree and choose *Open* to open the part *crank* in a new window. If you take a look at the model tree, as shown in Figure P1.42, you will see that the part is imported as a dumb block with no parametric solid feature, dimension, or sketch. Therefore, if you want to make design changes to the imported parts and assembly, you are recommended to use the FRT in Pro/ENGINEER to identify and modify solid features on the imported models.

P1.4.3 Feature recognition for SolidWorks part

With the FRT in Pro/ENGINEER, you can recognize features on an imported part, and edit the definition of recognized features and change their attributes or dimension parameters. For features that

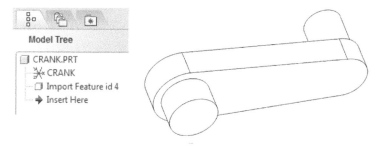

FIGURE P1.42 **Imported Crankshaft.**

FIGURE P1.43 **Enabling the Feature Recognition Tool (FRT).**

are based on sketches, you can edit the sketches to change the geometry of the features. In this lesson, we will use the crankshaft (Figure P1.42) in the slider-crank assembly discussed earlier as an example to illustrate the steps of using the FRT.

Before we start, we need to enable the FRT in Pro/ENGINEER, which is disabled by default. Choose from the pull-down menu *Tools > Options* to open the *Options* dialog box. Add a new option *frt_enabled* and set the value to *yes*, as shown in Figure P1.43. Click *Add/Change* and then click *OK* to close the dialog box.

FIGURE P1.44 Feature Recognition Tool (FRT) Toolbar.

Now repeat the steps discussed in the last section to import the SolidWorks part *crank.SLDPRT* into Pro/ENGINEER. Alternatively, you can import the complete assembly as we did in the last section and then open the part *crank* from the imported assembly. Once the crankshaft is imported or opened, you will see that the FRT toolbar appears in the toolchests area, as shown in Figure P1.44. These FRT options are also available in the *Edit* pull-down menu.

The FRT in Pro/ENGINEER supports six types of solid features:

- Protrusions or pockets on flat or curved surfaces
- Extruded slots
- Simple or sketched holes on flat or curved surfaces
- Constant radius rounds
- Chamfers
- Table pattern for holes

Now, we demonstrate how to use these capabilities to recognize features. First, recognize the larger pin on the crankshaft. Click the button in the FRT toolbar, pick the flat surface of the main body of the crankshaft, and then click *OK* in the *Select* dialog box (Figure P1.45). Pro/ENGINEER will search all protrusion features created from this basis surface.

As shown in Figure P1.46, the larger pin has been successfully identified and added to the model tree as feature *Extrude 1* with a sketch. At the same time, a *Remove 1* feature is created in the model tree prior to *Extrude 1*. In fact, during the feature recognition process, Pro/ENGINEER first removed the original geometry of the larger pin from the imported body, and then created a parametric feature to replace the removed geometry. If you right-click *Extrude 1* in the model tree and choose *Delete*, you should see that the larger pin on the crankshaft disappears, as shown in Figure P1.47. However, if you continue to delete the *Remove 1* feature as well, the pin will grow back again, but this time it is simply the original geometry rather than a parametric feature. Once both *Extrude 1* and *Remove 1* are deleted, the model returns to the status right after it was imported.

Now recognize the larger pin again; then, following the same steps, recognize the smaller pin, as shown in Figure P1.48. However, you will encounter problems when you proceed to recognize the last

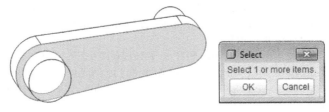

FIGURE P1.45 Recognizing the Larger Pin.

FIGURE P1.46 Larger Pin Recognized.

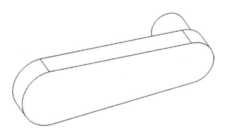

FIGURE P1.47 Larger Pin Removed from the Crankshaft.

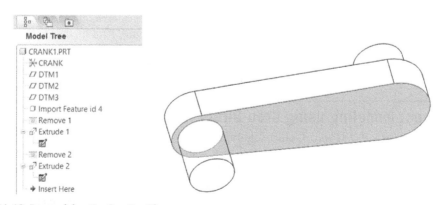

FIGURE P1.48 Recognizing the Smaller Pin.

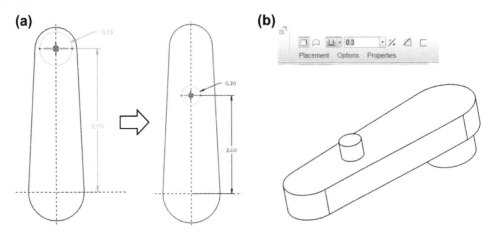

FIGURE P1.49 Making Changes to the Crankshaft. (a) Changing Diameter and Location of the Smaller Pin, and (b) Changing the Height of the Smaller Pin.

feature—the main body of the crankshaft. Using FRT, when recognizing an extrusion feature, we must identify a reference surface based on which the feature is created. Because the main body of the crankshaft is the first feature created when building the crankshaft, no reference surface for this feature can be found. Therefore, the main body is not able to be recognized using the FRT in Pro/ENGINEER, and the crankshaft has to be left as a partially parametric model.

Although the crankshaft is not completely converted into a parametric model, we can still make design changes to the recognized features (in this case, the two pins). For example, to change the location and diameter of the smaller pin, rightclick the sketch under *Extrude 2* in the model tree (Figure P1.48) and choose *Edit Definition*. In the graphics area, modify the radius and distance dimensions to 0.20 and 2.00, respectively, as shown in Figure P1.49a, and then click the *Done* ✔ button on the *Sketch* toolbar. The model of the crankshaft will be updated in the graphics area. To change the height of the smaller pin, right-click on *Extrusion 2* in the model tree and choose *Edit Definition*. In the text field of the toolbar on top of the graphics area, change the height of the extrusion feature to 0.3 and click *OK* ✔ to accept. As shown in Figure P1.49b, the smaller pin on the crankshaft has been successfully changed.

Save your models for future reference.

P1.5 Direct modeling using Creo Direct

As a member of the PTC Creo product family, Creo Direct is a stand-alone 3D CAD modeling application that supports creating, manipulating, and editing 3D geometries. Using a direct modeling approach, Creo Direct offers the most efficiency and flexibility when you are creating a concept design or editing models to defeature or simplify them for a downstream process, such as CAE analysis or numerical control and tooling design. The data created or edited in Creo Direct can be seamlessly shared among all other Creo applications used in the design process.

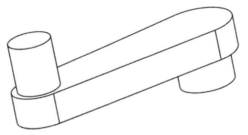

FIGURE P1.50 Geometry of the Crankshaft Example.

Creo Direct follows a history-free design approach. That is, geometric entities created during the design process are independent of each other. With Creo Direct, you will be able to make changes to a 3D CAD model that is imported without features or dimensions through direct interaction with geometry. For example, you can "push–pull" geometry to the desired position without losing the original design intent. Also, you can offset, mirror, and remove geometry, or modify rounds and analytic features such as fillets, cylinders, cones, and spheres. In addition, Creo Direct supports 3D surface modeling and assembly modeling.

In this tutorial lesson, we will learn how to directly edit the solid model of a crankshaft shown in Figure P1.50 using Creo Direct. Because SolidWorks files cannot be opened in Creo Direct, in this lesson, the crankshaft model will be imported as an IGES file.

It is worth mentioning that an add-on product named Flexible Modeling Extension (FMX), which also provides direct modeling capabilities, is embedded in Creo Parametric. This add-on contains part of the functionalities offered in Creo Direct. In this lesson, we will focus on Creo Direct (version 2.0) to show the steps in directly editing the imported geometry. However, most operations demonstrated in this lesson can also be accomplished using the FMX in Creo Parametric.

P1.5.1 Creo Direct user interface

As shown in Figure P1.51, the user interface of Creo Direct consists of a main window with a status bar at the bottom. The status bar displays prompts, filter settings, and the mode for the modeling operation. In the main window, the model tree to the left of the graphics area is similar to that in Pro/ENGINEER. Most tools and commands are grouped into two tabs—*Home* and *View*—in the ribbon above the graphics area. In part mode, the buttons in the *Home* tab include commands to create, modify, manipulate, dimension, and analyze geometry.

P1.5.2 Using Creo Direct

Choose from the pull-down menu *File > Open* (or click the 🖆 button on the *Quick Access Toolbar*). As shown in Figure P1.52, in the *File Open* dialog box, navigate to the folder *Direct Modeling Tutorial* (or the folder where you put the model), select *crank.igs*, and click *Open*. You should see the crankshaft appears in the graphics area as a dumb solid block without features. Note that the file formats supported in Creo Direct are listed in the *File Type* pull-down list (Figure P1.52).

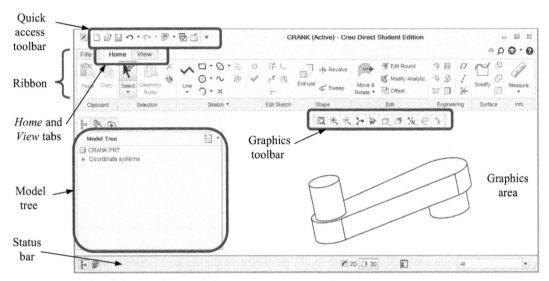

Quick access toolbar

Ribbon

Home and *View* tabs

Graphics toolbar

Model tree

Graphics area

Status bar

FIGURE P1.51 User Interface of Creo Direct.

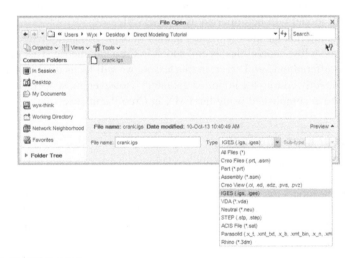

FIGURE P1.52 *File Open* **Dialog Box.**

First, we change the size and location of the smaller shaft on the crankshaft. Click the *Move & Rotate* button from the ribbon under the *Home* tab, as circled in Figure P1.53a, and then pick the front half-cylindrical surface of the smaller shaft (Figure P1.53b). You should see a graphical control tool—the dragger—appear near the selected surface. In the meantime, a live toolbar that includes most relevant commands for the selected entity shows up in the graphics area (Figure P1.53c). The arrows

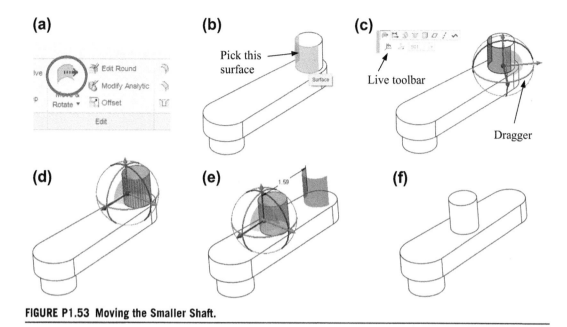

FIGURE P1.53 Moving the Smaller Shaft.

and rings of the dragger are called handles, which enable movement in one or more degrees of freedom. The handles that you can use for operations are highlighted in blue (Figure P1.53c).

As shown in Figure P1.53c, the origin of the dragger is placed at the location where you click to pick the surface, and the highlighted arrow is normal to the surface. Drag to move the origin of the dragger so that the blue arrow points approximately towards the larger shaft, as shown in Figure P1.53d. Next, pull the blue arrow along its direction (Figure P1.53e), and you should see the movement of the smaller shaft shown dynamically in the graphics area, with the relative distance from the starting point. Click the middle mouse button to accept the new location of the smaller shaft (Figure P1.53f).

Similarly, to adjust the length of the shaft, simply pick the end face of the shaft and pull the highlighted arrow upwards, as shown in Figure P1.54.

Now, edit the diameter of the smaller shaft. Choose *Modify Analytic* from the ribbon (Figure P1.55a), and then pick the cylindrical surface of the shaft. A red arrow will appear normal to the selected surface with the current diameter displayed. Drag the arrow towards the opposite direction of the arrow to reduce the diameter of the shaft, as shown in Figure P1.55b–d. Press the middle mouse button to accept the change.

Next, duplicate the modified smaller shaft along one edge of the main body of the crankshaft. Pick all three surfaces of the smaller shaft (two half cylindrical surfaces and one top surface) in the graphics area, and then click *Pattern* from ribbon (Figure P1.56a). Note that you need to hold the *Crtl* key when selecting multiple entities. The live toolbar switches automatically to the *Pattern* tab. Pick the straight edge on the crankshaft as shown in Figure P1.56b, and then drag the red arrow to adjust the location of the duplicated geometry. Click the middle mouse button to accept the change.

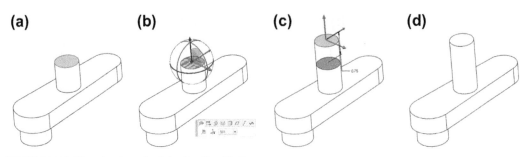

FIGURE P1.54 Changing Length of the Smaller Shaft.

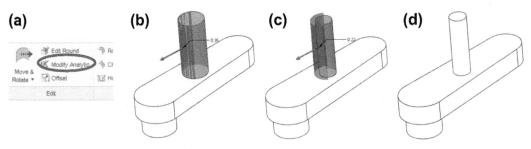

FIGURE P1.55 Changing Diameter of the Smaller Shaft.

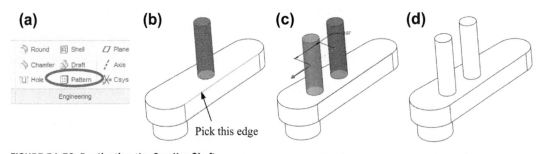

FIGURE P1.56 Duplicating the Smaller Shaft.

Finally, create a fillet on the duplicated shaft. Choose *Round* from the ribbon (Figure P1.57a) and select the top circular edge of the duplicated shaft (Figure P1.57b). A fillet will be created and you can adjust its radius by dragging either one of the two red arrows. As shown in Figure P1.57c, the fillet radius will be dynamically updated in the graphics area. Click the middle mouse button to accept the change.

Save the model for future reference.

(a) 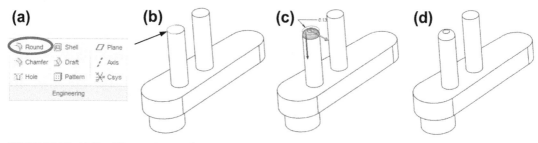 **(b)** **(c)** **(d)**

FIGURE P1.57 Adding Fillet to the Duplicated Shaft.

Exercises

P1.1. In this problem, you are asked to create the pulley assembly from parts provided on the book's companion website (http://booksite.elsevier.com/9780123985132). Note that the unit system mmNs is assumed. Use Set Up > Units to choose your unit system.

   a. Create the assembly like the one shown in Figures EP1.1 and EP1.2 (cut-out view) and print out four views (uncut, unexploded, with hidden lines) of the assembly: i.e. front, side, top, and isometric views.

   b. Create a matrix that shows component placement constraints, like the one in Table P1.5.

   c. Check interference and report the volume of interference.

   d. Create a cut in the base component to eliminate the interference problem.

   e. Create a key with proper dimensions between the hole of the pulley and shaft in assembly mode.

   f. Generate an exploded view to visualize the assembly like the one in Figure EP1.3. Print out and submit an isometric view.

FIGURE EP1.1 The Pulley Assembly.

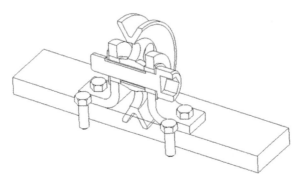

FIGURE EP1.2 The Cut-Out View of the Pulley Assembly.

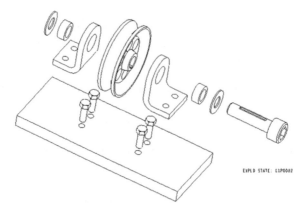

EXPLD STATE: L1P0002

FIGURE EP1.3 The Exploded View of the Pulley Assembly.

FIGURE EP1.4 The Block Part of IGES Model.

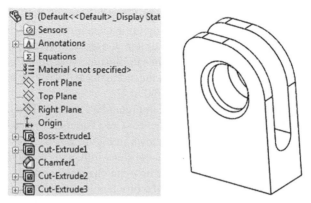

FIGURE EP1.5 The Block Part of SolidWorks.

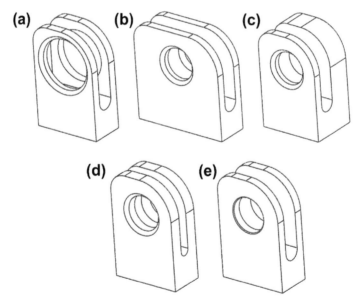

FIGURE EP1.6 Changes to Make for the Block Example. (a) Hole Size. (b) Base Width. (c) Base Depth. (d) Slot Width. and (e) Chamfer Size.

g. Define proper relationships among dimensions so that a physically valid model can be regenerated when diameter of the shaft is changed. Report these relations with associated screen captures for illustration.

Example files were extracted from Pro/ENGINEER Wildfire 5.0 Tutorial and MultiMedia CD by Roger Toogood and Jack Zecher, SDC Publications, ISBN: 978-1-58503-415-4.

P1.2. Open the block part of IGES shown in Figure EP1.4 in Pro/ENGINEER following the steps shown in Section P1.4. Use the feature recognition capability to recognize solid features embedded in the IGES model. Report features that you are able to recognize. Briefly state how they are recognized and attach screen captures to illustrate your approach. Describe problems encountered in the process, if any. Discuss the strengths and weaknesses of the feature recognition capability offered by Wildfire 5.0.

P1.3. Import the block part of SolidWorks shown in Figure EP1.5 to Pro/ENGINEER following the steps shown in Section P1.4. Use Creo Direct (or Creo Parametric) to edit the part, following steps shown in Section P1.5, for changes shown in Figure EP1.6. These changes include: (1) hole size, (2) base width, (3) base depth, (4) slot width, and (5) chamfer size, as shown in Figure EP1.6a–e, respectively. Briefly state how the changes are made and attach screen captures to illustrate your approach. Describe problems encountered in the process, if any. Discuss the strengths and weaknesses of the direct modeling capability offered by Creo 2.0.

Index

Printed and bound by CPI Group (UK) Ltd, Croydon, CR0 4YY

03/10/2024

01040329-0001